The Essential Guide to Passing
E:

A Reference Manual with 160 Key Questions

First Edition

Jacob Petro

PhD, PMP, CEng, PE

The Essential Guide to Passing the California Seismic Principles Exam

A Reference Manual with 160 Key Questions

First Edition

Jacob Petro
PhD, PMP, CEng, PE

PE Essential Guides
Hillsboro Beach, Florida

For the most up-to-date list of errata, or for any other information please contact us at
Errata@PEessentialguides.com

Or

Info@PEessentialguides.com

The Essential Guide to Passing the California Seismic Principles Exam
A Reference Manual with 160 Key Questions

First Edition Print 1.1

For written permissions contact: Permissions@PEessentialguides.com

For general inquiries contact: Info@PEessentialguides.com

Imprint name: PE Essential Guides

Company owning this imprint: Petro Publications LLC. Established in Florida, 2023. Company website accessed at https://peessentialguides.com/

ISBN: 979-8-9921358-0-0

Disclaimer

The information provided in this book is intended solely for educational and illustrative purposes. It is important to note that the technical information, examples, and illustrations presented in this book should not be directly copied or replicated in real engineering reports or any official documentation.

While there may be resemblances between the examples in this book and real structures, users must exercise caution and conduct comprehensive verification of all information before implementing it in any practical setting. The author and all affiliated parties explicitly disclaim any responsibility or liability arising from the misuse, misinterpretation, or misapplication of the information contained in this book.

Furthermore, it is essential to understand that this book does not constitute legal advice, nor can it be considered as evidence or exhibit in any court of law. It is not intended to replace professional judgment, and readers are encouraged to consult qualified experts or seek legal counsel for any specific legal or technical matters.

By accessing and utilizing the information in this book, readers acknowledge that they do so at their own risk and agree to hold the author and all affiliated parties harmless from any claims, damages, or losses resulting from the use or reliance upon the information provided herein.

This page is intentionally left blank

About the Author

Dr. Petro is a professional engineer and a business leader with over 20 years of experience in leading and growing engineering companies. Throughout his career, he worked with some of the most prestigious engineering firms. With a vast background in design and construction, he earned a reputation for delivering innovative and cutting-edge projects throughout his career. Dr. Petro is also the author of the series of the PE Essential Guides, which is a best seller series that coverages all civil engineering disciplines.

Dr. Petro is a civil engineer, he holds a Doctorate degree, he has earned a Professional Engineering (PE) license as well as Chartered Engineer (CEng) certification. Additionally, Dr. Petro has earned a Project Management Professional (PMP) certification, further demonstrating his expertise in managing complex projects. Over the years, he successfully led and managed teams of engineers, designers, and other professionals, overseeing complex projects from conception to completion.

Throughout his career, Dr. Petro designed and delivered numerous innovative and interesting projects that have contributed significantly to various industries he has worked in. His passion for engineering and business has driven him to publish several papers and articles in industry-leading journals and magazines. His work has been recognized as state-of-the-art and has been referenced by many industry professionals.

As an international civil engineer who worked across the globe, Dr. Petro brings an interesting perspective to the table. He has a deep understanding of how civil facilities and structures work and how to optimize them for maximum efficiency and safety. His ability to communicate complex engineering concepts to both technical and non-technical stakeholders has been key to his success.

This page is intentionally left blank

Important Information about this Book

General Information about the Book

This book is designed to help civil engineers who have passed the NCEES exam prepare for and pass the California Seismic Principles Exam, a prerequisite for obtaining the Professional Engineering (PE) license in California. The Seismic Principles Exam is a 2.5-hour test consisting of 55 specific questions. Along with the Engineering Surveying Exam, these two exams are important prerequisites for obtaining the PE license in California.

I understand that not all civil engineers aiming to obtain their engineering license in California have the same background. For instance, not all of them come from a structural engineering background and may lack the required seismic theory understanding. If you are a water resources engineer, a geotechnical engineer, or a traffic engineer, or even a structural engineer, you may not necessarily have the required information on Seismic Design Categories, risk categories, or how to distribute loads to diaphragms whether vertically or horizontally. You might also be unfamiliar with concepts like base shear, or the required structural details for wood, concrete, or steel, and many other topics. Therefore, I found it unfair to author practice questions without first providing a full background on the required topics on seismic design.

This book with its theory explanation is designed to be neither overly complex, which might be overwhelming, nor too simplistic, which might not achieve the intended purpose. Based on this, I decided to structure the book in a way that guides any engineer, regardless of their discipline or background, through the journey of learning the science of earthquakes. This approach is tailored to the Seismic Principles test plan as outlined on the Board for Professional Engineers, Land Surveyors, and Geologists (BPELSG) website.

Book Structure

This book is structured into seven chapters, each designed to align with the test plan from the Board for Professional Engineers, Land Surveyors, and Geologists (BPELSG) website. Each chapter begins with a theoretical explanation, providing the necessary equations, figures, and references to ensure a comprehensive understanding of the material in an easy-to-follow format. At the end of each chapter, you will find practice questions that mimic the exam style, along with detailed solutions to help you apply the theory you have just learned. These explanations, questions, and solutions are supplemented with detailed diagrams and equations to help you grasp the underlying principles and solve problems efficiently and accurately.

Below is a summary of the chapters and the number of questions provided per chapter:

Chapter	Title	Examples within Chapter	Questions at end of Chapter
1	Seismic Data and Seismic Design Criteria	3	10
2	Seismic Characteristics of Engineered System	4	15
3	Seismic Forces: Building Structures	9	25
4	Seismic Analysis Procedures	9	25
5	Seismic Forces: Non-Building Structures	2	25
6	Design Codes, Detailing and requirements	3	25
7	Seismic Vulnerability and Improvements	-	5

This book offers comprehensive coverage and understanding of the relevant exam topics and potential question scenarios you might encounter during the actual test. If certain topics or methods are not covered, the book's presentation style will guide you on where and how to find a certain topic or a solution to a question independently, showing you where to look and how to provide timely answers during the exam.

The questions included at the end of each chapter in this book are designed to be constructive and creative, neither too easy nor too difficult. They are crafted to help you remember the core concepts discussed in each chapter, challenging you to think critically and apply your knowledge to both exam and real-world scenarios. Some questions may require a deeper level of analysis and understanding than simple recall, often involving few steps or different perspectives.

Additionally, there are two non-technical references that are publicly available and necessary for the exam application process. These documents are not included in this book and should be reviewed at your own pace prior to applying to the exam as passing those is part of your application for this exam:

- The Professional Engineers Act (Business and Professions Code Sections 6700-6799)
- The Board Rules (Title 16, California Code of Regulations sections 400-476)

About the Exam
General information
The California Seismic Principles Exam is administered in one 2.5 hours session. The exam mostly consists of multiple-choice questions, sometimes point-and-click, drag-and-drop, or it could ask you to choose more than one option per question. The exam is designed to test not only one's knowledge and technical skills, but also their ability to think critically and work under pressure.

Dissecting the Exam
The exam at the date of publishing this book consists of 55 questions and candidates are given a total of 150 minutes to solve those questions. This means that each question is allotted an average time of less than three minutes. It is very important to keep in mind that some questions during the real exam will take only one minute to solve while others could take up to five minutes to complete. To prepare for the exam, it is crucial to practice solving as many questions as you can prior to the exam to familiarize yourself with the exam content and the locations of equations or tables needed to solve these questions.

Which References to Own
The California Seismic Principles Exam requires you to thoroughly study six references as noted below. It is also recommended to check the BPELSG website from time to time to verify updates to this list. Those references are:

- 2022 California Building Code (CBC) California Code of Regulations, Title 24, Part 2, Volumes 1 and 2.

  This publication can be accessed online through the following official website: https://codes.iccsafe.org accessed in January 2025

- American Society of Civil Engineers, ASCE/SEI 7-16 Minimum Design Loads for Buildings and Other Structures.

- AWC SDPWS 2021 Special Design Provisions for Wind and Seismic.

  This publication can be accessed online through the following official website: https://awc.org/publications/2021-sdpws accessed in January 2025

- TMS 402/602-16 Building Code Requirements and Specifications for Masonry Structures.

- ATC 20 Procedures for Postearthquake Safety Evaluation of Buildings and ATC 20-1 Field Manual, 2nd Edition, 1995.

- Design of Wood Structures ASD/LRFD 7th Edition, 2014 by Donald E. Breyer, Kenneth J. Fridley and Kelly E. Cobeen. McGraw-Hill, Inc. New.

For example, the ASCE 7 Standard – one of the main focuses of this book – contains critical information, and its presentation can be overwhelming. This book seeks to break down and explain the theory behind these details in a clear and easy-to-follow sequence. The CBC 2022 code has certain requirements and modifications to the ASCE 7 Standard. You will find some of these key requirements integrated into the chapters of this book when needed in *italics*. Same applied to TMS 402/602 and SDPWS 2021. You are required however to familiarize yourself with all the relevant chapters of the above referenced codes to be fully prepared for the exam.

This book has been tested prior to its publication in an exam setting and proven to be a valuable resource for preparation, also as a resource during the exam as it includes key summaries, bespoke diagrams, and some key ASCE 7 tables which have been copied here with permission from the American Society of Civil Engineers.

Additional Practice Material

If you are seeking additional practice materials with more examples that are similar to the actual exam, reach out to us at info@PEessentialguides.com. We always seek to prepare extra workbooks and practice content that may not be available through traditional retail channels.

Sticky Notes or Tabbing Advise for Quick Referencing

To help you save time during both your study sessions and the exam, I recommend marking key pages of this book with tabs or sticky notes. Focus on pages that feature important tables from specific standards or summaries I have carefully created for you. These pages will be easy to identify as you go through the guide. I have worked hard to condense the most crucial points from entire chapters of codes and standards, and you will find these summaries throughout the book. Additionally, be sure to highlight process diagrams, as they provide a clear, step-by-step walkthrough of complex design processes.

However, be careful not to overdo it, as too many marks can lead to confusion or hinder the maneuverability of the book during the exam. I recommend using no more than ten tabs: five at the top right edge and the other five at the very top edge of the book. This will leave the bottom right edge free, making it easier for you to navigate and turn pages smoothly during the exam.

Finally, to maximize your chances of success on the exam, I recommend reviewing this book and the relevant material several times before your exam date.

Tips for Passing the Exam

This exam is neither too hard nor too easy. Some questions can be answered in 30 seconds, while others may require more time-consuming calculations that could take up to 5 minutes, though these should be relatively few. Remember that every exam is different, so be prepared for anything.

The exam can be tricky, so make sure to review the "Common Mistakes" section at the end of this book and add to it as you go along. One thing you will notice is the importance of becoming familiar with the SFRS tables included in the book. Take the time to study these tables and the notes beneath them. Read each line carefully and ensure you know where to find key details like the pump seismic coefficient, a steel/concrete composite eccentric system, or coefficients for partitions, which you might not expect to be tested on.

As mentioned earlier, this book has been used in actual exam setting prior to publication and has proven highly effective as a reference guide during the exam. However, be aware that you may encounter only few questions that require you to reference the CBC (e.g., shear capacities for stabled walls or diaphragms and others), the SDPWS, or other references. The book will guide you on how to navigate the online service provided by the CBC or SDPWS. For a small fee for the CBC code, you can print the tables or pages you plan to bring with you during the exam, as long as they are ring-bound.

You will also encounter questions that test your knowledge directly – either you know the answer, or you don't. This book helps by explaining how seismic systems work so that when you face these types of questions, you can rely on your understanding to answer them correctly.

Many people describe this exam as a "speed test," but this is only the case if you do not know where to find the necessary equations or tables, and if you did not prepare well or practiced well enough prior to the test. If you are well-prepared, know the material thoroughly, and are familiar with this book and any other references you may need, you will be able to finish ahead of time leaving you with extra minutes to review any questions you flagged or marked for review.

Acknowledgement and Dedication

I would like to thank the readers of this book, who I hope will find it informative, engaging, and thought-provoking. It is my sincere hope that this book will inspire others to pursue their own passion, and that it will serve as a valuable resource for all those interested in the field of engineering and seismology.

Additionally, I would like to thank the engineers and readers of the original series (Structural, Construction, Geotechnical, WRE, and Transportation disciplines) who trusted my work and the new method of delivery. Their support led to the series being categorized as a bestseller, which encouraged me to continue this journey and author more technical books with high-quality content, aiming to reach a broader audience of engineers and engineering students.

CONTENT

Table of Contents

List of Figures

List of Tables

List of Examples

List of Problems

This page is intentionally left blank

CHAPTER
1

Seismic Data and Seismic Design Criteria

1.1. Purpose of this Chapter

This chapter offers you a brief introduction to the causes of earthquakes, the seismology specific to California, and the methods used to measure seismic activity. It delves into the seismic hazards by utilizing geotechnical data, which is important for influencing design considerations. The chapter explains how this data can be employed to classify sites for design purposes, ensuring safety and compliance with relevant standards.

Additionally, the chapter discusses site-related coefficients and the applicable codes and manuals that guide the design process. It defines Seismic Design Categories (SDCs), which are built on the varying levels of seismic risks associated with different building types, and they help in deciding the methods that shall be used for designing and detailing structures. Furthermore, it outlines building risk categories and their practical applications, providing a clear framework for designing structures that can withstand seismic events.

By integrating this information, the chapter provides you with sufficient knowledge that is required to assess and mitigate seismic risks effectively, ensuring that buildings are designed to be resilient in the face of earthquakes.

1.2. Causes of Earthquakes and the San Andreas Fault

While human activities like using explosives or causing buildings to collapse can sometimes cause earthquakes, this book focuses on those that occur naturally due to the movement of earth's tectonic plates.

The earth is made up of several large plates, known as tectonic plates (see Figure 1-1), which float on a sea of molten lava. As these plates try to move relative to each other, their movement is often hindered by friction at their boundaries, known as faults. For example, the Pacific Plate and the North American Plate interact at the San Andreas Fault in California (see Figure 1-2).

When the movement of these plates is restricted by friction, energy builds up in the plates and at the fault, much like compressing a spring stores kinetic energy. Eventually, the fault gives way, releasing this energy in the form of elastic rebounds. This release generates seismic waves that travel in all directions. Depending on their characteristics, these waves can be classified as either surface waves or body waves.

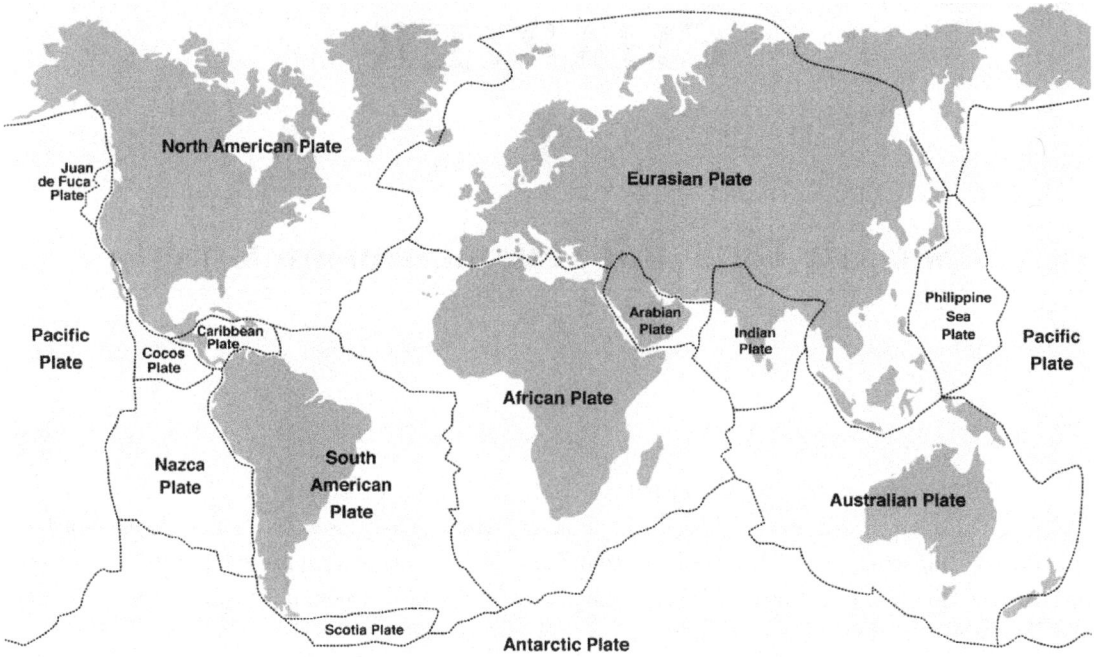

Figure 1-1 Earth Tectonic Plates

Figure 1-2 San Andreas Fault in California

The San Andreas Fault marks the boundary between the Pacific Plate and the North American Plate as identified in Figure 1.1. It spans approximately 1,200 kilometers from Cape Mendocino to the Mexican border. The fault line is visible on the Earth's surface, showcasing the plates' movements through features such as scarps and pressure ridges (See Figure 1-3).

Figure 1-3 Aerial Photo for San Andreas Fault

The plates around the San Andreas Fault move past each other at an average rate of a few inches per year. However, this movement is not smooth; it occurs in sudden bursts when the built-up strain is released, leading to earthquakes. The San Andreas Fault is part of a complex network of faults that makes California one of the most seismically active regions in the world.

Due to the importance of these active fault zones, the Alquist-Priolo Earthquake Fault Zones were established in California. These regulatory zones surround the surface traces of active faults. In areas where an active fault exists, especially if it has the potential for surface rupture, no structures for human occupancy are allowed directly over the fault. Additionally, there must be a minimum distance – typically 50 feet – between the nearest point of the building to the closest edge of the Alquist-Priolo fault zone.

Given California's seismic characteristics, which are deeply influenced by its historical faults and significant past earthquakes, understanding seismic design is essential. As an aspiring professional engineer in California, it is crucial that you understand seismic risks, safety measures, and resilience needed when designing or supervising the construction of any structure in the state. This knowledge ensures that you can effectively address potential seismic events and contribute to the safety and durability of buildings or any other type of structure in California.

1.3. Earthquake Measurements

Earthquakes can be measured using two main criteria: *intensity* and *magnitude*. Intensity measures the degree of destruction caused by an earthquake and varies inversely with the distance from the epicenter – The epicenter is the point on the Earth's surface directly above the earthquake's origin, known as the hypocenter or focus. Intensity is measured using the Modified Mercalli Intensity (MMI) Scale, which ranges from I (least felt) to XII (most destructive). There are correlations between the Mercalli scale and ground acceleration that can be used for design purposes. This subject is not covered in this book.

Earthquake magnitude refers to the size or amount of energy released by an earthquake. It is a single value that does not depend on the distance from the epicenter. Historically, earthquake magnitude was measured using the Richter scale, which is now rarely used except for small earthquakes. Today, the more accurate Moment Magnitude Scale is used. This scale is based on the physical properties of the earthquake, derived from an analysis of all the seismic waves recorded during an earth-shaking event. Another way to measure the size of an earthquake is by calculating the energy it releases, which indicates the potential damage to structures and buildings.

1.4. Earthquake Propagation Medium and Site Classification
1.4.1.   Shear Wave Velocity

This section explains how earthquakes travel through different types of ground or soil, which is essential for site classification. Understanding this process is crucial for predicting the impact of earthquakes on buildings and structures. Knowing the type of medium through which an earthquake moves helps determine various seismic site and design coefficients. It also aids in establishing the Seismic Design Category (SDC) for structures, calculating base shear, and selecting appropriate design methods for analyzing or designing structures at those sites.

In seismic design, it is important to understand the speed at which earthquake waves move through different mediums, determined by measuring shear wave velocity. Shear wave velocity indicates the speed at which shear waves – waves that cause soil particles to move perpendicular to the direction of earthquake wave propagation – travel through the ground. This velocity is influenced by soil density and stiffness, making it a key factor in determining soil site classification for seismic design.

Several methods can be used to determine shear wave velocity through geophysical tests. One such method is the Spectral Analysis of Surface Waves (SASW), which involves applying impact loads to the ground surface while measuring wave propagation using geophones placed at specific locations. Another method, considered the most accurate, is the cross-borehole method (Figure 1-4). This technique uses two boreholes: a wave source "S" is inserted into one hole, and a receiver "R" is inserted into the other. The speed of wave propagation is measured while knowing the distance between the two boreholes.

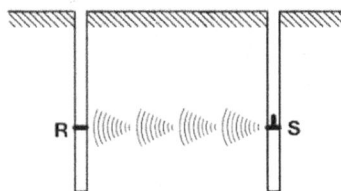

Figure 1-4 Cross Hole Method for Determining the Shear Wave Velocity

Sites are classified using Table 1-1, which is reproduced here with permission from ASCE/SEI 7-16, Chapter 20, page 204. The soil tests mentioned above are typically conducted for the upper 100 ft only. This section in this book does not intend to summarize this chapter or any other from the ASCE 7 Standard; rather, it is recommended that you familiarize yourself with the ASCE 7 Standard at your own pace.

Table 1-1 Site Classification

Site Class	Shear wave velocity $\bar{v}_s$	Average standard penetration resistance $\bar{N}$ or $\bar{N}_{ch}$	Average undrained shear strength $\bar{s}_u$
A. Hard Rock	> 5,000 ft/s	NA	NA
B. Rock	2,500 to 5,000 ft/s	NA	NA
C. Very dense soil and soft rock	1,200 to 2,500 ft/s	> 50 $blows/ft$	> 2,000 lb/ft^2
D. Stiff soil	600 to 1,200 ft/s	15 to 50 $blows/ft$	1,000 to 2,000 lb/ft^2
E. Soft clay soil	< 600 ft/s	< 15 $blows/ft$	< 1,000 lb/ft^2
	Also soil with more than 10 ft with Plasticity Index PI > 20, Moisture content $w \geq 40\%$ and undrained shear $\bar{s}_u$ < 500 lb/ft^2		
F. Soil requiring site response analysis using ASCE Section 21.1	See description below		

Important information about this table and site classification:

- Classes A and B shall not be assigned if there is more than 10 ft of soil between bedrock and the bottom of a spread footing or a mat foundation.

- The following equations, which are provided in Chapter 20 of the ASCE/SEI 7-16 shall be used to calculate the average shear wave velocity ($\bar{v}_s$), or the average standard penetration resistance ($\bar{N}$) where (N_i) is the number of blows with a maximum of 100 blows, [$\bar{N}_{ch}$ is used for cohesionless soil layers], or undrained shear strength ($\bar{s}_u$). Each soil layer is assumed to have a thickness of (d_i):

$$\bar{N} = \frac{\sum_{i=1}^{n} d_i}{\sum_{i=1}^{n} \frac{d_i}{N_i}}$$

$$\bar{N}_{ch} = \frac{\sum_{i=1}^{m} d_i}{\sum_{i=1}^{m} \frac{d_i}{N_i}} \quad \ldots\ldots\ldots \ (m) \ \textit{here is the number of cohesionless layers}$$

$$\bar{v}_s = \frac{\sum_{i=1}^{n} d_i}{\sum_{i=1}^{n} \frac{d_i}{v_{si}}}$$

$$\bar{s}_u = \frac{\sum_{i=1}^{k} d_i}{\sum_{i=1}^{k} \frac{d_i}{s_{ui}}} \quad \ldots\ldots\ldots \ (k) \ \textit{here is the number of cohesive layers}$$

- Class F is assigned in the following cases – a site response analysis as indicated in Table 1-1 shall be performed as well unless there is an exception:

  ➢ Liquifiable soils, quick and highly sensitive clays and collapsible weak cemented soils where the structure has a fundamental period[1] > 0.5 *sec*.

  ➢ Peats or highly organic clays with layer thickness > 10 *ft*.

  ➢ Very high plasticity clays with thickness > 25 *ft* and Plasticity Index *PI* > 75 in a soil profile that should have been classified as D or E:

    ▪ An exception to this, values for F_a and F_v[2] obtained for Site Class D or E can be multiplied by a sliding scale factor that varies from 1.0 (for *PI* = 75) to 1.3 (for *PI* = 125) provided that the resulting values for S_{DS} and S_{D1}[3] do not exceed the upper value for Design Category B (i.e., 0.33 for S_{DS} and 0.133 for S_{D1}[4]).

  ➢ Very thick – soft or medium – stiff clay > 120 *ft* with s_u < 1,000 *psf*.

    ▪ An exception to this, F_a and F_v[3] can be calculated treating the class as E while ensuring that the resulting values for S_{DS} and S_{D1}[4] do not exceed the upper value for Design Category B as explained in the previous condition.

- Class E is assigned when conditions do not satisfy Class F and there is a total thickness of soft clay > 10 *ft* and s_u < 500 *psf* with moisture content ≥ 40% and *PI* > 20.

- Class D shall be used when soil properties are not sufficiently known. If this was the case, value for F_a and F_v per Table 1-3 and 1-4 here should be > 1.2.

Example 1.1 – Average Shear Wave Velocity and Site Class

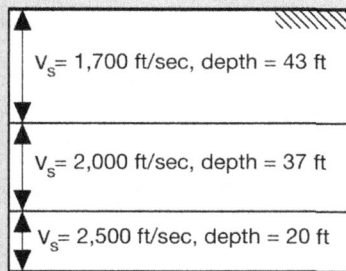

V_s= 1,700 ft/sec, depth = 43 ft

V_s= 2,000 ft/sec, depth = 37 ft

V_s= 2,500 ft/sec, depth = 20 ft

The average shear wave velocity for the above soil layers is calculated using the previously discussed equations as follows:

$$\bar{v}_s = \frac{\sum_{i=1}^{n} d_i}{\sum_{i=1}^{n} \frac{d_i}{v_{si}}} = \frac{43 + 37 + 20}{\frac{43}{1,700} + \frac{37}{2,000} + \frac{20}{2,500}} \cong 1,931 \text{ ft/sec}$$

Based on this value, the site classification is C – very dense soil and soft rock

---

[1] Buildings or structures fundamental period is defined in footnote No. 5 of section 1.5 here (page 10).
[2] Values for F_a and F_v are key seismic design coefficients discussed in Section 1.5.2 here.
[3] S_{DS} and S_{D1} are explained in more details in Section 1.5.2 here.
[4] Upper bound value can be found in Table 11.6-1 and 11.6-2 in ASCE 7 Standard, and this table is explained in Section 1.7 Seismic Design Categories here.

1.4.2. Standard Penetration Test

The Standard Penetration Test (SPT) is a widely used in-situ geotechnical test conducted during drilling operations. A $140\ lb$ ($63.8\ kg$) hammer is used to drive a split-spoon sampler into the soil. The sampler is hammered down in increments of $6\ inches$ ($15\ cm$) over three attempts, with each attempt measuring the number of blows required to penetrate the soil by $6\ inches$. Consequently, in any borehole log where SPT has been performed, you will find three numbers representing the blow counts from these three attempts. The blow count for the initial $6\ inches$ is typically disregarded, as the soil may be disturbed from the boring operation; only the penetration counts for the second and third increments (which totals to $1\ ft$) are considered for analysis.

The Standard Penetration Test (SPT) provides several benefits in geotechnical investigations. One significant advantage is its ability to quickly assess soil properties, such as density and strength, which are essential for understanding subsurface conditions. SPT results can also be correlated to determine site classification as indicated in Table 1-1, or liquefaction potential as explained in Section 1.4.3 and shown in Figure 1-5.

1.4.3. Liquefaction Potential

Liquefaction occurs when saturated soil loses its strength and stiffness in response to applied stress, often during an earthquake or other ground shaking events. In such cases, the soil behaves like liquid. This happens in loose, water-saturated, unconsolidated sediments. Liquefiable soils present a significant hazard to transportation infrastructure during earthquakes. It occurs as the seismic shaking generates pore water pressure, reducing the effective stress in the ground to the point that they can no longer hold in place.

The US Department of Transport and the Federal Highway Administration in their publications FHWA NHI-16-072 provided the following conditions for liquefaction to occur:

1. The soil is saturated and below the groundwater level.
2. The soil is predominantly coarse-grained (typically less than about 20% fines).
3. The soil is loose/cohesionless (relative density less than about 40%).
4. The ground shaking is significantly strong.

Another FHWA publication "NHI-11-032" defines highly sensitive soils which are susceptible to liquefaction as those with Liquid limit < 0.4, Liquidity index > 0.6, and $(N_1)_{60}$ < 5, where $(N_1)_{60}$ is the adjusted blows for overburden depth.

During an earthquake, the shaking increases pore water pressure within the soil, reducing the friction between soil particles. As a result, the soil can no longer support loads, leading to issues such as sinking, lateral spreading, or even the failure of foundations. Liquefaction poses significant risks to structures and infrastructure built on susceptible soils, making it a critical consideration in geotechnical engineering and seismic design.

Liquefaction potential is typically assessed using charts, such as the one presented in Figure 1-5. This chart illustrates the relationship between the Cyclic Stress Ratio (CSR) versus corrected SPT in situ test measurements $(N_1)_{60}$. The curves on the diagram represent empirical boundaries that differentiate conditions likely to produce liquefaction from those where liquefaction is unlikely to occur.

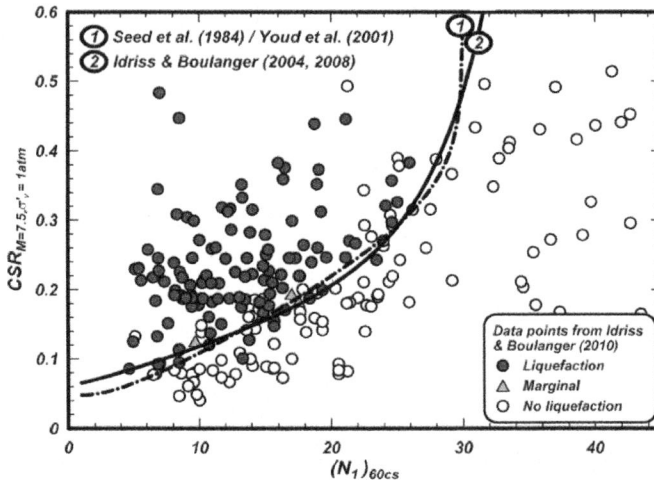

Figure 1-5 Liquefaction Potential Chart for SPT Measurements, Idriss & Boulanger (2008)

The Cyclic Stress Ratio (CSR) represents the ratio of the cyclic shear stress (τ_{ave}) induced by seismic loading (i.e., the shear stress that causes liquefaction), to the effective vertical stress (σ'_{vo}) in the soil. See below:

$$CSR = \frac{\tau_{ave}}{\sigma'_{vo}}$$

$$\tau_{ave} = CSR \times \sigma'_{vo}$$

The above value indicates that any shear stress caused by seismic excitement exceeding (τ_{ave}) can cause liquefaction. Based on this information, a Factor of Safety for liquefaction can be calculated and further used to improve the design of foundations and structures.

Example 1.2 – Liquefaction Factor of Safety

A 20 ft deep clay layer with density of 110 pcf, a Cyclic Stress Ratio of CSR = 0.3, and an expected lateral shear stress due to seismic activity of 350 psf can be assessed for liquefaction potential at the bottom of the clay layer as follows:

First calculate the effective stress at the bottom of the layer by calculating the vertical stress and deducting the pore water pressure 62.4 pcf from it as follows:

$$\sigma'_{vo} = (110 - 62.4)\,pcf \times 20\,ft = 952\,psf$$

$$\tau_{ave} = CSR \times \sigma'_{vo} = 0.3 \times 952\,psf = 285.6\,psf$$

The Factor of Safety can therefore be calculated as follows:

$$FS = \frac{285.6}{350} = 0.82$$

This indicates that this soil is susceptible to liquefaction at the given shear stress with a Factor of Safety < 1.0.

The Cone Penetration Test (CPT) can also be used to assess soils for liquefaction potential by measuring cone resistance and pore water pressure as the cone is pushed into the ground.

1.4.4.   Slope Seismic Stability Analysis

Earth lateral pressures acting on retaining walls or slopes should be adjusted to account for seismic activity. Slopes may fail due to excessive lateral earth pressure, typically resulting in a circular failure mechanism, as illustrated in Figure 1-6. This lateral pressure is calculated by multiplying the weight of the potential failing wedge (W_{wedge}) multiplied by the seismic coefficient (K_S) as shown in the figure using a method known as Pseudostatic analysis. This method uses static loads to approximate the effects of dynamic seismic loads. Although seismic vertical loads can also be calculated, they are typically ignored due to their minor impact on slope failures.

When evaluating slope failures, it is essential to assess all potential circular and sliding wedges, which can be several hundreds. In such cases, computer programs are often used to perform this analysis. For each evaluated circular slope failure, a Capacity to Demand (C/D) ratio is calculated by determining the seismic loads or moments and their corresponding resisting movements, similar to a retaining wall overturning assessment. Overturning and resisting moments are calculated around the centroid of the failing circular surface as shown in Figure 1.6 – this figure depicts only one of the failure surfaces.

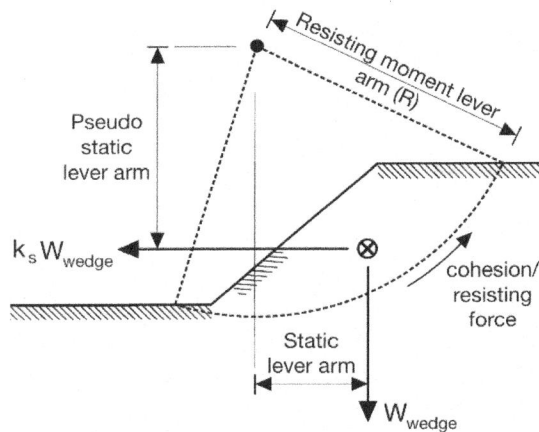

Figure 1-6 Slope Seismic Failure Analysis

Pseudostatic analysis is a widely used method for assessing the stability of slopes, retaining walls, and other geotechnical structures during seismic events. It simplifies the analysis by assuming that ground motion occurs instantaneously (statically) rather than over time, disregarding dynamic effects such as soil inertia.

The Pseudostatic analysis and the calculation of Pseudostatic coefficients are not included in the exam and are outside the scope of this book. It is sufficient to understand the concepts of slope failures and the associated resisting mechanisms. However, if you wish to learn more about these calculations, a valuable resource is the FHWA publication NHI-11-032, "LRFD Seismic Analysis and Design of Transportation Geotechnical Features and Structural Foundations Reference Manual," specifically Section 6.2.2 on Limit Equilibrium Pseudostatic Stability Analysis. This resource is publicly available.

1.5. Response Spectrum

1.5.1.   Understanding the Response Spectrum

You may have come across the term "response spectrum." It is a fundamental concept in seismic design. The site coefficients, which will be discussed later, are factors derived for different site conditions (classes A, B, C, D, E, and F, as mentioned earlier). These coefficients are applied to the response spectra to tailor them to the specific site or building in question. Buildings are identified by their fundamental periods (T)[5]. It is important to note that the response spectrum can be used for the design of both regular and irregular simple buildings. Nonelastic models should be developed as part of the design process of either complex or irregular complex structures in high-risk areas.

Explaining the response spectrum and its derivation can be complex, this book will simplify the concepts as much as possible. While you may not need this information for the exam, understanding how the response spectrum is constructed and what it signifies will enhance your understanding of seismic analysis and design throughout this book.

We will begin by discussing how earth shaking is measured. Ground shaking is recorded using an accelerogram, which captures ground acceleration – also the rate of ground displacement during shaking. Figure 1-7 shows an example of an accelerogram record from a random earthquake.

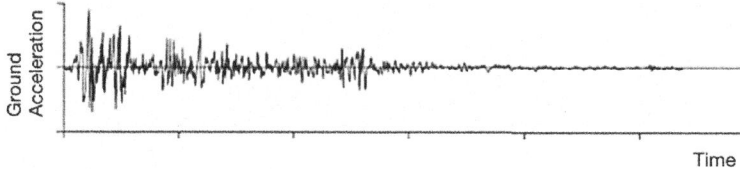

Figure 1-7 Random Earthquake Record

Now, imagine a simple one-story building placed on this shaking ground (see Figure 1.8). If we assume the building's columns are infinitely rigid, its fundamental period (T) would be zero in this case. This means the building top slab would experience the same ground acceleration during an earthquake. In contrast, if we test a less rigid building with a fundamental period greater than zero $(T > 0)$, the acceleration experienced at the top slab of the building (i.e., the diaphragm) would be lower.

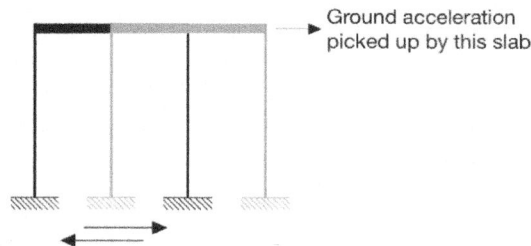

Figure 1-8 Single Story Building Subjected to Earth Shaking

---

[5] The fundamental period of a building is the time it takes for the building to sway back and forth in response to seismic excitation. A fundamental period of 2 seconds for example means that it takes the building 2 seconds to complete one full cycle of swaying back and forth in response. Buildings with a longer fundamental period are typically taller or more flexible, which can affect how they respond to ground motion. Refer to the note section note No. 7 to understand the relationship between fundamental period (T) and frequency (f).

This reduced acceleration at the slab level is known as the Spectral Response Acceleration (S_a), typically measured as a percentage of ground acceleration (g), as shown in Figure 1-9 where the x-axis represents the building or the structure fundamental period. For design purposes, numerous experiments are conducted using various fundamental periods and damping[6] levels (see how this is depicted in Figure 1-9). The results are then combined and smoothed to create the shape shown in Figure 1-10.

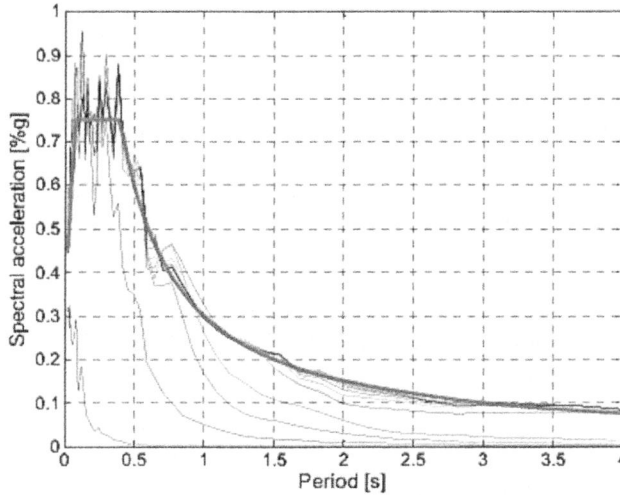

Figure 1-9 Building a Spectral Response Spectrum, Suarez & Montejo (2005)

Remember that force equals mass times acceleration. Therefore, to understand the lateral forces generated by the shaking of the slab or the diaphragm, we multiply the mass of the slab by the acceleration it experiences in a similar fashion to the mass of a wedge (W_{wedge}) multiplied by the seismic coefficient (K_S) as discussed in Section 1.4.4 and Figure 1-6 for a failing slope. It is not as simple as this description, therefore more details to follow.

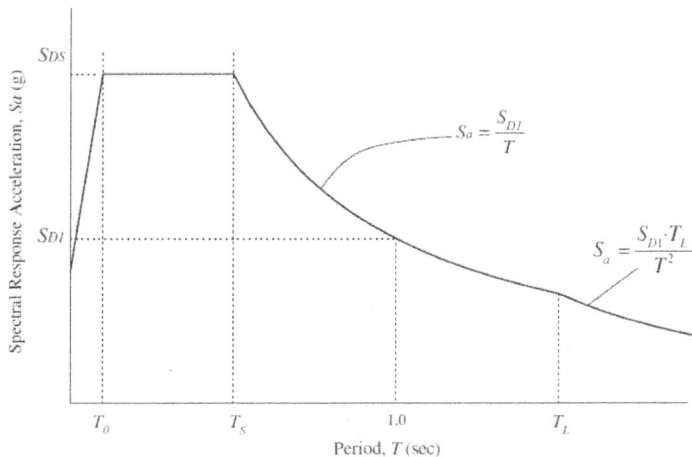

Figure 1-10 Design Response Spectrum copied from ASCE with Permission

---

[6] To keep the description of the response spectrum simple, damping is not discussed here. For more information on damping, refer to Note No. 6 in the notes section. Generally, damping refers to the mechanism for dissipating energy. By reducing the amplitude of vibrations, damping helps prevent excessive swaying or oscillation, contributing to the building's overall stability.

1.5.2.   Building a Response Spectrum

In the previous section, you learned about response spectra. A response spectrum shows the maximum acceleration that can affect a specific building or a structure at a certain location, considering its fundamental period. It gives the spectral acceleration, which indicates the maximum acceleration a structure may experience during an earthquake, based on its natural frequency.

As shown previously in Figure 1-10, there are several seismic coefficients you need to understand in order to build a response spectrum and perform seismic analysis and design. The main ones are the **design spectral response accelerations** at short periods – i.e., at 0.2 sec, labeled as (S_{DS}), and the at one second, labeled as (S_{D1}). Later in this book, you will see how these coefficients will be put to use to conduct seismic analysis and design. In this section, we will discuss how to calculate those and how to build a response spectrum accordingly.

To understand how to derive (S_{DS}) and (S_{D1}), we first need to find the maximum spectral response acceleration parameters at short periods (S_S) and at a period of one second (S_1) for the Maximum Considered Earthquake (MCE). These parameters are specific to geographic locations. You can find these values in the Seismic Hazard Maps provided in Chapter 22 of the ASCE 7 standard. Additionally, there is an online tool from ASCE (see Reference [2] at chapter end) that allows you to pinpoint the exact location, select the relevant seismic risk category[7] and soil site class, and obtain the required parameters. This tool makes the process automated and more precise when it comes to location.

Figure 1-11 provides a screenshot of the online ASCE Hazard Tool, where Los Angeles Airport was selected for evaluation, along with Risk Category IV[7] and Site Soil Class C. At this point, we are focusing on how to derive (S_S) and (S_1), which do not require site class or risk category at this stage.

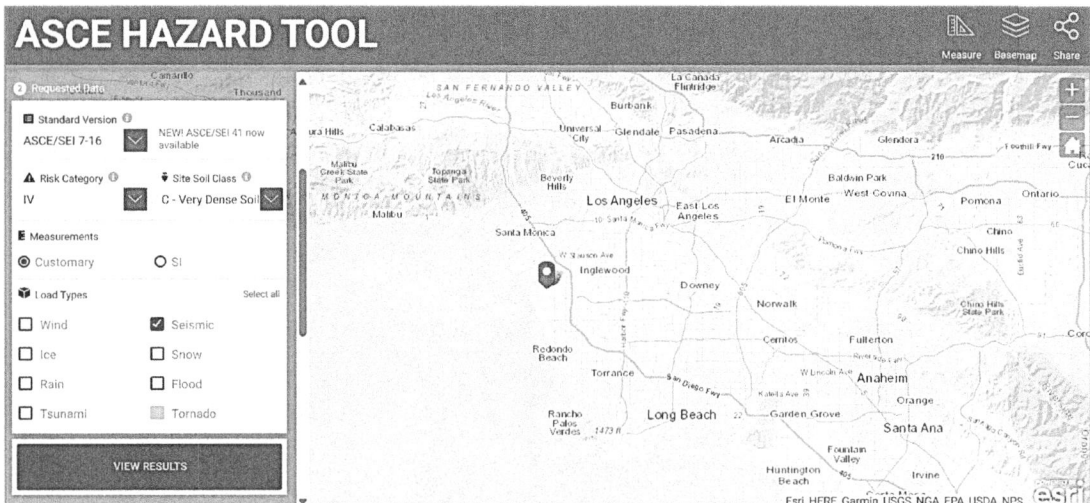

Figure 1-11 ASCE Hazard Tool Screen Shot with Los Angeles Airport Pinpointed

Figure 1-12 presents the results of the Los Angeles Airport assessment in a table format copied from the ASCE Hazard Tool. You will see that the short-term spectral acceleration (S_S) is 1.832 and the one-second spectral acceleration (S_1) is 0.644 (both expressed as portions of ground acceleration, i.e., 1.832g and 0.644g respectively). These values will later be used to

---

[7] Risk categories are explained in Section 1.6 here and are listed in Table 1.5-1 of the ASCE 7 Standard.

calculate other seismic design coefficients. While the Hazard Tool provides these parameters, you will learn how to derive them independently and create the relevant response spectrum. In the meantime, you are encouraged to attempt to run the same analysis for Los Angeles Airport using the charts provided in Chapter 22 of the ASCE 7 standard.

REPORT SUMMARY	✕
Seismic	
S_S	1.832
S_1	0.644
F_a	1.2
F_v	1.4
S_{MS}	2.198
S_{M1}	0.902
S_{DS}	1.466
S_{D1}	0.601
T_L	8
PGA	0.786
PGA_M	0.943
F_{PGA}	1.2
I_e	1.5
C_v	1.266
Seismic Design Category	D

Figure 1-12 ASCE Hazard Tool Summary Results for the Los Angeles Airport

So far, we have defined and explained how to determine the spectral response parameters for the Risk-Targeted Maximum Considered Earthquake (MCE_R) at short periods (S_S) and at one second (S_1). Next, these values need to be adjusted to account for site conditions (i.e., Site Class A, B, C, D, E, or F). This adjustment involves converting them to (S_{MS}) and (S_{M1}) by multiplying (S_S) and (S_1) with the appropriate site coefficients (F_a and F_v), as described in equations 11.4-1 and 11.4-2 of the ASCE 7 Standard as follows:

$$S_{MS} = F_a S_S$$

$$S_{M1} = F_v S_1$$

The short-period site coefficient (F_a) and long-period site coefficient (F_v) depend on the site class discussed in Section 1.4.1 and the spectral acceleration parameters (S_S and S_1) outlined earlier. You can find the values for (F_a) in Table 1-2 here and for (F_v) in Table 1-3 here. These tables are located in the ASCE 7 Standard Tables 11.4-1 and 11.4-2 respectively and are copied here with ASCE permission. Similar tables are also available in other FHWA publications (see reference [4] in the reference section in this chapter if interested).

Table 1-2 Short Period Site Coefficient (F_a)

Site Class	$S_S \leq 0.25$	$S_S = 0.5$	$S_S = 0.75$	$S_S = 1.0$	$S_S = 1.25$	$S_S \geq 1.5$
A	0.8	0.8	0.8	0.8	0.8	0.8
B[b]	0.9	0.9	0.9	0.9	0.9	0.9
C	1.3	1.3	1.2	1.2	1.2	1.2
D	1.6	1.4	1.2	1.1	1.0	1.0
E	2.4	1.7	1.3	1.2^a from CBC[c]	1.2^a from CBC[c]	1.2^a from CBC[c]
F	a	a	a	a	a	a

a: Site-specific geotechnical investigations and dynamic site response analysis is recommended to evaluate the influence of local soil conditions on the site per ASCE Chapter 21. See ASCE 7 Standard Section 11.4.8 for exceptions.
b: If determined per ASCE Chapter 20 revealing rock conditions with no site-specific velocity measurements, F_a = 1.0.
c: Values collected from CBC 2022 Section 1613 specifically for OSHPD (abbrev. in Chapter 6 Appendix).

Table 1-3 Long Period Site Coefficient (F_v)

Site Class	$S_1 \leq 0.1$	$S_1 = 0.2$	$S_1 = 0.3$	$S_1 = 0.4$	$S_1 = 0.5$	$S_1 \geq 0.6$
A	0.8	0.8	0.8	0.8	0.8	0.8
B[b]	0.8	0.8	0.8	0.8	0.8	0.8
C	1.5	1.5	1.5	1.5	1.5	1.4
D	2.4	2.2^a	2.0^a	1.9^a	1.8^a	1.7^a
E	4.2	3.3^a From CBC[c]	2.8^a from CBC[c]	2.4^a from CBC[c]	2.2^a from CBC[c]	2.0^a from CBC[c]
F	a	a	a	a	a	a

a: Site-specific geotechnical investigations and dynamic site response analysis is recommended to evaluate the influence of local soil conditions on the site per ASCE Chapter 21. See ASCE 7 Standard Section 11.4.8 for exceptions.
b: If determined per ASCE Chapter 20 revealing rock conditions with no site-specific velocity measurements, $F_v = 1.0$.
c: Values collected from CBC 2022 Section 1613.

Finally, the design values needed for seismic analysis and design – the design earthquake spectral response acceleration parameters for short (S_{DS}) and one-second-long periods (S_{D1}) – are calculated per below. This excludes the simplified method discussed in the next section. The below equations can be found in the ASCE 7 Standard equations number 11.4-3 and 11.4-4:

$$S_{DS} = \frac{2}{3} S_{MS}$$

$$S_{D1} = \frac{2}{3} S_{M1}$$

Example 1.3 – Building a Design Response Spectrum for Los Angeles Airport

Using the example in Figure 1-11 and the results in Figure 1-12, we will apply the values of $S_S = 1.832$ and $S_1 = 0.644$, along with the definition of Site Class C, to verify the design values (S_{DS}) and (S_{D1}) shown in Figure 1-12. We will also build the response spectrum in a manner similar to that presented in Figure 1-10 for this location.

Refer to Table 1-2 here (or Table 11.4-1 in the ASCE 7 Standard) for $F_a = 1.2$ and Table 1-3 (or Table 11.4-2 in ASCE) for $F_v = 1.4$.

$$S_{DS} = \frac{2}{3} S_{MS} = \frac{2}{3} F_a S_S = \frac{2}{3} \times 1.2 \times 1.832 = 1.466 \ (\approx 1.5)$$

$$S_{D1} = \frac{2}{3} S_{M1} = \frac{2}{3} F_v S_1 = \frac{2}{3} \times 1.4 \times 0.644 \approx 0.6$$

Based on the above information, and information provided in the ASCE 7 Standard Section 11.4.6, the following table can be constructed:

T	S_a	Equation used per ASEC Section 11.4.6
0.0	0.6	$S_a = S_{DS}(0.4 + 0.6\,T/T_o) = 1.466(0.4 + 0.6 \times 0/0.08) \approx 0.6$
$T_o = 0.08$	$S_{DS} = 1.5$	$T_o = 0.2 S_{D1}/S_{DS} = 0.2 \times 0.6/1.5 = 0.08$
$T_S = 0.4$	$S_{DS} = 1.5$	$T_S = S_{D1}/S_{DS} = 0.6/1.5 = 0.4$
0.5	1.2	$S_a = S_{D1}/T = 0.6/0.5 = 1.2$
0.6	1.0	$S_a = S_{D1}/T = 0.6/0.6 = 1.0$
0.7	0.86	$S_a = S_{D1}/T = 0.6/0.7 = 0.86$
0.8	0.75	$S_a = S_{D1}/T = 0.6/0.8 = 0.75$
0.9	0.66	$S_a = S_{D1}/T = 0.6/0.9 = 0.67$
1.0	$S_{D1} = 0.6$	$S_a = S_{D1}/T = 0.6/1.0 = 0.6$
1.1	0.55	$S_a = S_{D1}/T = 0.6/1.1 = 0.55$

Based on the above table, the following is the **Design Response Spectrum** for Los Angeles Airport:

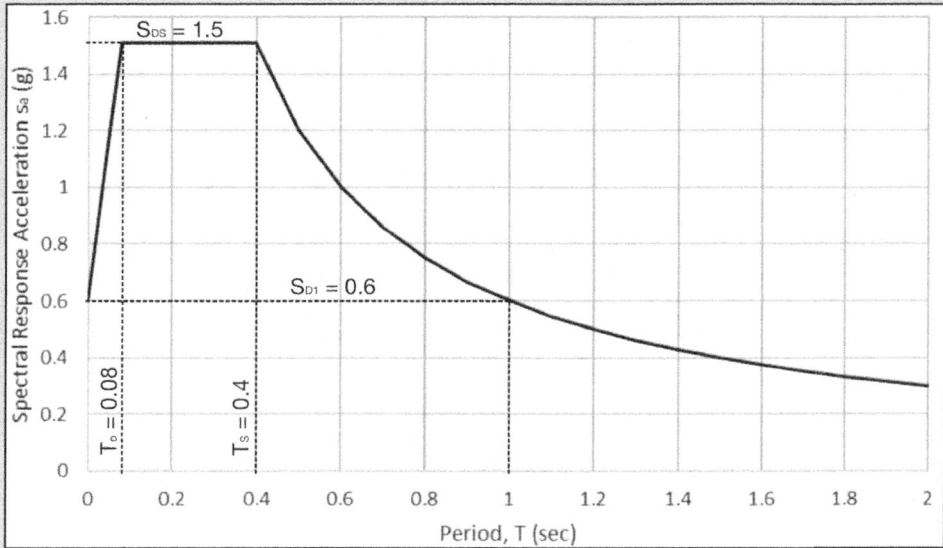

It is important to recognize that the above Spectrum is the design one which equals to two-thirds the Risk-Triggered Maximum Considered Earthquake (MCE_R) which is why equation 11.4-3 & 4 of the ASCE 7 Standard were multiplied by (2/3).

1.5.3.   The Simplified Method

The ASCE 7 Standard provides a simplified method for calculating spectral acceleration coefficients and the associated seismic loads in its Section 12.14. However, this method has certain limitations. Below are some of the key requirements for its application, but for a comprehensive understanding of all requirements and qualifications, refer to Section 12.14.1.1 of the ASCE 7 Standard.

To use the simplified method (refer to Figure 1-13 for an illustrative sketch), the structure must be classified as either Risk Category I or II (risk categories are discussed in Section 1.6 here or Section 1.5.1 of the ASCE 7 Standard). Additionally, the soil site class must not be classified as E or F. The structure should not exceed three stories and must include two lateral lines of resistance positioned on either side of the geometric centroid of the diaphragm for its two major axes. The force-resisting system must consist of either a shear wall or a building frame system. Furthermore, the eccentricity between the center of mass in the slab and its center of rigidity should not deviate more than 10% the length of the slab – Chapter 4 explains how to calculate the center of rigidity. It is also important to note that there are restrictions on overhangs for buildings utilizing this method as described in the ASCE and shown in Figure 1-13. Additionally, irregularities are not permitted when using this method.

Although this information is illustrated in Figure 1-13, you are encouraged to consult Section 12.14.1.1 of the ASCE 7 Standard for a more comprehensive understanding for this method. Also, it is important to know that forces generated from the simplified method should be applied to the two orthogonal directions.

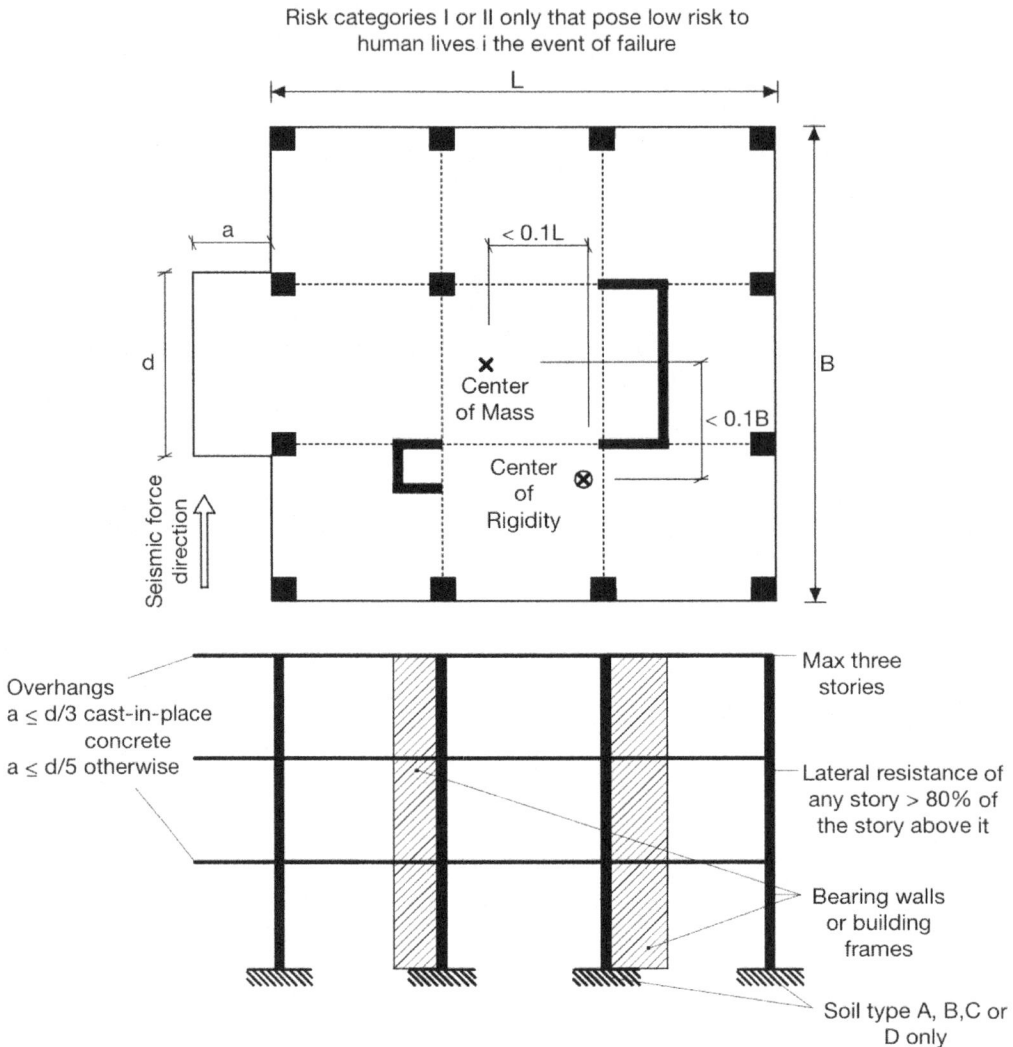

Figure 1-13 The Simplified Method Acceptable System

Provided the above conditions apply, the design spectral acceleration for short periods can be calculated using the ASCE 7 Standard Section 12.14.8.1 as follows:

$$S_{DS} = \frac{2}{3} F_a S_S$$

For this purpose, the following assumptions can be used:

- S_{D1} is not required.
- S_{DS} should be a maximum of 1.5.
- F_a can be taken as 1.0 for rock and 1.4 for soil.
- Sites can be considered as rock if the distance from the bedrock layer to the bottom of foundations is less than 10 ft.

The "Simplified Method" offers variations for applying lateral load equations, drift calculations, and other procedures. These are discussed at the end of each relevant section in this book.

1.5.4.   Peak Ground Acceleration and Spectral Acceleration

In the previous section, we briefly discussed Peak Ground Acceleration (PGA) before moving on to spectral acceleration (S_a). Spectral acceleration is widely used because it accounts for building frequencies or their fundamental periods (T). There is a distinction between short ($0.2 - second$) spectral acceleration and $one - second$ spectral acceleration, each serving different purposes. For example, short spectral acceleration is used to evaluate seismic forces in short, rigid buildings, while one-second spectral acceleration is used for taller buildings. You can see how these spectral parameters are used to derive seismic forces in Chapter 2 here or Chapter 12 of the ASCE 7 Standard.

PGA denotes the maximum ground acceleration, as shown in Figure 1-14. Although it has applications which are beyond the scope of this book or the California exam, it is important to understand that PGA is mostly used in slopes, bridges, and other non-building structures, or, as input to geotechnical investigations and dynamic site response analysis for specific site classes and risk and design categories. PGA can be obtained similarly to spectral parameters, either from the ASCE online Hazard Tool shown previously in Figure 1-11 here, or from Chapter 22, Figures 22-9 to 22-13 of the ASCE 7 Standard.

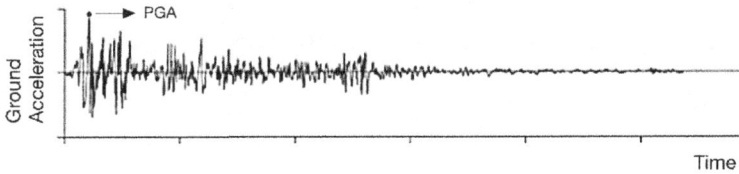

Figure 1-14 Peak Ground Acceleration

To give an example for one of its applications, recall our discussion on slope stability and Pseudostatic forces in Section 1.4.4. The pseudo-seismic force is calculated using a factor (K_S) multiplied by the weight of the wedge (W_{wedge}). The site factor, or the seismic coefficient for this wedge, taken from FHWA publication reference [4] here, is calculated as follows:

$$K_{max} = F_{PGA}PGA$$

The above is the (K_{max}), while (K_S) is derived from it using methods not covered here. This equation can also be found in Section 11.8.3 of the ASCE 7 Standard in another form which is as follows:

$$PGA_M = F_{PGA}PGA$$

In this context, (PGA_M) is the (MCE_G) peak ground acceleration adjusted for site effects, and (F_{PGA}) is the site factor, which can be found in Table 1-4 here or ASCE Table 11.8-1.

Table 1-4 Site Coefficient (F_{PGA})

Site Class	$PGA \leq 0.1$	$PGA = 0.2$	$PGA = 0.3$	$PGA = 0.4$	$PGA = 0.5$	$PGA \geq 0.6$
A	0.8	0.8	0.8	0.8	0.8	0.8
B[b]	0.9	0.9	0.9	0.9	0.9	0.9
C	1.3	1.2	1.2	1.2	1.2	1.2
D	1.6	1.4	1.3	1.2	1.1	1.1
E	2.4	1.9	1.6	1.4	1.2	1.1
F	a	a	a	a	a	a

a: Site-specific geotechnical investigations and dynamic site response analysis is recommended to evaluate the influence of local soil conditions on the site per ASCE Chapter 21. See ASCE 7 Standard Section 11.4.8 for exceptions.
b: If determined per ASCE Chapter 20 revealing rock conditions with no site-specific velocity measurements, $F_v = 1.0$.

1.6. Risk Categories and the Importance Factor

In earthquake design (also for the design to withstand wind load, snow load and ice load) , structures are classified into different risk categories based on their intended use and the potential consequences of their failure. These categories help determine the level of seismic resistance required by assigning a certain importance factor as shown in Table 1-5 here – also Table 1.5-2 of the ASCE 7 Standard.

The ASCE 7 Standard defines four primary risk categories in Section 1.5.1, which are summarized below.

- **Risk Category I:** Low hazard to human life (e.g., agricultural buildings, temporary facilities, small warehouses).

- **Risk Category II:** General buildings (e.g., residential, single family, small commercial, industrial). Failure of buildings in this category do not impose significant life-threatening risks. This category includes structures or facilities that do not belong to risk category I, III or IV.

  Additionally, buildings containing toxic, highly toxic, or explosive substances can be classified as Risk Category II if a hazard assessment shows no public threat from potential releases.

- **Risk Category III:** Buildings where their failure can pose substantial risk to human life and are not included in category IV. Examples to this includes high occupancy building, schools, assemblies such as theaters or stadiums, large commercial centers. **See amendments to this category in Section 6.2.1 of Chapter 6 in this book.**

- **Risk Category IV:** Essential facilities (e.g., hospitals, major airports, fire stations, emergency response centers and data centers and buildings that should maintain functionality during a catastrophic event), also facilities where their failure can pose substantial risk to the community buildings and other structures (e.g., facilities that store fuel or other hazardous substances that exceed a certain threshold and in case of failure could spill beyond their boundary into the communities around it). **See amendments to this category in Section 6.2.1 of Chapter 6 in this book.**

The importance factor, denoted as (I_e), is a multiplier used in seismic design to account for the varying levels of importance and risk associated with different structures. This factor increases the design seismic forces for buildings that are critical to safety and functionality during and after an earthquake. For example, essential facilities like hospitals and emergency response centers have a higher importance factor, typically 1.5, to ensure they remain operational during seismic events. See Table 1-5 here for more information about the values assigned to the different risk categories. This table is copied here with permission from ASCE.

Table 1-5 Seismic Importance Factors

Risk Category	Seismic Importance Factor (I_e)
I	1.00
II	1.00
III	1.25
IV	1.50

1.7. Seismic Design Categories (SDCs)

1.7.1.   Importance and Use of Seismic Design Categories

In this chapter, so far, we have discussed site coefficients and other seismic parameters, which eventually help determine the Seismic Design Categories (SDCs). While determining how SDCs will be explained in the next section, this section briefly covers some of their applications.

Throughout this book – and in any resource on seismic design, codes, or manuals – you will find that different designs, methods, and details are recommended for various SDCs. Table 1-6 here provides a high-level overview with few examples from various codes as to the application of some of the SDCs. More details on this are covered as needed throughout this book. Those are mostly summarized and provided in chapter 6.

Table 1-6 Examples on the use of SDCs in Different Codes, Manuals or Standards

Code, Manual or Standard	Relevant Section	SDCs Applications Examples
ASCE 7 Standard	Section 12.2.1 Selection and Limitations, Table 12.2-1.	This table provides limitations on building heights for various SDCs.
	Table 12.3-1 Horizontal Structural Irregularities.	This table defines the allowed type of irregularity for several SDCs.
	Section 12.5 Direction of loading.	Defines which direction of seismic loading should be applied for each SDC. For example, in buildings assigned to SDC B, seismic loading can be applied in each direction independent of the other.
	Table 12.6-1 Permitted Analytical Procedures.	This is an important table as it points you to the method that you should use to analyze the building in question per SDC.
	Table 15.4-1 Seismic Coefficients for Non-building Structures.	This table provides limitations on heights for various SDCs, and discussed in further detail in Chapter 5 of this book.
TMS 402/602	Chapter 7, Table CC 7.3.2-1 Requirements for Masonry Shear Walls Based on Shear Wall Designation.	This is an important table as it points you to the applicable section for each masonry element (whether participating to the seismic load resisting system or nonparticipating) for various SDCs – e.g., the table tells you that you cannot use unreinforced masonry in an SDC F.
	Chapter 7, Section 7.4 Seismic Design Category Requirements.	This section provides you with various details for different SDCs – e.g., Section 7.4.5.1, specifies that for SDC E, nonparticipating masonry not laid in running bond, the area of horizontal reinforcement should be 0.0015 of A_{gross}.
2021 SDPWS	Section 4.1.5.1	Provides requirements for anchoring concrete or masonry to wooden diaphragms.
	Section 4.1.10	Provides the requirements for toe-nailed connection for specific SDCs.
	Section 4.2.5	This section provides several torsional irregularity requirements for SDC B, C, D, E and F.

Table 1-6 is not meant to be comprehensive; rather, it offers a very small selection of important examples for the various uses and applications of SDCs. To prepare effectively, you are advised to review the relevant codes or manuals associated with this exam and their use of SDCs. Be sure to note the pages or sections that contain important information on SDCs, so you can easily find them during the exam. A good summary of those is provided in Chapter 6.

1.7.2. Determining the Seismic Design Categories

A Seismic Design Category (SDC) is a classification assigned to a structure based on its Risk Category and the severity of the design earthquake ground motion at the site.

The severity of the ground motion is assessed using the design spectral acceleration at short periods (S_{DS}) and at long periods (S_{D1}) which were discussed in Section 1.5.2 here. Based on this information, Seismic Design Categories can be determined using Tables 1-7, 1-8 and 1-9 below which are also discussed in Section 11.6 of the ASCE 7 Standard. These tables are copied here with permission from ASCE.

Table 1-7 SDC Based on (S_{DS})

	Risk Categories	
Value of S_{DS}	I or II or III	IV
$S_{DS} < 0.167$	A	A
$0.167 \leq S_{DS} < 0.33$	B	C
$0.33 \leq S_{DS} < 0.50$	C	D
$0.5 \leq S_{DS}$	D	D

Table 1-8 SDC Based on (S_{D1})

	Risk Categories	
Value of S_{D1}	I or II or III	IV
$S_{D1} < 0.067$	A	A
$0.067 \leq S_{D1} < 0.133$	B	C
$0.133 \leq S_{D1} < 0.20$	C	D
$0.2 \leq S_{D1}$	D	D

Table 1-9 SDC Based on (S_1)

	Risk Categories	
Value of S_1	I or II or III	IV
$0.75 \leq S_1$	E	F

In the event of a conflict between the above tables and/or other guidelines, the more severe outcome prevails. Additionally, you will find in other chapters of the ASCE 7 Standard that the spectral design acceleration (S_{DS}) is permitted to be adjusted based on Soil Structure Interaction (SSI) as noted in Chapter 19. It is crucial to use the spectral parameters provided above to determine the Seismic Design Category (SDC) before making any adjustments, following the guidance in this chapter or in Chapter 11 of the ASCE 7 Standard.

Moreover, if the simplified method outlined in Section 1.5.3 is applicable, you may use either Table 1-7 or Table 1-9, selecting the one that results in the more severe outcome.

1.8. Vertical Acceleration Component

In previous sections we discussed horizontal acceleration components for an earthquake, don't forget however that earthquakes excite the ground in all possible directions and hence in some cases, depending on the geometry of the structure or building, vertical acceleration shall be considered. For instance, if you have a balcony, or any other cantilever element or component. You will see that this component is included in load combinations in Chapter 4 of this book, also in nonbuilding structures with cantilever elements in Chapter 5 here.

The vertical response spectral acceleration (S_{av}) is calculated as follows:

$$S_{av} = \frac{2}{3} S_{aMv}$$

Where (S_{aMv}) and the relevant factor (C_v) can be calculated using Table 1-10 and Table 1-11 as shown below:

Table 1-10 Determining Vertical Response Spectrum

Vertical period (T_v)	Equation
$T_v \leq 0.025\ sec$	$S_{aMv} = 0.3C_vS_{MS}$
$0.025 < T_v \leq 0.05$	$S_{aMv} = 20C_vS_{MS}(T_v - 0.025) + 0.3C_vS_{MS}$
$0.05 < T_v \leq 0.15$	$S_{aMv} = 0.8C_vS_{MS}$
$0.15 < T_v \leq 2.0$	$S_{aMv} = 0.8C_vS_{MS} \left(\dfrac{0.15}{T_v}\right)^{0.75}$

Table 1-11 Values of Vertical Coefficient (C_v)

	Site Class A, B	Site Class C	Site Class D, E, F
$S_S \geq 2.0$	0.9	1.3	1.5
$S_S = 1.0$	0.9	1.1	1.3
$S_S = 0.6$	0.9	1.0	1.1
$S_S = 0.3$	0.8	0.8	0.9
$S_S \leq 0.2$	0.7	0.7	0.7

1.9. Maximum Considered Earthquake (MCE_R) and (MCE_G)

In this chapter and in ASCE 7, you will notice references to two important terms in earthquake engineering: the Risk-Targeted Maximum Considered Earthquake (MCE_R) and the Maximum Considered Earthquake Geometric Mean (MCE_G). These terms are used in different applications, and it is important to understand their distinct definitions and roles. In this section, we will clarify both terms and explain when each is used.

MCE_R represents the most severe earthquake effects adjusted for risk-targeting, ensuring that structures are designed to withstand the maximum expected ground shaking while minimizing risk to life safety and property. This value considers factors such as building importance and occupancy, and accounts for the orientation that generates the largest response to horizontal ground motions.

MCE_G refers to the geometric mean of Peak Ground Acceleration (PGA) during an earthquake. It represents the most severe earthquake effects without any risk-based adjustments. MCE_G is typically used as a baseline measure of ground shaking, calculated by averaging the peak accelerations in two horizontal directions. It is mainly applied in geotechnical evaluations to assess site-specific hazards like soil behaviour, liquefaction potential, or lateral spreading.

In summary, (MCE_R) is used to derive spectral accelerations for structural design, ensuring that buildings are safe and resilient to strong ground motions. On the other hand, (MCE_G) is used to evaluate site-specific and geotechnical hazards, providing important data for understanding soil behaviour and earthquake-induced risks.

1.10. Chapter References

The following references were used in this chapter. The reference(s) in italics is/are required for the exam, as noted in the preface of this book.

[1] *American Society of Civil Engineers, ASCE/SEI 7-16 Minimum Design Loads for Buildings and Other Structures.*

[2] American Society of Civil Engineers. ASCE Hazard Tool. https://ascehazardtool.org/ (Accessed February 1, 2025).

[3] EM 1110-2-1902: USACE Engineering and Design: Slope Stability, 2003, U.S. Army Corp of Engineers, Washington D.C.

[4] FHWA NHI-11-032 GEC No. 3: FHWA LRFD Seismic Analysis and Design of Transportation Geotechnical Features and Structural Foundations Reference Manual, 2011, Geotechnical Engineering Circulars, U.S. Department of Transportation, Federal Highway Administration, Washington, D.C.

[5] FHWA NHI-16-072 GEC No. 10: FHWA Geotechnical Site Characterization, 2017, Geotechnical Engineering Circulars, U.S. Department of Transportation, Federal Highway Administration, Washington, D.C.

[6] Idriss, I. M., and Boulanger, R. W. (2008). Soil Liquefaction During Earthquakes. Monograph MNO-12, Earthquake Engineering Research Institute, Oakland, CA.

[7] Idriss, I. M., and Boulanger, R. W. (2010). SPT-Based Liquefaction Triggering Procedures. Report UCD/CGM-10/02, Department of Civil and Environmental Engineering, University of California, Davis, CA.

[8] Kramer (1996). Geotechnical Earthquake Engineering. Prentice-Hall International (UK) Limited, London.

[9] Mononobe & Matsuo (1929). On The Determination of Earth Pressures During Earthquakes. in Proceedings of the World Engineering Congress, p. 9, Tokyo, Japan, 1929.

[10] NCHRP Report 611. Seismic Analysis and Design of Retaining Walls, Buried Structures, Slopes, and Embankments. National Cooperative Highway Research Program.

[11] Suarez, L. & Montejo, L. (2005). Generation of Artificial Earthquake via the Wavelet Transform. International Journal of Solids and Structures. 42(21): 5905-5919.

This page is intentionally left blank

Problems & Solutions

The following questions have been carefully designed to provide comprehensive coverage of the material presented in this book, while also being educational in nature, as you will notice. Additionally, you may come across a few questions not directly covered in this chapter; these have been intentionally included to ensure broader coverage of the content.

- ➤ Problem 1.1: Site Classification
- ➤ Problem 1.2: Resultant of Vertical Pseudostatic Forces
- ➤ Problem 1.3: Site Classification with Limited Site Information
- ➤ Problem 1.4: Site Classification for Clay with High Organic Content
- ➤ Problem 1.5: Vertical Acceleration
- ➤ Problem 1.6: SDC For a Tall Building
- ➤ Problem 1.7: Liquefaction Potential
- ➤ Problem 1.8: Short Period Site Coefficient
- ➤ Problem 1.9: Geotechnical Investigation Report
- ➤ Problem 1.10: Spectral Response Acceleration

PROBLEM 1.1 *Site Classification*
The below is a borehole log for a borehole taken at a certain location where it is required to classify the site to carry out seismic analysis and design.

BOREHOLE LOG SHEET				
Depth (ft)	Groundwater Level	Graphic Log	Material Description	Samples Test Remarks
0			SANDY SILTY CLAY, medium plasticity, well graded with traces of poorly graded gravel. *Density = 112 pcf*	SPT N = 8 blows per ft
10				
20				
30			GRAVELLY SANDY CLAY, medium plasticity, motted brown, moist to wet and fully submerged at water level at 4 ft depth. *Density = 118 pcf*	SPT N = 23 blows per ft
40				
50				
60				
70				
80			VERY DENSE SOIL, Low plasticity, poorly graded with traces of well graded gravel, cohesionless *Density = 120 pcf*	SPT N = 32 blows per ft
90				
100				

Based on the above information, the site classification for this location is:

(A) Site C – Dense soil or soft rock

(B) Site D – Stiff soil

(C) Site E – Soft clay soil

(D) Site F – Vulnerable to failure

PROBLEM 1.2 *Resultant of Vertical Pseudostatic Forces*
The weight of the failure wedge behind a retaining wall is $W_s = 30.0\ kip$ per linear foot and the coefficient of vertical acceleration is $k_v = 0.1$. The vertical resultant for the static and the Pseudostatic forces behind this wall per linear foot when the vertical acceleration portion is acting upwards is:

(A) 27.0 kip/ft ↓

(B) 30.0 kip/ft ↑

(C) 57.0 kip/ft ↓

(D) 3.0 kip/ft ↓

PROBLEM 1.3 *Site Classification with Limited Site Information*
The site soil classification that can be assigned for a soil that has limited geotechnical information is:

(A) Site Class C

(B) Site Class D

(C) Site Class E

(D) Site Class F

PROBLEM 1.4 *Site Classification for Clay with High Organic Content*
The site soil classification that can be assigned for a 20 ft thick layer of clay that has high organic content is:

(A) Site Class C

(B) Site Class D

(C) Site Class E

(D) Site Class F

PROBLEM 1.5 *Vertical Acceleration*
What is the vertical response spectral acceleration for a building's cantilever horizontal extension that is 5 ft long with a vertical fundamental period of 0.1 $seconds$, Soil Class C, along with the following seismic coefficients: $S_S = 1.5$, $S_1 = 0.6$ & $T_L = 6\ sec$?

(A) 0.65g

(B) 0.85g

(C) 1.15g

(D) 1.85g

PROBLEM 1.6 *SDC for a Tall Building*
A retail building with risk category II located in San Francisco (*). The building is founded on soil type C and is built using

steel and concrete composite ordinary braced frames.

The Seismic Design Category SDC for this building is:

(A) SDC A

(B) SDC B

(C) SDC C

(D) SDC D

(*) Use the following spectral parameters for this building in San Francisco:

$$S_S = 1.5$$
$$S_1 = 0.6$$
$$T_L = 6 \; sec$$

PROBLEM 1.7 *Liquefaction Potential*
Which of the following soils with the adjusted SPT blow counts $(N_1)_{60}$ showing below has the highest liquefaction potential knowing that all soils are fully saturated:

(A) Soil 1 with $(N_1)_{60} = 4$

(B) Soil 2 with $(N_1)_{60} = 7$

(C) Soil 3 with $(N_1)_{60} = 11$

(D) Soil 4 with $(N_1)_{60} = 27$

PROBLEM 1.8 *Short Period Site Coefficient*
The short period site coefficient F_a for a 30 ft thick soil type D with a plasticity index of $PI = 100$ and a short period spectral response acceleration $S_s = 0.2$ is:

(A) 1.45

(B) 1.65

(C) 1.85

(D) Further geotechnical studies are required to determine it.

PROBLEM 1.9 *Geotechnical Investigation Report*
The below are some of the additional geotechnical investigation requirements that shall be included a geotechnical study

for a Site Class D for a building with a basement, except for:

(A) Slope instability.

(B) Assessment of potential consequences of soil liquefaction.

(C) Peak ground acceleration that may cause liquefaction.

(D) Mitigation measures such as ground stabilization.

PROBLEM 1.10 *Spectral Response Acceleration*
What is the spectral response acceleration for a tall building located in a Site Class B with $S_s = 0.5$ and $S_1 = 0.2$ and $T_L = 4$, while the building has a fundamental period of 5 sec:

(A) 0.017g

(B) 0.021g

(C) 0.048g

(D) 0.060g

SOLUTION 1.1
Reference is made to the ASCE 7-16 Standard, Section 20.4 Definitions of Site Class Parameters, and Table 20.3-1 Site Classification (also Section 1.4.1 here).

Per the above, the following equation is used and is only applied to the first $100\,ft$ while the given borehole is $105\,ft$ deep:

$$\bar{N} = \frac{\sum_{i=1}^{n} d_i}{\sum_{i=1}^{n} \frac{d_i}{N_i}}$$

Where is d_i the layer thickness and N_i is number of blows per foot.

$$\bar{N} = \frac{30 + 45 + 25}{\frac{30}{8} + \frac{45}{23} + \frac{25}{32}} = 15.4$$

In reference to Table 20.3-1 of the ASCE 7 Standard (Table 1-1 here), with $\bar{N} = 15.4$, this site is classified as **Site D – Stiff Soil**.

Correct Answer is (B)

SOLUTION 1.2
As explained in Section 1.4.4 in this book., Pseudostatic force is calculated by multiplying the relevant coefficient of acceleration by the mass that will be affected (or moved) due to seismic activity.

For a clearer understanding, please refer to the following sketch:

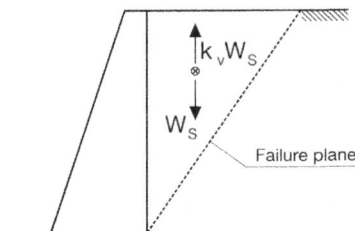

$$R = W_S - k_v W_S$$
$$= 30 - 0.1 \times 30$$
$$= 27.0\,kip \downarrow$$

Correct Answer is (A)

SOLUTION 1.3
Reference is made to the ASCE 7-16 Standard, Section 11.4.3 states that where soil properties are not known in sufficient detail to determine the site class, **Site Class D** can be used unless the authority having jurisdiction determines otherwise.

This information is also included in Section 1.4.1 of this book.

Correct Answer is (B)

SOLUTION 1.4
Reference is made to the ASCE 7-16 Standard, Section 20.3.1, point No. 2 assigns soil layers with peats or highly organic clays with layer thickness $> 10\,ft$ a **Site Class F**.

This information is also included in Section 1.4.1 of this book.

Correct Answer is (D)

SOLUTION 1.5
Reference is made to the ASCE 7-16 Standard Section 11. Also, you can find the same information in this chapter in Sections 1.5.2 and 1.8 along with Tables 1-2, 1-10 and 1-11 as follows:

$$S_{MS} = F_a S_S$$
$$= 1.2 \times 1.5$$
$$= 1.8$$
$$S_{aMv} = 0.8\,C_v\,S_{MS}$$
$$= 0.8 \times 1.2 \times 1.8$$
$$= 1.73$$
$$S_{av} = \frac{2}{3} S_{aMv}$$
$$= \frac{2}{3} \times 1.73$$
$$= 1.15$$

Correct Answer is (C)

SOLUTION 1.6

Use Sections 11.4.4 and 11.4.5 of the ASCE 7-16 Standard where $F_a = 1.2$ and $F_v = 1.4$ as collected from Tables 1-2 and 1-3 in this book. The spectral parameters are:

$$S_{DS} = \frac{2}{3} S_{MS}$$

$$= \frac{2}{3} F_a S_S$$

$$= \frac{2}{3} \times 1.2 \times 1.5$$

$$= 1.2$$

$$S_{D1} = \frac{2}{3} S_{M1}$$

$$= \frac{2}{3} F_v S_1$$

$$= \frac{2}{3} \times 1.4 \times 0.6$$

$$= 0.56$$

The Seismic Design Category (SDC) can now be determined from Section 1.7.2 of this book with $S_{DS} > 0.5$ and $S_{D1} > 0.2$ and a risk category II (retail), as **SDC D**.

Correct Answer is (D)

SOLUTION 1.7

To assess a soil's potential for liquefaction, additional parameters are needed, as outlined in Section 1.4.3. However, we can make an initial assessment based on the Standard Penetration Test (SPT) blow count, which may require further verification through lab testing later.

The SPT blow count reflects the hardness or stiffness of the soil – i.e., a lower blow count means the soil is softer.

Since all the soils are already saturated, **Soil 1**, with the lowest blow count, is the softest and therefore most susceptible to liquefaction.

Correct Answer is (A)

SOLUTION 1.8

The method of verifying this site classification is explained in Section 1.4 of this book or Section 20.3.1 point 3 of the ASCE 7 Standard.

Since this site is classified as D and $S_s < 0.25$, the short period coefficient $F_a = 1.6$ per Table 11.4-1 of the ASEC 7 Standard.

However, since this soil has a $PI > 75$ and thickness $> 25\ ft$, F_a should be multiplied by a sliding scale factor that varies from 1.0 (for $PI = 75$) to 1.3 (for $PI = 125$):

$$F_a = 1.15 \times 1.6 = 1.84$$

The above is valid and will not require any further geotechnical studies provided that $S_{DS} < 0.33$:

$$S_{DS} = \frac{2}{3} S_{MS}$$

$$= \frac{2}{3} F_a S_S$$

$$= \frac{2}{3} \times 1.84 \times 0.2$$

$$= 0.25 < 0.33 \rightarrow OK$$

As a conclusion, the short period coefficient can be used as concluded above as $F_a = 1.84$ with no further geotechnical investigations required to determine it.

Correct Answer is (C)

SOLUTION 1.9

Geotechnical hazards and investigations are discussed in ASCE 7 Standard Section 11.8.2 and 11.8.3.

In this section you will observe that most of the mentioned studies in this question are required for this study, However, **slope instability** is not relevant because there are no slopes at this site. Section 11.8.3 only addresses applicable conditions.

Correct Answer is (A)

SOLUTION 1.10

Reference in this question is made to Section 11.4.4 of the ASCE 7 Standard and Figure 11.4-1 of the same section – you can also refer to Example 1.3 or Figure 1-10 of this book)

Given that the building fundamental period $T = 5\ sec$ is larger than the long-term transition period $T_L = 4\ sec$ (*), the following equation can be used to determine the spectral response acceleration:

$$S_a = \frac{S_{D_1} T_L}{T^2}$$

Where:

$$S_{D1} = \frac{2}{3} S_{M1}$$

$$= \frac{2}{3} F_v S_1$$

$$= \frac{2}{3} \times 0.8 \times 0.2$$

$$= 0.107$$

Where F_v is collected from Table 1-3 of this book (or Table 11.4-2 of ASCE 7 Standard) as 0.8.

$$S_a = \frac{0.107 \times 4}{5^2} = 0.017$$

Correct Answer is (A)

(*) The Mapped Long-Period Transition Period (T_L) refers to a specific time frame that concerns how buildings respond to long-period ground motions generated by large earthquakes. The long-period ground motions are seismic waves with longer wavelengths, typically associated with significant earthquakes. They can cause prolonged shaking and affect taller structures more than shorter ones.

Geologic data on (T_L) can be collected from Figures 22.14 to 22.17 of the ASCE 7 Standard.

This page is intentionally left blank

CHAPTER

2

Seismic Characteristics of Engineered System

2.1. Purpose of this Chapter

Building on the first chapter's introduction of key facts and seismic parameters that are essential for the seismic design process, this chapter focuses on how structures, particularly buildings, withstand earthquakes and prevent collapse. It explores various structural systems that enhance a building's lateral stiffness, outlining design parameters and their limitations.

Additionally, this chapter addresses how irregularly shaped buildings may lack the necessary lateral support during ground shaking. It categorizes these irregularities as either horizontal, related to the plan geometry of the building's slabs (diaphragms), or vertical. The discussion includes the challenges posed by these irregularities and the additional connections or horizontal support needed to address them to increase the lateral stiffness when irregularly shaped.

Finally, the chapter examines buildings drift, which refers to horizontal deflections during earthquakes. It highlights the importance of implementing setbacks to prevent buildings from colliding and to minimize lateral displacement during seismic events.

2.2. Seismic Force Resisting System (SFRS)

2.2.1. Introduction and the Need for SFRS

If a building is constructed without a Seismic Force Resisting System (SFRS), it becomes significantly more vulnerable to earthquake damage. Lacking the necessary structural components to resist lateral forces generated by ground shaking, the building cannot effectively counteract these forces, leading to inadequate lateral support. This can result in shear failure of structural elements such as walls and columns, as the lateral forces may exceed the material strength, compromising structural integrity. The absence of an SFRS can also cause excessive drift, which is the horizontal movement of the structure during an earthquake, potentially leading to colliding with adjacent buildings if they were too close.

In high seismic zones, and with the lack of an SFRS, the building may experience catastrophic failure, resulting in partial or total collapse due to an inability to distribute seismic forces. Additionally, increased lateral forces can negatively impact the foundation, causing tilting, sliding, or failure. Even if the main structure remains standing, non-structural elements such as windows, walls, and interior finishes may sustain significant damage, resulting in costly repairs and safety hazards.

In buildings constructed with a column-beam support system, the absence of a Seismic Force Resisting System can have serious implications during an earthquake. While column-beam systems are designed to carry vertical loads, they may not adequately address the lateral forces generated by seismic activities. Beams in this case may experience excessive bending and shear forces, as they try to accommodate lateral movements. This can cause the beams to deform or even fail, particularly at their connections to the columns. The columns themselves can be subjected to increased axial loads and lateral forces, which may exceed their design capacity, leading to buckling or shear failure – see Figure 2-1.

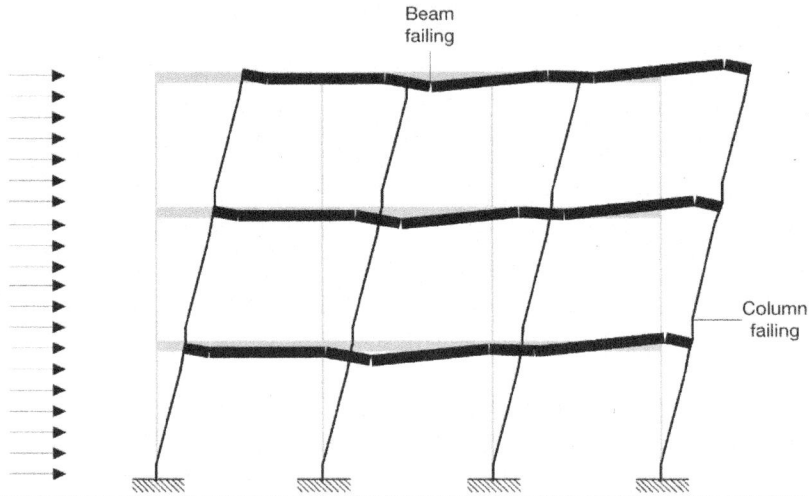

Figure 2-1 Beam and Column System Failure

As a result of the vulnerabilities outlined earlier, every structure must be designed and constructed to withstand lateral and horizontal loads. To achieve this, structures employ various techniques to counteract these forces. A bracing system is one of these techniques typically installed perpendicular to the direction of the anticipated force. Since seismic forces can originate from multiple directions, bracing is often implemented in all directions. Figure 2-2 below provides some of the bracing techniques.

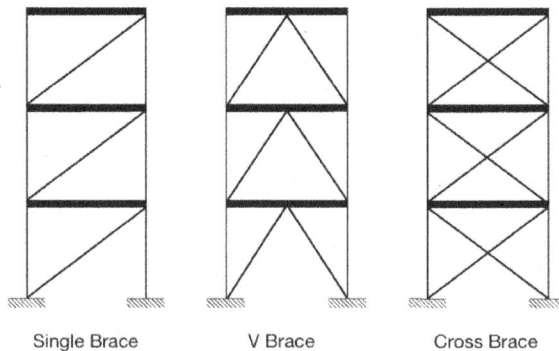

Figure 2-2 Bracing Techniques – The Rightmost is Concentrically Braced

The ASCE 7 Standard outlines different Seismic Force Resisting Systems in Chapter 12, specifically in Table 12.2-1. Each system has its own limitations concerning height and other factors, which will be explored in greater detail later in this book, which are also tabulated in the Table 12.2-1 of ASCE, and provided at this chapter's appendix with permission from ASCE.

In the following sections, the below SFRS systems which form part of ASCE 7 Standard are discussed:

A. Bearing wall systems.
B. Building frame systems.
C. Moment-resisting frame systems.
D. Dual systems with special moment frames capable of resisting at least 25% of prescribed seismic forces.
E. Dual systems with intermediate moment frames capable of resisting at least 25% of prescribed seismic forces.
F. Shear wall-frame interactive system with ordinary reinforced concrete moment frames and ordinary reinforced concrete shear walls.
G. Cantilevered column systems.
H. Steel system not specifically detailed for seismic resistance, excluding cantilever column system.

The above-mentioned systems are included in Table 12.2-1 of the ASCE 7 Standard – the first two are included in Table 12.14-1 which only deals with the simplified method. The simplified method is explained in Section 1.5.3 here.

The most important factors that are discussed here are the height limitations for each of the above systems and the response modification coefficient (R). This coefficient represents the building's ability to dissipate seismic energy through its design and materials, enhancing its resilience and reducing the forces it needs to withstand during an earthquake. The higher the response modification coeffect (R) the lower the shear force would be – See Section 3.3.1 here or Section 12.8.1.1 of the ASCE 7 Standard.

2.2.2. Bearing Wall Systems

The bearing wall system is designed to support the building's gravity loads and resist lateral forces using shear walls (see Figure 2-3). There are 18 wall systems classified under this category, as outlined in Table 12.2-1 of the ASCE 7 Standard. These walls transfer lateral forces from the floors directly to the foundation. However, this system lacks redundancy; if one shear wall fails, others may not be able to carry the load, jeopardizing the entire structure. Moreover, the system is less ductile, this means that the response modification coefficient (R) assigned to these 18 systems are generally lower than those for systems in other categories. Additionally, this system is often classified as Not Permitted (NP) in most Seismic Design Category (SDC) C and above.

You will notice in the systems under this category, and some of the other categories which have shear walls as either a separate or a dual system, that there is a distinction between three main systems: Ordinary, intermediate, and special walls. Ordinary, intermediate, and special shear walls differ primarily in their design, detailing, and seismic performance. Ordinary shear walls are more suitable for low seismic risk areas; they have simpler reinforcement and lower ductility, making them less effective in significant seismic events. For instance, you will notice under ASCE 7 Standard Table 12-2.1 that the use is not permitted in most applications under some of the high Seismic Design Categories, and their ductility, or their response modification factor (R) is low compared to its peers.

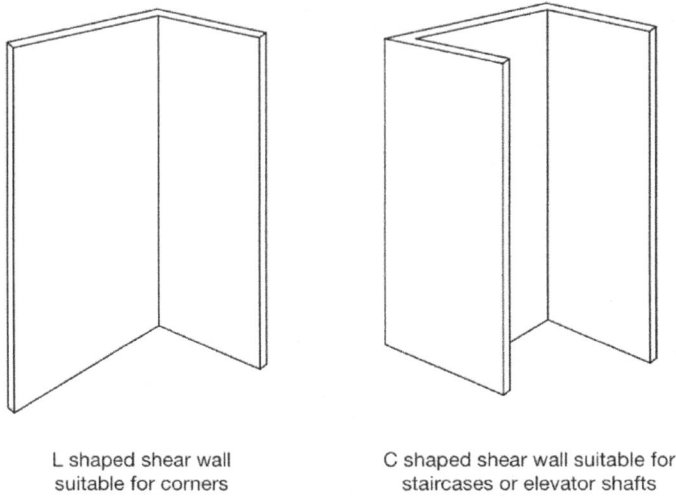

L shaped shear wall   C shaped shear wall suitable for
suitable for corners   staircases or elevator shafts

Figure 2-3 Different Shear Wall Systems

Intermediate shear walls, on the other hand, are designed for moderate seismic zones, with enhanced detailing that allows for better lateral load resistance and moderate ductility. Moreover, special shear walls are used for high seismic risk areas as they include detailing and high ductility to absorb energy and withstand substantial lateral forces during earthquakes – you can refer to Section 6.5.2 here for more information on this.

2.2.3. Building Frame Systems

There are two systems in this category, but this does not qualify them as dual systems as they are independent from each other. The first system relies primarily on braced frames to resist lateral loads, functioning like trusses. In this setup, the braced frames coexist with a traditional gravity-resisting system, which is not designed to handle lateral loads (see Figure 2.4). Similarly, the second system consists of shear walls that resist lateral loads alongside another system, which may be of a beam-column type, to support vertical gravity loads (see Figure 2-5).

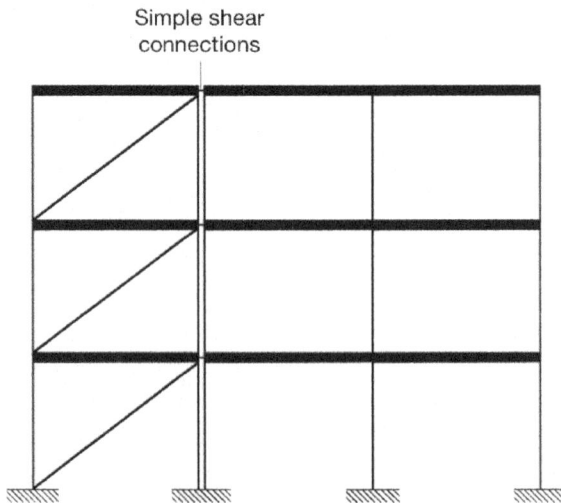

Simple shear
connections

Figure 2-4 Building Frame System - Braced System

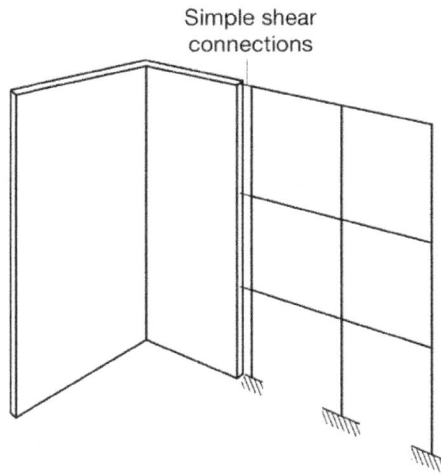

Simple shear
connections

Figure 2-5 Building Frame System - Shear Walls

2.2.4.   Moment-Resisting Frame Systems

Moment-resisting frame systems support both vertical and lateral loads through flexural action at the connections between beams and columns (see Figure 2-6). This design eliminates the need for diagonal bracing or shear walls. There are several types under this category, for instance the ordinary and intermediate moment-resisting frames have some height limitations, and sometimes they are Not Permitted (NP) for specific Seismic Design Categories. The special moment-resisting frames on the other hand can be designed for any height.

These frames are highly flexible, which can lead to significant inelastic deformation during seismic events. Such deformation may cause notable damage to nonstructural elements and architectural features. These systems demonstrate higher ductility, resulting in a higher response modification coefficient (R), consequently, lower lateral design forces. To maintain stability and prevent P-delta effects (P-delta is discussed later in Section 2.4.2 here), it is crucial to achieve a uniform distribution of stiffness, strength, and mass along the height of the frame.

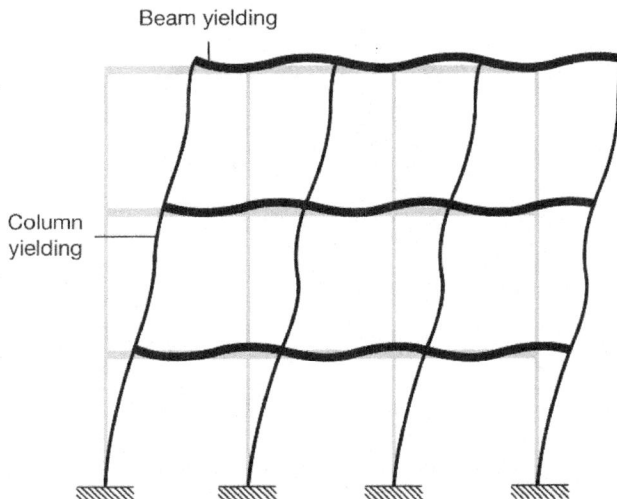

Beam yielding

Column
yielding

Figure 2-6 Moment Resisting Frame System

Seismic Characteristics of Engineered System

Steel moment frames are typically composed of beams and columns connected through bolts and welds. They provide more flexibility compared to shear walls and braced frames, or even concrete frames. They allow large openings and smaller wall sections.

Moment frames are classified into three main types: Ordinary Moment Frames (OMFs), Intermediate Moment Frames (IMFs), and Special Moment Frames (SMFs) each suited for different seismic zones (see below) – also, see Figure 2-7 for an example of a high-rise construction using special moment frames.

- OMFs are used in areas with low or no seismic activity.
- IMFs are appropriate for low- to mid-seismic activity regions.
- SMFs are designed for mid- to high-seismic activity areas.

The key distinctions among OMFs, IMFs, and SMFs relate to their ductility, influenced by member sizes and detailing. OMFs possess the lowest energy dissipation capacity due to the absence of designated plastic hinges, meaning there are no specific areas designed for yielding. This lack of designated failure points results in fewer restrictions on connections and member sizes.

In contrast, IMFs and SMFs both have enhanced detailing, such as reduced beam sections or specialized brackets, allowing for controlled yielding. IMFs can utilize moderately ductile materials, while SMFs require highly ductile members, evaluated based on their cross-sectional properties.

Figure 2-7 Highrise Construction using Special Moment Frames

To clarify, it is crucial to understand that both columns and beams in Special Moment Frames (SMFs) must possess sufficient strength and ductility to resist significant lateral forces. Typically, columns are designed to be stronger than beams to prevent their failure. One effective strategy to achieve this in steel structures is by carefully determining the locations of potential plastic hinge formation, a topic explored in more detail in Section 6.6.1 of Chapter 6 here. In these locations, sections may be intentionally weakened to allow for controlled energy dissipation. For example, Figure 2-8 below illustrates a Reduced Beam Section (RBS), which is an AISC-approved connection design. This section offers several advantages, which are further discussed in Chapter 6 here. One of the important advantages is preventing the column from bulking and failing due to a strong beam that refuses to rotate.

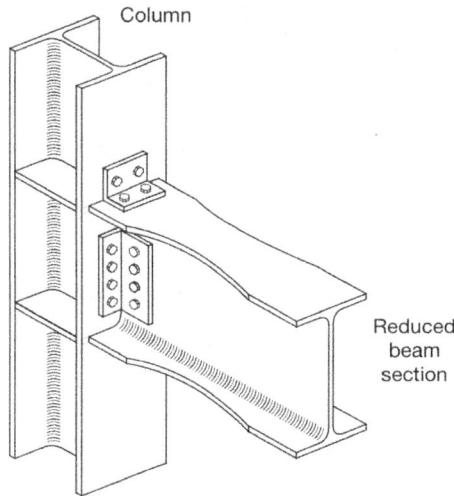

Figure 2-8 Reduced Beam Section in Special Moment Frames

As an example of reinforced concrete design, Figure 2-9 illustrates an exterior joint in a reinforced concrete Special Moment Frame (SMF). This figure for instance highlights a requirement that hooks must extend beyond the mid-depth of the joint to facilitate the development of a joint diagonal compression strut. More important information on SMFs is provided in Chapter 6 here.

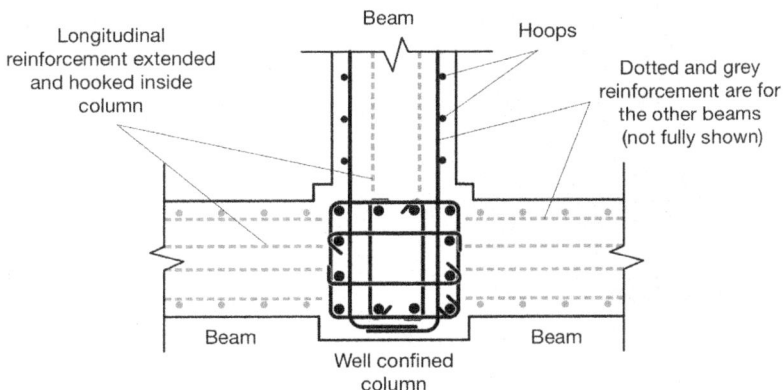

Figure 2-9 Exterior RC Joint in Special Moment Frames

2.2.5.   Dual System with SMF or IMF Moment Resisting Frames

This category covers items D and E in ASCE 7 Standard Table 12.2-1. It indicates that this system combines with either the Special Moment Frame (SMF) or the Intermediate Moment Frame (IMF). For example, item 2 under category D refers to "steel special concentrically braced frames" indicating that these braced frames work in conjunction with the SMF system (see Figure 2-10). Similarly, item 2 under category E refers to "special reinforced concrete shear walls" meaning that these shear walls are combined with the IMF system (see Figure 2-11). It is important to ensure that the IMF or SMF is capable of resisting at least 25% of the seismic forces as mentioned in the title of this category. Additionally, be mindful of the height limitations for each system and their corresponding coefficients during the assessment or design process.

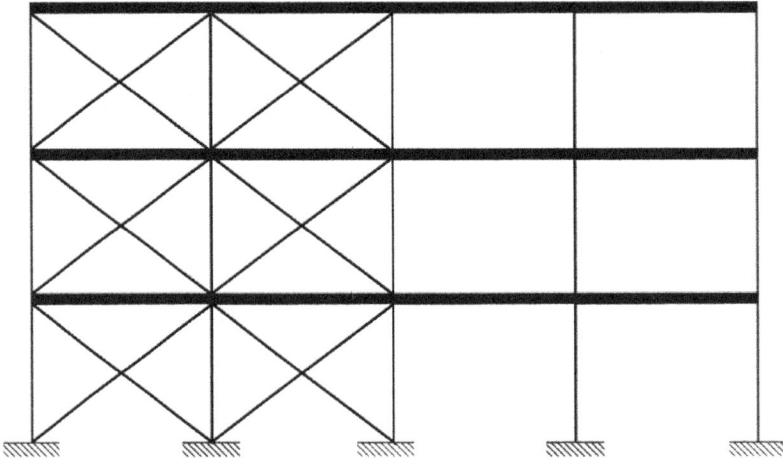

Figure 2-10 Steel Concentrically Braced Frames in a Dual System

Figure 2-11 Special Reinforced Concrete Shear Walls in a Dual System

### 2.2.6.	Shear Wall-Frame Interactive System

In a shear wall-frame interactive system, ordinary reinforced concrete moment frames (OMFs) work alongside ordinary reinforced concrete shear walls, forming a dual system suitable for structures in seismic design categories A and B only. According to ASCE 7 Standard Section 12.2.5.8, the shear strength of the shear wall must be at least 75% of the design story shear at each story, and the frames must resist at least 25% of the design story shear at every story.

### 2.2.7.	Cantilevered Column System

As the name suggests, this system is commonly found in low-rise residential buildings, where vehicles are parked on the ground floor beneath open columns and apartments occupy the upper levels (see Figure 2-12). However, this system has several limitations. For most Seismic Design Categories (SDCs), the maximum height is typically capped at $35\,ft$, and the system may not be permitted in some other SDCs. Additionally, its ductility is low, leading to a very low response modification coefficient (R), which results in higher lateral forces and base shear. Consequently, the drift or lateral deflection during seismic events can be significant.

Figure 2-12 Cantilevered Column System – Residential Complex

Figure 2-12 illustrates a ten-story building located in SDC A in Florida, using a cantilevered column system. Although the building exceeds the $35\,ft$ height limit, this is permitted due to the absence of a height restriction for SDC A in Table 12.2-1 of ASCE 7. However, if this building was located elsewhere, and constructed in SDC B or a higher seismic design category, it would not be permitted then.

2.2.8.   Steel Systems Not Detailed for Seismic Resistance

These systems, often including ordinary moment frames (OMFs) and other conventional steel framing methods, generally exhibit lower ductility and energy dissipation capabilities compared to systems designed for seismic resistance. Consequently, they may not be suitable for buildings in high seismic zones, as they do not meet the necessary criteria for resisting lateral forces. Typically, these systems have a lower response modification coefficient (R), indicating a reduced ability to absorb seismic energy, which leads to higher base shear and lateral forces during seismic events.

2.3. Buildings Irregularities

2.3.1.   Overview of Irregularities

Building designs often move away from conventional square shapes, with elevations that can be anything but straight and uniform. To meet various architectural needs or to stand out with innovative concepts, architects frequently embrace irregularly shaped structures. These irregularities can appear in the floor plan, creating **horizontal irregularities**, or in the elevation, leading to **vertical irregularities**. While these creative choices can enhance visual appeal and functionality, they also pose challenges, especially during seismic events.

For example, structures with heavily asymmetrical floor plans may risk twisting and buckling under pressure (see Figure 2-13), while weaker floors could lead to the dangerous collapse of entire floors (see Figure 2-14). Understanding these complexities is essential for ensuring the safety and stability of building unconventional designs which is what the following sections aim to discuss.

Figure 2-13 illustrates an example of a building's collapse due to **horizontal irregularities**, captured during the devastating earthquake in Nepal in April 2017, which tragically claimed the lives of 9,000 people and injured over 17,000. This earthquake caused extensive damage, destroying hundreds of thousands of homes, schools, businesses, healthcare facilities, and critical infrastructure, as well as historic landmarks.

Figure 2-13 shows three buildings that are marked for analysis:

1. Building No.1 exhibits an irregular floors shape, with columns and support structures not centered around the floors' center of mass. This misalignment caused the floor to twist during the seismic shaking, as indicated by the curved black arrow. The twisting motion likely led to the cracked columns on the ground floor and the crack on the first floor, highlighted by white arrows.

2. Building No.2 shows an out-of-plane offset, where there is a discontinuity in the lateral force-resisting path, marked by a dotted line. In this structure, the walls of the first floor do not extend to the second floor, creating a gap where the second-floor walls begin elsewhere. This discontinuity caused the diaphragm to twist during the earthquake, and as shown in the photo, this twisting may have resulted in the failure of the ground floor columns, ultimately causing the building to collapse onto its side.

3. Building No.3, in contrast, is regular in design and remained unaffected by the earthquake.

Figure 2-13 Horizontal Irregularity Collapse Example

Figure 2-14 depicts a building collapse caused by **vertical irregularity** in Sichuan, China, during the earthquake in 2008. This earthquake ruptured a fault over 240 kilometers (150 miles), resulting in surface displacements claiming 69,225 lives and injuring 374,176 individuals.

Figure 2-14 Vertical Irregularity Collapse Example

In this image, the building has a soft first floor, which contributes to the vertical irregularity. The structure appears to be built on a very rigid shear wall system beneath, transitioning to a less stiff support system on the first floor, as indicated by the white arrow in the photo. This discrepancy in stiffness likely weakened the building's stability, leading to its collapse during the seismic event.

2.3.2. Diaphragms

Prior to exploring building irregularities, it is important to understand what a diaphragm is, its function, and its components. We mentioned this term several times in this book, typically in relation to floors, slabs, or roofs, and this is discussed in further detail in Chapter 4 here.

A diaphragm primarily consists of a slab, floor, or roof designed to carry vertical or gravity loads. These diaphragms are some of the heaviest elements in a structure, and during an earthquake, they exert a force on the vertical support system that is proportional to their mass and the spectral acceleration. If diaphragms are irregularly shaped, they can twist and turn during earth shaking relative to the center of mass and the location of the center of stiffness of the vertical support system.

Diaphragms, during an earthquake, function like a deep beam and tend to bend, as illustrated in Figure 2-15, depending on their flexibility. To counteract flexural forces, **chords** are typically provided. Without these, the entire diaphragm acts as a deep beam, requiring additional reinforcement if made of concrete at its ends to address in-plane moments. **Collectors** are also included to transfer shear forces from laterally unsupported areas to vertical resisting elements, as shown in the figure below.

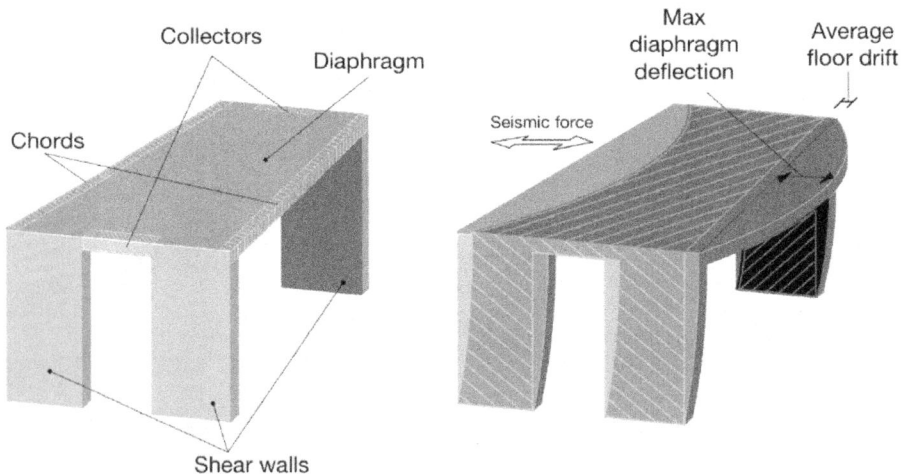

Figure 2-15 Diaphragm Components and Deflection

Typically, there are three diaphragm types: rigid, flexible, and semi-rigid. Diaphragms can be assumed to be rigid (e.g., concrete slabs or concrete-filled metal decks) if their aspect ratio is 3 or less, and provided the structure has no significant horizontal irregularities. Diaphragms are considered flexible (e.g., wood panels) when their maximum in-plane deflection under lateral load exceeds twice the average story drift of adjacent vertical elements (see Figure 2-15). Semi-rigid diaphragms sit in between and should be modeled to reflect their actual in-plane stiffness and behavior when significant deformation occurs or when required by the code. You can check some of the flexible diaphragm types in ASCE 7 Standard Section 12.3.1.

Additionally, it is important to understand that the internal forces in diaphragms are calculated differently depending on the type of diaphragm used. For instance, flexible diaphragms, particularly when combined with shear walls, can often be modeled as simply supported deep beams – see Example 2.1 for clarification.

In contrast, rigid diaphragms require other considerations during analysis and design. Rigid diaphragms distribute lateral forces more uniformly across their surface, leading to different load transfer mechanisms compared to flexible diaphragms taking into account the diaphragm's stiffness and the rigidity of the lateral supports like walls and columns. Chapter 4 delves deeper into the principles of diaphragms load distribution providing guidelines and examples that illustrate these key considerations.

Example 2.1 – Calculating Diaphragm Forces for a Flexible Diaphragm

Find the chord forces for the below flexible diaphragm which was presented in Figure 2-15 with a seismic load of 1,000 lb/ft as shown below:

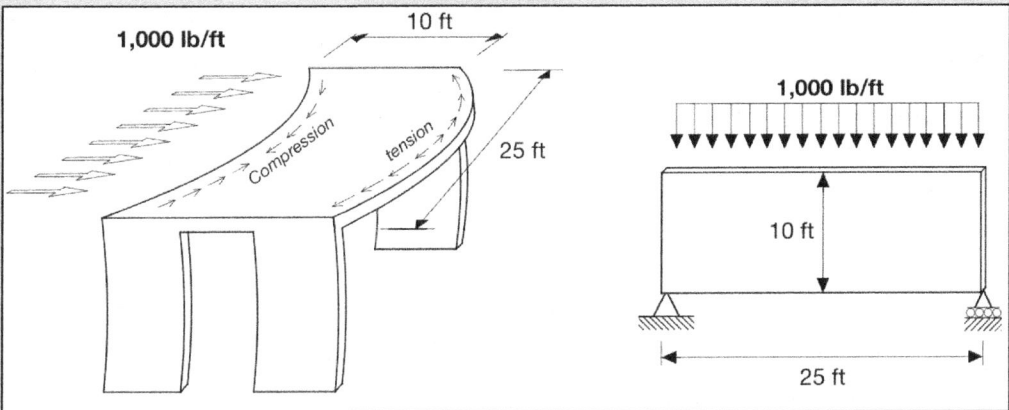

As discussed in this section, this diaphragm can be modelled as a deep beam on two supports. The moment generated from this arrangement is as follows – refer to Section 4.4.2 here for moment and shear equations for simple beams:

$$M = \frac{Wl^2}{8} = \frac{1\ kip/ft \times (25\ ft)^2}{8} = 78.1\ kip.ft$$

This moment can be converted to a tension/compression couple in the two chords as shown in the below figure:

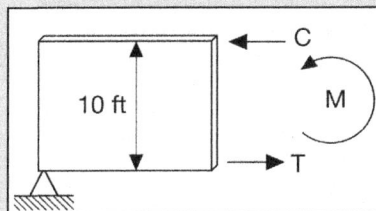

$$\mp C = \mp T = \frac{78.1\ kip.ft}{10\ ft} = 7.8\ kip$$

2.3.3.   Horizontal Irregularities

There are five horizontal irregularities included in the ASCE 7 Standard listed below in Table 2-1 and defined in Table 12.3-1 of ASCE 7 Standard. Figure 2-16 here provides a graphical representation for each of those irregularities. Instead of providing a narrative we opted to provide a sketch in the aforementioned figure for clarity. Additionally, Table 12.3-1 of ASCE 7 Standard points you to the design treatment for each of the mentioned horizontal irregularities. Those treatments are summarized in Table 2-1 here. This summary is intended to be of a high level only. If you seek more details, you can refer to the relevant sections in the ASCE 7 Standard.

Table 2-1 Horizontal Irregularity Design Treatments (*italics* check Section 2.3.5 or relevant CBC section)

Irregularity	Applicable SDC	Design Treatment (ASCE 7 relevant section between brackets)
Type 1a: Torsional Irregularity *Applicable to rigid or semirigid diaphragms only.*	D, E & F	Diaphragm design force for connectors to vertical elements and collectors to be increased by 25%. (12.3.3.4)
	B, C, D, E & F	3D model shall be built with 3 degrees of freedom and 2 orthogonal directions analysis. (12.7.3)
	C, D, E & F	*Accidental torsional moment* magnified by $1 \leq A_x = \left(\frac{\delta_{max}}{1.2\delta_{avg}}\right)^2 \leq 3$. (12.8.4.3)
	C, D, E & F	Story drift (Δ) < allowable drift (Δ_a) of Table 12.12-1 of ASCE 7 Standard.
	B, C, D, E & F	Consider 5% accidental eccentricity of diaphragm's center of mass. (16.3.4)
Type 1b: Extreme Torsional Irregularity *Applicable to rigid or semirigid diaphragms only.*	E & F	Prohibited. (12.3.3.1)
	D, E &F	*Prohibited per CBC 2022 Section 1617A.1.10 for state regulated DSA & OSHPD.*
	D	Diaphragm design force for connectors to vertical elements and collectors to be increased by 25%. (12.3.3.4)
	D	*Redundancy factor* (ρ) shall be taken as 1.3. (12.3.4.2)
	B, C, D	3D model shall be built with 3 degrees of freedom and 2 orthogonal directions analysis. (12.7.3)
	C & D	*Accidental torsional moment* magnified by $1 \leq A_x = \left(\frac{\delta_{max}}{1.2\delta_{avg}}\right)^2 \leq 3$. (12.8.4.3)
	C & D	Story drift (Δ) < allowable drift (Δ_a) per Section 12.12.1 of ASCE 7.
	B, C & D	Consider 5% accidental eccentricity of diaphragm's center of mass. *(16.3.4)*
Type 2: Reentrant Corner Irregularity	D, E & F	Diaphragm design force for connectors to vertical elements and collectors to be increased by 25%. (12.3.3.4)
Type 3: Diaphragm Discontinuity	D, E & F	Diaphragm design force for connectors to vertical elements and collectors to be increased by 25%. (12.3.3.4)
Type 4: Out-of-Plane Offset Irregularity	B, C, D, E & F	Seismic loads shall include *overstrength.* (12.3.3.3)
	D, E & F	Diaphragm design force for connectors to vertical elements and collectors to be increased by 25%. (12.3.3.4)
	B, C, D, E & F	3D model shall be built with 3 degrees of freedom and 2 orthogonal directions analysis. (12.7.3)
	B, C, D, E & F	Consider 5% accidental eccentricity of diaphragm's center of mass. *(16.3.4)*
Type 5: Nonparallel System Irregularity	C, D, E & F	*Analysis* shall either be conducted with: (1) simultaneous application of seismic load in orthogonal directions using linear or nonlinear response history. Or (2) using the equivalent analysis method or the linear response or MRSPA, loads can be applied per direction independent of the other provided the other direction can resist 30% of the load. (Standard 12.5.3)
	B, C, D, E & F	3D model shall be built with 3 degrees of freedom and 2 orthogonal directions analysis. (12.7.3)
	B, C, D, E & F	Consider 5% accidental eccentricity of diaphragm's center of mass. (16.3.4)

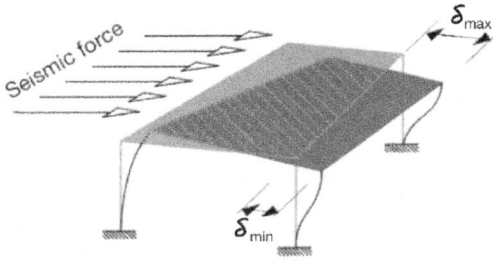

$$\delta_{avg} = \frac{\delta_{min} + \delta_{max}}{2}$$

$$\frac{\delta_{max}}{\delta_{avg}} > 1.2$$

Type 1.a
Torsional Irregularity

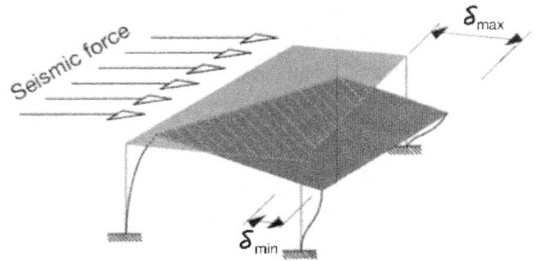

$$\delta_{avg} = \frac{\delta_{min} + \delta_{max}}{2}$$

$$\frac{\delta_{max}}{\delta_{avg}} > 1.4$$

Type 1.b
Extreme Torsional
Irregularity

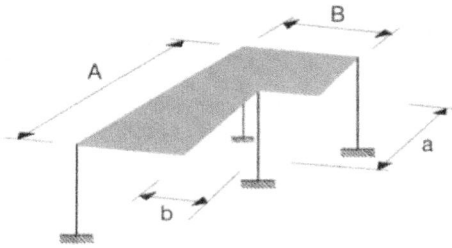

a/A > 0.15
and
b/B > 0.15

Type 2
Reentrant Corner
Irregularity

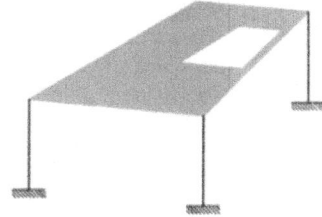

$A_{opening} = 50\% A_{diaphragm}$
or
Change in diaphragm stiffness more
than 50% from story to next

Type 3
Diaphragm Discontinuity
Irregularity

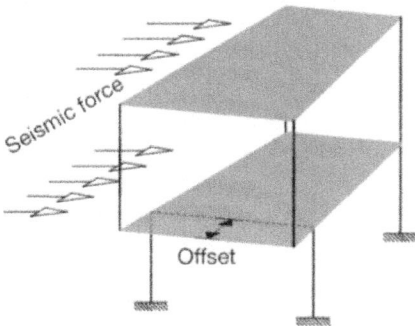

Offset

Type 4
Out-of-Plane Offset
Irregularity

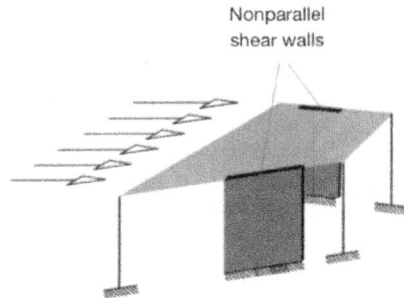

Nonparallel
shear walls

Type 5
Nonparallel System
Irregularity

Figure 2-16 ASCE 7 Standard Horizontal Irregularity Types

Seismic Characteristics of Engineered System

2.3.4. Vertical Irregularities

In this section, we discuss vertical irregularities, as summarized in Figure 2-17, with a detailed narrative available in Table 12.3-2 of the ASCE 7 Standard. In these summaries, you may have noticed that Type 1 vertical irregularity refers to a soft story, which emphasizes story stiffness, while Type 5 pertains to a weak story, focusing on story strength. Story strength refers to the capacity of the structural elements to withstand the lateral applied loads without experiencing failure, such as yielding or buckling. It is measured in terms of force and is critical in ensuring that structural components, like columns and walls, can support the loads imposed on them during an earthquake. In contrast, stiffness measures a structural element's resistance to deformation under load, indicating how much it will deflect when a force is applied. Stiffness is quantified as a ratio of force to displacement, reflecting how a structure deforms in response to lateral forces.

Table 2-2 here provides you with the summary of design treatments needed per vertical irregularity. This summary is intended to be of a high level only. If you seek more details, you can refer to the relevant sections in the ASCE 7 Standard.

Table 2-2 Vertical Irregularity Design Treatments (*italics* check Section 2.3.5 or relevant CBC section)

Irregularity	Applicable SDC	Design Treatment (ASCE 7 relevant section between brackets)
Type 1a: Story Stiffness Irregularity - Soft Story Irregularity	D, E & F	Check method of *analysis* (i.e., Equivalent load, MRSPA, Linear response history or nonlinear response history) in Table 12.6-1 of ASCE 7 Standard.
Type 1b: Story Stiffness Irregularity – Extreme Soft Story Irregularity	E & F	Prohibited. (12.3.3.1)
	D, E & F	*Prohibited per CBC Section 1617A.1.10 for state regulated DSA & OSHPD.*
	D, E & F	Check method of analysis in Table 12.6-1 of ASCE 7 Standard.
Type 2: Weight (Mass) Irregularity	D, E & F	Check method of *analysis* in Table 12.6-1 of ASCE 7 Standard.
Type 3: Vertical Geometry Irregularity	D, E & F	Check method of *analysis* in Table 12.6-1 of ASCE 7 Standard.
Type 4: In-Plane Discontinuity in Vertical Lateral Force-Resisting Element Irregularity	B, C, D, E & F	Seismic loads shall include *overstrength*. (12.3.3.3)
	D, E & F	Diaphragm design force for connectors to vertical elements and collectors to be increased by 25%. (12.3.3.4)
	D, E & F	Check method of analysis in Table 12.6-1 of ASCE 7 Standard.
Type 5a: Discontinuity in Lateral Strength Irregularity - Weak Story Irregularity	E & F	Prohibited. (12.3.3.1)
	D, E & F	*Prohibited per CBC Section 1617A.1.10 for state regulated DSA & OSHPD.*
	D, E & F	Check method of *analysis* in Table 12.6-1 of ASCE 7 Standard.
Type 5b: Discontinuity in Lateral Strength Irregularity – Extreme Weak Story Irregularity	D, E & F	Prohibited. (ASCE 7 12.3.3.1 *also CBC 1617A.1.10*)
	B & C	Weak story cannot be more than 30 *ft* or two stories except if they can resist total seismic force multiplied by *overstrength*. (12.3.3.2)
	D, E & F	Check method of *analysis* in Table 12.6-1 of ASCE 7 Standard.

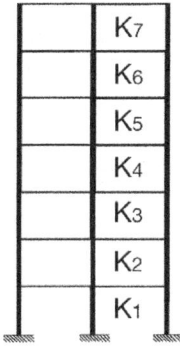

Type 1.a (*)
$K_1 < 0.7K_2$
or
$K_1 < 0.8av(K_2, K_3, K_4)$

Type 1.b (*)
$K_1 < 0.6K_2$
or
$K_1 < 0.7av(K_2, K_3, K_4)$

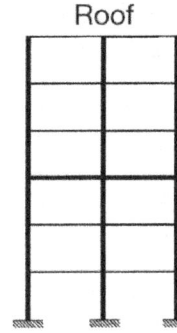

Type 1.a Soft Story
Type 1.b Extreme Soft Story
Soft Story Irregularity

(*) K: Stiffness

$M_3 > 1.5M_4$
or
$M_3 > 1.5M_2$
A lighter roof need not to be considered

Type 2
Weight (Mass)
Irregularity

$L_1 > 1.3L_2$

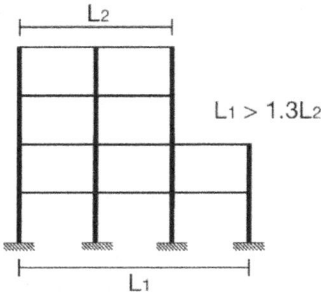

Type 3
Vertical Geometric
Irregularity

This causes overturning demands

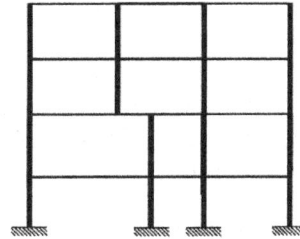

Type 4
In-Plane Discontinuity
Irregularity

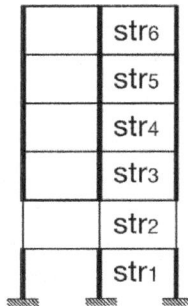

Type 5.a (*)
$str_2 < 0.8str_3$

Type 5.b (*)
$str_2 < 0.65str_3$

Type 5.a Weak Story
Type 5.b Extreme Weak Story
Discontinuity in Lateral
Strength Irregularity

(*) str: Story lateral supports
strength

Figure 2-17 ASCE 7 Standard Vertical Irregularity Types

Seismic Characteristics of Engineered System

2.3.5.   Clarifications on Diaphragm Irregularities Design Treatments

In Section 2.3.3 and Section 2.3.4 here, we discussed the design treatment for both horizontal and vertical irregularities in seismic design. Within those treatments, several key terminologies were utilized, which are explained here to provide a clearer understanding of their implications.

Overstrength factor (Ω_o) refers to the additional capacity that a structure possesses beyond the nominal strength required by the design loads and is included in ASCE 7 Standard Table 12.2-1. When seismic loads include overstrength, it acknowledges that structures should perform better than expected under extreme conditions, allowing for some margin of safety. This is particularly relevant in ductile systems where energy dissipation occurs.

Conversely, when seismic loads do not account for overstrength, the design is based strictly on expected load conditions without considering this inherent capacity, potentially leading to less resilient structures in real-world seismic events.

Accidental torsional moment (M_{ta}) accounts for unintended rotations that may occur due to asymmetries in the structure or loading conditions. Normally by moving the center of mass of the concerned diaphragm by 5% in relation to the affected length. This factor is essential because it helps ensure that the structure can withstand the additional forces generated by these unexpected moments, which could lead to uneven distributions of forces and potential failure in irregularly shaped buildings.

Redundancy factor (ρ) is a consideration in seismic design that addresses the reliability of a structural system. It is used when a structure's load path contains multiple elements that can carry loads, ensuring that if one element fails, others can redistribute the loads.

Section 16.3.4 of ASCE 7 mentioned in the design treatment Table 2-1 for Type 1b and Type 4 horizontal irregularities, and the requirement for 5% accidental eccentricity for the diaphragm's center of mass, refers to the nonlinear response history analysis.

Finally, the analytical methods summarized in Table 12.6-1 of the ASCE 7 Standard provide various approaches to assess a structure's response to seismic loads. These methods include:

1. *Equivalent Lateral Force Procedure:* A simplified approach that uses equivalent static forces based on building height and mass to estimate seismic loads.

2a. *Modal Response Spectrum Analysis (MRSPA):* A more sophisticated method that evaluates the response of a structure using its natural frequencies and mode shapes, providing a more accurate representation of dynamic behaviors.

2b. *Linear Response History Analysis:* This method examines the response of a structure to specific ground motion records, assuming linear elastic behavior during real seismic events.

3. *Nonlinear Response History Analysis:* A comprehensive approach that considers material nonlinearity and inelastic behavior under seismic loading, providing detailed potential damages and performance during extreme conditions.

Example 2.2 – Weak Story

A 100 ft tall building consists of a composite Seismic Force Resisting System (SFRS) made of steel and concrete special concentrically braced frames. The second story is classified as an "extremely weak story" with a Type 5b irregularity due to its height of 40 ft, designed as a large reception area with a high ceiling. Total seismic forces are illustrated in the accompanying sketch and the building is assigned a Seismic Design Category B.

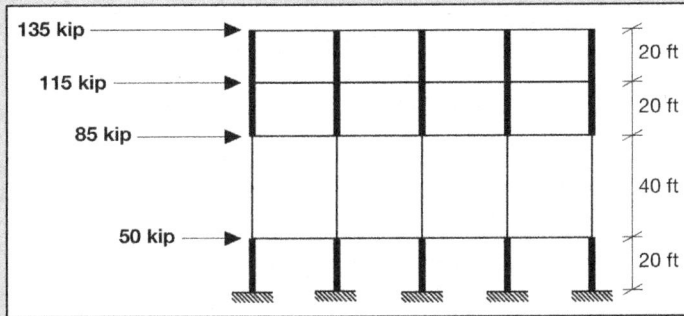

What should be the total lateral force that the lateral resisting system for this weak floor needs to be designed to withstand?

The total forces resisted by the lateral resisting system on that floor are the sum of all the above forces, multiplied by the overstrength factor (Ω_o) of the described SFRS system as indicated in Table 2-2 here and Section 12.3.3.2 of the ASCE 7 Standard. This factor can be found in Table 12.2-1 of the ASCE 7 Standard, under item D.6, where Ω_o is specified as 2.5. See the calculation below:

$$= 2.5 \times (85 + 115 + 135) = 837.5 \ kip$$

Otherwise, this design is prohibited, and the second story height should be lowered to a maximum of 30 ft as indicated in the ASCE 7 Standard.

2.4. Story Drift and P-Delta Effect
2.4.1.   Story Drift

Story drift refers to the lateral deflection experienced by each story of a building during an earthquake, measured in relation to the story below it. In seismic design, story drift is typically calculated using the strength design method, which differs from standard practice where allowable service loads are used to determine deflections in non-seismic designs.

When analyzing buildings' deflections and their stories' drifts, it is important to differentiate between the two. Story drift is represented by the symbol (Δ_x), while the overall building deflection is denoted by (δ_x) or (δ_{xe}) – the difference is explained below. For clarity, refer to Figures 2-18 and 2-19: Figure 2-18 illustrates building deflection/displacement, which does not reflect the relationship between stories.

Building deflection is generally calculated through elastic analysis, often performed using computer software. This analysis yields the deflection corresponding to seismic loads, typically measured at the center of the diaphragm and denoted as (δ_{xe}).

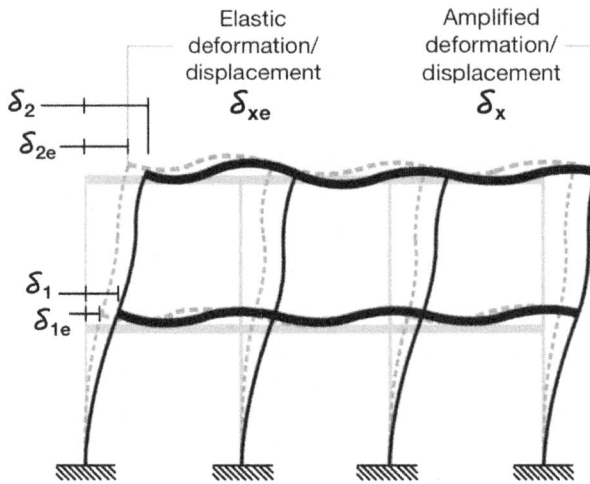

Figure 2-18 Building Elastic and Amplified Displacement

To account for the full impact of these loads, the elastic deflection (δ_{xe}) is amplified to obtain (δ_x), as shown by the solid line of Figure 2-18, using the following equation:

$$\delta_x = \frac{C_d \delta_{xe}}{I_e}$$

Where (I_e) is the importance factor discussed in Section 1.6 and Table 1-5 of this book, while (C_d) is the deflection amplification factor which is SFRS system related and can be collected from Table 12.2-1 of the ASCE 7 Standard.

Story drift (Δ_x) on the other hand is determined by subtracting the amplified elastic deflection (δ_x) of the floor below, as illustrated in Figure 2-19, (e.g., $\Delta_2 = \delta_2 - \delta_1$). The calculated story drift is then compared to the allowable limits (Δ_a) specified in Section 12.12 of the ASCE 7 Standard – See problems 2.8 and 2.9 at the end of this chapter, also Example 2.3 here for more details on this.

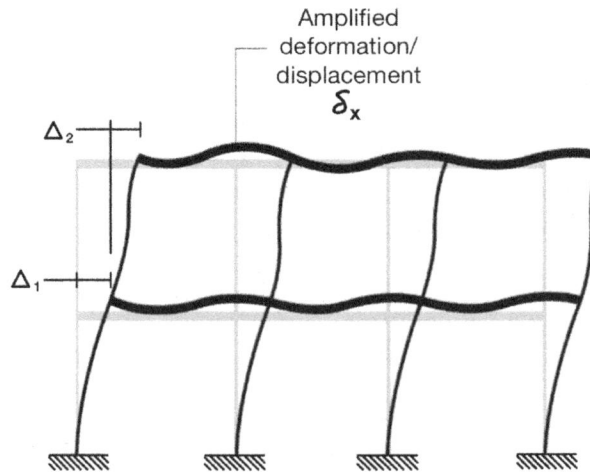

Figure 2-19 Story Drift Determination

Example 2.3 – Story Drift

Drawing from the discussions of Section 2.4.1 of this book, the below building has three stories with a total height of 46 ft. The total elastic displacements for floors 1, 2 and 3 calculated using a structural analysis software for a certain seismic load are 1.7 in, 2.5 in and 3.3 in respectively.

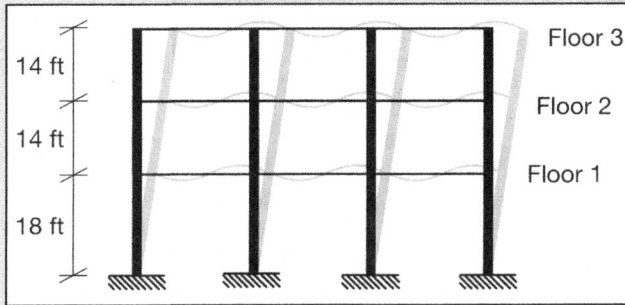

The building SFRS system is made of steel ordinary moment frames and is assigned a Seismic Design Category B, with a risk category III and an importance factor of $I_e = 1.25$.

Calculate the drift for each floor and compare it with the allowable drift per ASCE 7 Standard.

First, determine the deflection amplification factor (C_d) from Table 12.2-1 of ASCE 7 Standard for the given SFRS system. Line-item C.4 of this table specifies($C_d = 3.0$). Using this information along with the equation presented in Section 2.4.1 here (taken from Equation 12.8-15 of ASCE 7 Standard), the amplified displacement (δ_x) can be calculated as follows – floor 1 only shown in the calculation below.

$$\delta_1 = \frac{C_d \delta_{1e}}{I_e} = \frac{3.0 \times 1.7}{1.25} = 4.08 \ in$$

Allowable story drift (Δ_a) is determined using Section 12.12 and Table 12.12-1 of ASCE 7 Standard. Given this building is assigned to an SDC B, Section 12.12.1.1 with the use of redundancy factor (ρ) does not apply – See Problem 2.9 here for more details on this.

Floor 1 allowable drift is calculated only. The rest of results are presented in the table that follows:

$$\Delta_{a1} = 0.02 h_{sx} = 0.020 \times 18 \ ft \times 12 \frac{in}{ft} = 4.32 \ in$$

Floor	Elastic Displacement (δ_{xe})	Amplified Displacement (δ_x)	Story Drift ($\Delta_x = \delta_x - \delta_{x-1}$)	Allowable Drift (Δ_a)
Floor 1	1.7 in	4.08 in	4.08 in	4.32 in
Floor 2	2.5 in	6.00 in	1.92 in	3.36 in
Floor 3	3.3 in	7.92 in	1.92 in	3.36 in

2.4.2.   P-Delta Effect

When a building experiences significant lateral displacements, particularly under high gravity loads on their floors, these displacements can be exacerbated while the vertical loads are pushing down. This occurs because the lateral deflection introduces a secondary moment, potentially worsening the situation. See Figure 2-20 for better understanding.

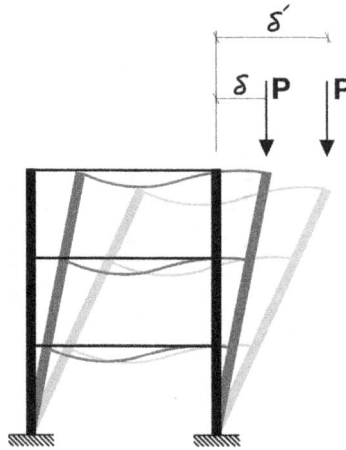

Figure 2-20 P-Delta Effect Representation

Figure 2-20 represents a building that displaced by (δ), and due to vertical loads P, this displacement increased to (δ'). This effect is called the P-Delta effect where P represents the vertical load, while Delta is the lateral deflection. The interaction between these two factors leads to what is known as a second-order analysis.

Analyzing a structure for P-Delta effects can be quite complex. Prior to the introduction of the advanced analysis software, codes and standards mandated the use of stiffer elements to reduce potential P-Delta effects.

Including the effects of P-Delta analysis is not always required especially if the delta was small enough. The ASCE 7 Standard for instance, in Section 12.8.7, addresses this matter with the use of the stability coefficient (θ). If this coefficient was found to be $\theta \leq 0.1$, this is an indication that the story drift (Δ) is sufficiently small that, when combined with vertical loads above level x (P_x) and base shear (V_x) – where the height above level x is (h_{sx}) – the P-Delta effects can be ignored. See equation below:

$$\theta = \frac{P_x \Delta I_e}{V_x h_{sx} C_d}$$

Conversely, the stability coefficient cannot exceed a maximum value of (θ_{max}). If the design of a particular structure yields a stability factor greater than (θ_{max}), it indicates instability, necessitating a redesign.

$$\theta_{max} = \frac{0.5}{\beta C_d} \leq 0.25$$

Where (β) in this case is the ratio of shear demand (i.e., applied shear) to shear capacity for the story between levels x and $x - 1$ and is permitted to be taken as 1.0 as a conservative measure.

In light of this, if $\theta > 0.1$, P-Delta effects must be integrated into the analysis and design. Several methods for doing this are discussed in the literature (see Reference [3] for further reading). Some examples are below:

1. *Amplification of Horizontal Loads:* This method requires the increase of horizontal loads to mimic a Delta effect.

2. *Amplification of Displacements:* Similar to the first method but requires increasing displacements.

3. *Iterative Method:* Developed by Chen and Liu (1991), this method involves multiple iterations for refining the analysis. Further details can be found in Reference [3].

4. *Non-Linear Analysis:* This approach utilizes the Full Newton-Raphson method for a comprehensive analysis.

You are not required to know the above methods for the California seismic exam. However, you must understand how ASCE 7 Standard addresses this effect, which is dependent on the analytical procedure discussed in Section 2.3.5 of this book and Table 12.6-1 of the ASCE 7 Standard. Specifically, when $0.1 \leq \theta \leq \theta_{max}$ ASCE 7 recommends employing a rational method that involves mathematical modeling for the Equivalent Lateral Procedure (ELP) as well as the Modal Response and Linear Response methods. As an alternative approach that only applies to ELP, it allows amplifying displacements by a factor of $1/(1-\theta)$.

2.4.3.   Recommended Setbacks

When two adjacent buildings are in close proximity, there is a significant risk of structural pounding during a seismic event, particularly if the buildings' response is out of phase. Figure 2-21 illustrates few scenarios of structural pounding between buildings.

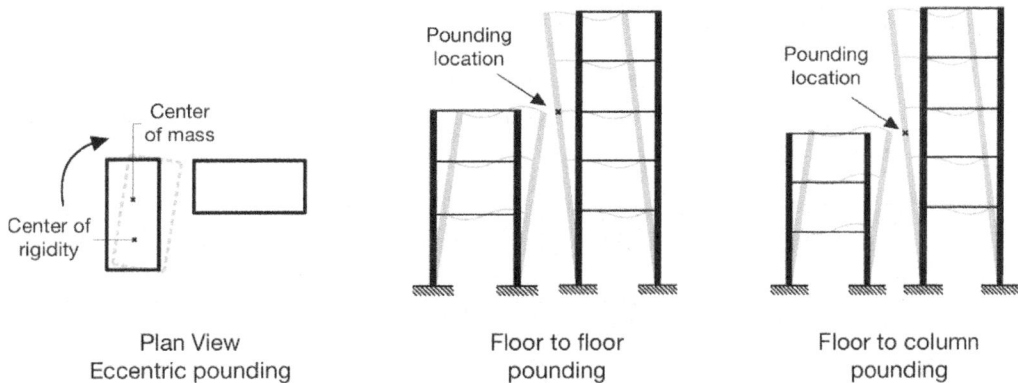

Plan View	Floor to floor	Floor to column
Eccentric pounding	pounding	pounding

Figure 2-21 Types of Buildings' Pounding

According to ASCE 7 Standard Section 12.12.3, buildings should be separated by a distance that accommodates the maximum inelastic response displacement (δ_M). This separation distance should consider the heights of the buildings, potential locations for pounding, torsional irregularities, and diaphragm deflection where (δ_{max}) is the maximum elastic displacement at the critical location.

$$\delta_M = \frac{C_d \delta_{max}}{I_e}$$

In determining the required separation distance between adjacent buildings (δ_{MT}), we typically do not add the maximum displacements ($\delta_{M1,M2,etc}$) of each structure. Instead, we use the statistical method known as the Square Root of the Sum of the Squares (SRSS) to calculate this distance. See below:

$$\delta_{MT} = \sqrt{(\delta_{M1})^2 + (\delta_{M2})^2}$$

It is also essential to consult the commentary section of the ASCE 7 Standard. In C12.12.3, it states that for rigid shear wall structures with rigid diaphragms, where lateral deflections cannot be reasonably estimated, the NEHRP provisions (FEMA 2009) recommend implementing structural separations of at least 1 *in*, plus an additional ½ *in* for every 10 *ft* of height above 20 *ft* – See reference [5] at chapter end for more details.

Finally, when designing supports for members that span two buildings – a common feature in many high-rise structures such as a connecting bridge – the requirements are more stringent, as outlined in ASCE 7 Section 12.12.4. This is crucial because if the support on one side fails during an earthquake, it could lead to the complete collapse of the entire bridge, resulting in catastrophic consequences. In this case, ASCE 7 Section 12.12.4 recommends designing the supports for those connecting members between buildings to withstand a displacement of:

$$\frac{1.5R}{C_d} \times \delta_x$$

Where (δ_x) is calculated according to Section 2.4.1 of this book or ASCE 7 Equation 12.8-15. Most importantly, the final setback for this case only should be determined by absolute addition rather than using the Square Root of the Sum of the Squares (SRSS) method.

Example 2.4 – Buildings Separation

What is the minimum distance that should be implemented to separate two concrete buildings in a buildings complex both with equal heights of 140 *ft*. The two buildings' SFRS system is that of special reinforced concrete shear walls with rigid diaphragms?

Since those two buildings are made of rigid shear wall structures with rigid diaphragms, Section C12.12.3 of the ASCE 7 Standard in the commentary section is consulted, and as mentioned in this section of this book, the separation in this case is calculated as follows:

$$= 1\ in + (0.5\ in) \times \frac{140 - 20}{10} = 7\ in$$

2.4.4.   The Simplified Method

In the "Simplified Method" discussed in Section 1.5.3 in this book, drift calculation is not required. However, if drift needs to be determined for specific purposes, such as for cladding, 1% of the structural height can be used for drift unless computed to be less – See ASCE 7 Standard Section 12.14.8.5 for more details on this.

2.5. Effects of Ductility and Damping on Seismic Performance

This section expands on concepts that were briefly discussed in earlier sections in this book to provide some more understanding of their importance in assessing seismic performance.

We will begin with the concept of ductility. Seismic ductility is important for the design and performance of structures subjected to seismic loading, as it refers to the structure's ability to undergo significant deformation without failure. This property allows structures to absorb and dissipate energy during seismic events. According to ASCE 7, structures must be designed with adequate ductility to accommodate the inelastic behavior expected during strong ground shaking. This is particularly crucial for systems that rely on yielding mechanisms, such as moment-resisting frames, which were discussed earlier in this chapter.

When examining ductility and deflection through the deflection amplification factor (C_d), it is important to understand why these values differ among various structural systems. For example, as shown in line-item B.2 of Table 12.2-1 of ASCE 7, steel special concentrically braced frames, has a $C_d = 5$, while a more rigid system, like the detailed plain concrete shear wall in line-item B.6, has a $C_d = 2$. These differences reflect how each system responds to seismic forces and their inherent ductility.

In addition to ductility, damping is another important characteristic that enhances seismic performance. Damping refers to the mechanisms by which vibrational energy is dissipated, thereby reducing the amplitude of oscillations during an earthquake. Various damping systems, including base isolation techniques, or innovative building damping devices such as the viscoelastic dampers, can significantly improve a building's resilience to seismic forces. Figure 2-22 below illustrates examples of building dampers and foundation isolation systems for demonstration purposes. For further information on damping systems or foundation isolation, refer to ASCE 7, Chapters 17 and 18.

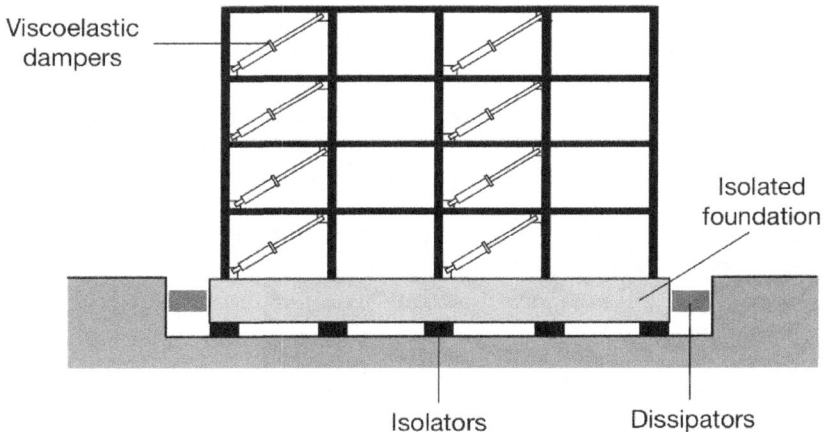

Figure 2-22 Example of a Building's Isolation and Damping System

Damping plays a critical role in seismic design, influencing how structures react to ground motion. ASCE 7 Chapter 11, along with the relevant risk coefficient maps presented in Chapter 22 and discussed in Section 1.5.2 of this book, assumes a critical damping ratio of 5%. This assumption is based on empirical data and past research, indicating that many conventional structural systems exhibit this level of inherent damping during seismic events. The 5% damping ratio serves as a benchmark, simplifying design calculations and providing a baseline for comparing different structural systems.

While engineers can use other damping values other than the standard 5%, they must conduct a detailed assessment of the specific damping characteristics of the structure. For structures designed with innovative damping techniques for instance – such as viscoelastic dampers or energy dissipating devices illustrated in Figure 2-22 – the actual damping ratio can be significantly higher, potentially exceeding 20%.

2.6. Seismic Base Isolation

As discussed in the previous section, seismic isolation for buildings and bridges can significantly reduce earthquake damage. As illustrated in Figure 2-22, this is achieved through the installation of layers of rubber. These rubber layers are vertically stiff to support substantial gravity loads while remaining laterally flexible, allowing the structure to move horizontally and thus protecting it from potential damage.

One of the most commonly used seismic isolators is the Lead-Rubber Bearing (LRB). This system comprises layers of elastomeric material, similar to those used in bridge supports, sandwiched between two pads, as shown in Figure 2-23. The rubber is reinforced with steel sheets, and a circular lead core is positioned in the center of the rubber pad. While the rubber effectively supports vertical loads, it does not dissipate the energy generated by ground motion. The lead core, however, yields under stress, allowing the rubber to accommodate displacement while simultaneously dissipating energy.

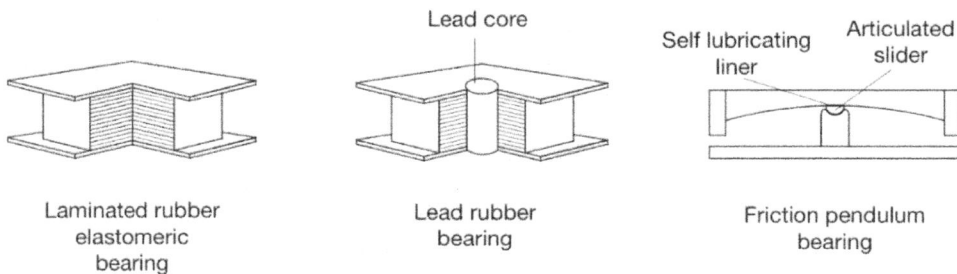

Figure 2-23 Types of Seismic Base Isolation

Other base isolation systems are shown in the same figure, each with its specific applications. The Laminated Rubber Bearings are suitable for minor earthquakes and can be employed in both bridges and buildings, as well as beneath large industrial equipment to reduce vibrations especially when installed over floors – or underneath a floor to create a floating floor for equipment. The Lead-Rubber Bearings (LRB) provide enhanced support for larger earthquakes and are particularly beneficial in critical structures such as nuclear power plants due to their high damping capabilities. The Friction Pendulum Bearing (FPB) is another option, designed for high-rise buildings and essential facilities like hospitals that must remain operational during seismic events.

2.7. Chapter References

The following references were used in this chapter. The reference(s) in italics is/are required for the exam, as noted in the preface of this book.

[1] *American Society of Civil Engineers, ASCE/SEI 7-16 Minimum Design Loads for Buildings and Other Structures.*

[2] *California Building Code (CBC 2022) California Code of Regulations, Title 24, Part 2, Volumes 1 and 2. Accessed through https://codes.iccsafe.org in January 2025.*

[3] Chen W. F. & Lui E. M. (1991). Stability Design of Steel Frames. CRC Press.

[4] Direct Relief. (2017, April 25). As Nepal Recovers from 2015 Quake, Direct Relief support continues.   https://www.directrelief.org/2017/04/as-nepal-recovers-from-2015-quake-direct-relief-support-continues/

[5] FEMA (2009). NEHRP Recommended Seismic Provisions for New Buildings and Other Structures. Federal Emergency Management Agency, Washington, DC.

[6] FEMA (2018). Assessing Seismic Performance of Buildings with Configuration Irregularities. Calibrating Current Standard and Practices. FEMA P-2012, September 2018, The National Earthquake Hazards Reduction Program (NEHRP).

[7] Hamburger, R. Krawinkler, H., Malley, J. & Adan, S. (2009). Seismic Design of Steel Special Moment Frames: A Guide for Practicing Engineers. NEHRP Seismic Design Technical Brief No.2. National Institute of Standards and Technology. U.S. Department of Commerce.

[8] Moehle, J., Hooper, J. & Lubke, C. (2008). Seismic Design of Reinforced Concrete Special Moment Frames: A Guide for Practicing Engineers. NEHRP Seismic Design Technical Brief No.1. National Institute of Standards and Technology. U.S. Department of Commerce.

[9] Ravikumar, C. M., Babu Narayan, K. S., Sujith, B. V., & Venkat Reddy, D. (2012). Effect of Irregular Configurations on Seismic Vulnerability of RC Buildings. Architecture Research, 2(3), 20-26.

[10] Sonawane, Y. & Patil, M. (2016). Base Isolation for Multistoried Buildings with Lead Rubber bearing. International Journal of New Innovations in Engineering and Technology. 4(4).

[11] Zabihullah, Singh, P. & Aryan, M. (2020). Effect of (Vertical & Horizontal) Geometric Irregularities on the Seismic Response of RC Structures. International Journal on Emerging Technologies, 11(3).

[12] Watts, Jonathan (August 14, 2008). "Sichuan Quake: China's Earthquake Reconstruction to Cost $150bn". The Guardian. Archived from the Original on October 8, 2017. Retrieved October 8, 2017.

This page is intentionally left blank

Chapter Appendix

This appendix contains the following information:

➤ Table 12.2-1 of ASCE 7 Standard
Design Coefficients and Factors for Seismic Force-Resisting Systems

> In this table, we shaded some systems in two different shades of grey. These shadings indicate limitations imposed by the California Building Code (CBC 2022), Sections 1617 and 1617A.1.4, as follows:
>
> **The light grey shading**
> *Not permitted in OSHPD (hospitals and other health care facilities).*
>
> **The darker grey shading**
> *Not permitted in both OSHPD and DSA (District of the State Architect) Community Colleges.*
>
> Refer to Appendix of Chapter 6 for more understanding of the above abbreviated authorities.

➤ Table 12.14-1 of ASCE 7 Standard
Design Coefficients for Seismic Force-Resisting Systems for the Simplified Design Method

➤ Table C12.2-1 of ASCE 7 Standard
Summary of Conditions for OMFs and IMFs in Structures Assigned to Seismic Design Category D, E, or F

➤ Table 12.12-1 of ASCE 7 Standard
Allowable Story Drift

All tables provided in this appendix are copied here with permission from ASCE.

ASCE Table 12.2-1 Design Coefficients and factors for SFRSs

Table 12.2-1 Design Coefficients and Factors for Seismic Force-Resisting Systems

Seismic Force-Resisting System	ASCE 7 Section Where Detailing Requirements Are Specified	Response Modification Coefficient, R^a	Overstrength Factor, $\Omega_0{}^b$	Deflection Amplification Factor, $C_d{}^c$	Structural System Limitations Including Structural Height, h_n (ft) Limits[d] Seismic Design Category B	C	D[e]	E[e]	F[f]
A. BEARING WALL SYSTEMS									
1. Special reinforced concrete shear walls[g,h]	14.2	5	2½	5	NL	NL	160	160	100
2. Ordinary reinforced concrete shear walls[g]	14.2	4	2½	4	NL	NL	NP	NP	NP
3. Detailed plain concrete shear walls[g]	14.2	2	2½	2	NL	NP	NP	NP	NP
4. Ordinary plain concrete shear walls[g]	14.2	1½	2½	1½	NL	NP	NP	NP	NP
5. Intermediate precast shear walls[g]	14.2	4	2½	4	NL	NL	40[i]	40[i]	40[i]
6. Ordinary precast shear walls[g]	14.2	3	2½	3	NL	NP	NP	NP	NP
7. Special reinforced masonry shear walls	14.4	5	2½	3½	NL	NL	160	160	100
8. Intermediate reinforced masonry shear walls	14.4	3½	2½	2¼	NL	NL	NP	NP	NP
9. Ordinary reinforced masonry shear walls	14.4	2	2½	1¾	NL	160	NP	NP	NP
10. Detailed plain masonry shear walls	14.4	2	2½	1¾	NL	NP	NP	NP	NP
11. Ordinary plain masonry shear walls	14.4	1½	2½	1¼	NL	NP	NP	NP	NP
12. Prestressed masonry shear walls	14.4	1½	2½	1¾	NL	NP	NP	NP	NP
13. Ordinary reinforced AAC masonry shear walls	14.4	2	2½	2	NL	35	NP	NP	NP
14. Ordinary plain AAC masonry shear walls	14.4	1½	2½	1½	NL	NP	NP	NP	NP
15. Light-frame (wood) walls sheathed with wood structural panels rated for shear resistance	14.5	6½	3	4	NL	NL	65	65	65
16. Light-frame (cold-formed steel) walls sheathed with wood structural panels rated for shear resistance or steel sheets	14.1	6½	3	4	NL	NL	65	65	65
17. Light-frame walls with shear panels of all other materials	14.1 and 14.5	2	2½	2	NL	NL	35	NP	NP
18. Light-frame (cold-formed steel) wall systems using flat strap bracing	14.1	4	2	3½	NL	NL	65	65	65
B. BUILDING FRAME SYSTEMS - shear wall (non-bearing)									
1. Steel eccentrically braced frames	14.1	8	2	4	NL	NL	160	160	100
2. Steel special concentrically braced frames	14.1	6	2	5	NL	NL	160	160	100
3. Steel ordinary concentrically braced frames	14.1	3¼	2	3¼	NL	NL	35[j]	35[j]	NP[j]
4. Special reinforced concrete shear walls[g,h]	14.2	6	2½	5	NL	NL	160	160	100
5. Ordinary reinforced concrete shear walls[g]	14.2	5	2½	4½	NL	NL	NP	NP	NP
6. Detailed plain concrete shear walls[g]	14.2 and 14.2.2.7	2	2½	2	NL	NP	NP	NP	NP
7. Ordinary plain concrete shear walls[g]	14.2	1½	2½	1½	NL	NP	NP	NP	NP
8. Intermediate precast shear walls[g]	14.2	5	2½	4½	NL	NL	40[i]	40[i]	40[i]
9. Ordinary precast shear walls[g]	14.2	4	2½	4	NL	NP	NP	NP	NP
10. Steel and concrete composite eccentrically braced frames	14.3	8	2½	4	NL	NL	160	160	100
11. Steel and concrete composite special concentrically braced frames	14.3	5	2	4½	NL	NL	160	160	100
12. Steel and concrete composite ordinary braced frames	14.3	3	2	3	NL	NL	NP	NP	NP
13. Steel and concrete composite plate shear walls	14.3	6½	2½	5½	NL	NL	160	160	100
14. Steel and concrete composite special shear walls	14.3	6	2½	5	NL	NL	160	160	100
15. Steel and concrete composite ordinary shear walls	14.3	5	2½	4½	NL	NL	NP	NP	NP
16. Special reinforced masonry shear walls	14.4	5½	2½	4	NL	NL	160	160	100
17. Intermediate reinforced masonry shear walls	14.4	4	2½	4	NL	NL	NP	NP	NP

Table 12.2-1 (Continued) Design Coefficients and Factors for Seismic Force-Resisting Systems

Seismic Force-Resisting System	ASCE 7 Section Where Detailing Requirements Are Specified	Response Modification Coefficient, R^a	Overstrength Factor, Ω_0^b	Deflection Amplification Factor, C_d^c	Structural System Limitations Including Structural Height, h_n (ft) Limits^d Seismic Design Category				
					B	C	D^e	E^e	F^f
18. Ordinary reinforced masonry shear walls	14.4	2	2½	2	NL	160	NP	NP	NP
19. Detailed plain masonry shear walls	14.4	2	2½	2	NL	NP	NP	NP	NP
20. Ordinary plain masonry shear walls	14.4	1½	2½	1¼	NL	NP	NP	NP	NP
21. Prestressed masonry shear walls	14.4	1½	2½	1¾	NL	NP	NP	NP	NP
22. Light-frame (wood) walls sheathed with wood structural panels rated for shear resistance	14.5	7	2½	4½	NL	NL	65	65	65
23. Light-frame (cold-formed steel) walls sheathed with wood structural panels rated for shear resistance or steel sheets	14.1	7	2½	4½	NL	NL	65	65	65
24. Light-frame walls with shear panels of all other materials	14.1 and 14.5	2½	2½	2½	NL	NL	35	NP	NP
25. Steel buckling-restrained braced frames	14.1	8	2½	5	NL	NL	160	160	100
26. Steel special plate shear walls	14.1	7	2	6	NL	NL	160	160	100
C. MOMENT-RESISTING FRAME SYSTEMS									
1. Steel special moment frames	14.1 and 12.2.5.5	8	3	5½	NL	NL	NL	NL	NL
2. Steel special truss moment frames	14.1	7	3	5½	NL	NL	160	100	NP
3. Steel intermediate moment frames	12.2.5.7 and 14.1	4½	3	4	NL	NL	$35^{h,k}$	$NP^{h,k,l}$	$NP^{i,l}$
4. Steel ordinary moment frames	12.2.5.6 and 14.1	3½	3	3	NL	NL	NP^i	NP^i	NP^i
5. Special reinforced concrete moment frames^m	12.2.5.5 and 14.2	8	3	5½	NL	NL	NL	NL	NL
6. Intermediate reinforced concrete moment frames	14.2	5	3	4½	NL	NL	NP	NP	NP
7. Ordinary reinforced concrete moment frames	14.2	3	3	2½	NL	NP	NP	NP	NP
8. Steel and concrete composite special moment frames	12.2.5.5 and 14.3	8	3	5½	NL	NL	NL	NL	NL
9. Steel and concrete composite intermediate moment frames	14.3	5	3	4½	NL	NL	NP	NP	NP
10. Steel and concrete composite partially restrained moment frames	14.3	6	3	5½	160	160	100	NP	NP
11. Steel and concrete composite ordinary moment frames	14.3	3	3	2½	NL	NP	NP	NP	NP
12. Cold-formed steel—special bolted moment frame^n	14.1	3½	3°	3½	35	35	35	35	35
D. DUAL SYSTEMS WITH SPECIAL MOMENT FRAMES CAPABLE OF RESISTING AT LEAST 25% OF PRESCRIBED SEISMIC FORCES	12.2.5.1								
1. Steel eccentrically braced frames	14.1	8	2½	4	NL	NL	NL	NL	NL
2. Steel special concentrically braced frames	14.1	7	2½	5½	NL	NL	NL	NL	NL
3. Special reinforced concrete shear walls^{g,h}	14.2	7	2½	5½	NL	NL	NL	NL	NL
4. Ordinary reinforced concrete shear walls^g	14.2	6	2½	5	NL	NL	NP	NP	NP
5. Steel and concrete composite eccentrically braced frames	14.3	8	2½	4	NL	NL	NL	NL	NL
6. Steel and concrete composite special concentrically braced frames	14.3	6	2½	5	NL	NL	NL	NL	NL
7. Steel and concrete composite plate shear walls	14.3	7½	2½	6	NL	NL	NL	NL	NL
8. Steel and concrete composite special shear walls	14.3	7	2½	6	NL	NL	NL	NL	NL
9. Steel and concrete composite ordinary shear walls	14.3	6	2½	5	NL	NL	NP	NP	NP
10. Special reinforced masonry shear walls	14.4	5½	3	5	NL	NL	NL	NL	NL
11. Intermediate reinforced masonry shear walls	14.4	4	3	3½	NL	NL	NP	NP	NP
12. Steel buckling-restrained braced frames	14.1	8	2½	5	NL	NL	NL	NL	NL
13. Steel special plate shear walls	14.1	8	2½	6½	NL	NL	NL	NL	NL

E. DUAL SYSTEMS WITH INTERMEDIATE MOMENT FRAMES CAPABLE OF RESISTING AT LEAST 25% OF PRESCRIBED SEISMIC FORCES — 12.2.5.1

Seismic Force-Resisting System	Section	R	Ω_0	C_d	B	C	D	E	F
1. Steel special concentrically braced frames[p]	14.1	6	2½	5	NL	NL	35	35	NP
2. Special reinforced concrete shear walls[g,h]	14.2	6½	2½	5	NL	NL	160	160	100
3. Ordinary reinforced masonry shear walls	14.4	3	3	2½	NL	160	NP	NP	NP
4. Intermediate reinforced masonry shear walls	14.4	3½	3	3	NL	NL	NP	NP	NP
5. Steel and concrete composite special concentrically braced frames	14.3	5½	2½	4½	NL	NL	160	160	100
6. Steel and concrete composite ordinary braced frames	14.3	3½	2½	3	NL	NL	NP	NP	NP
7. Steel and concrete composite ordinary shear walls	14.3	5	3	4½	NL	NL	NP	NP	NP
8. Ordinary reinforced concrete shear walls[g]	14.2	5½	2½	4½	NL	NL	NP	NP	NP

F. SHEAR WALL-FRAME INTERACTIVE SYSTEM WITH ORDINARY REINFORCED CONCRETE MOMENT FRAMES AND ORDINARY REINFORCED CONCRETE SHEAR WALLS[g]

Seismic Force-Resisting System	Section	R	Ω_0	C_d	B	C	D	E	F
	12.2.5.8 and 14.2	4½	2½	4	NL	NP	NP	NP	NP

G. CANTILEVERED COLUMN SYSTEMS DETAILED TO CONFORM TO THE REQUIREMENTS FOR: — 12.2.5.2

Seismic Force-Resisting System	Section	R	Ω_0	C_d	B	C	D	E	F
1. Steel special cantilever column systems	14.1	2½	1¼	2½	35	35	35	35	35
2. Steel ordinary cantilever column systems	14.1	1¼	1¼	1¼	35	35	NP[j]	NP[j]	NP[j]
3. Special reinforced concrete moment frames[m]	12.2.5.5 and 14.2	2½	1¼	2½	35	35	35	35	35
4. Intermediate reinforced concrete moment frames	14.2	1½	1¼	1½	35	35	NP	NP	NP
5. Ordinary reinforced concrete moment frames	14.2	1	1¼	1	35	NP	NP	NP	NP
6. Timber frames	14.5	1½	1½	1½	35	35	35	NP	NP

H. STEEL SYSTEMS NOT SPECIFICALLY DETAILED FOR SEISMIC RESISTANCE, EXCLUDING CANTILEVER COLUMN SYSTEMS

Seismic Force-Resisting System	Section	R	Ω_0	C_d	B	C	D	E	F
	14.1	3	3	3	NL	NL	NP	NP	NP

[a] Response modification coefficient, R, for use throughout the standard. Note that R reduces forces to a strength level, not an allowable stress level.

[b] Where the tabulated value of the overstrength factor, Ω_0, is greater than or equal to 2½, Ω_0 is permitted to be reduced by subtracting the value of 1/2 for structures with flexible diaphragms.

[c] Deflection amplification factor, C_d, for use in Sections 12.8.6, 12.8.7, and 12.9.1.2.

[d] NL = Not Limited, and NP = Not Permitted. For metric units, use 30.5 m for 100 ft and use 48.8 m for 160 ft.

[e] See Section 12.2.5.4 for a description of seismic force-resisting systems limited to buildings with a structural height, h_n, of 240 ft (73.2 m) or less.

[f] See Section 12.2.5.4 for seismic force-resisting systems limited to buildings with a structural height, h_n, of 160 ft (48.8 m) or less.

[g] In Section 2.3 of ACI 318. A shear wall is defined as a structural wall.

[h] In Section 2.3 of ACI 318. The definition of "special structural wall" includes precast and cast-in-place construction.

[i] An increase in structural height, h_n, to 45 ft (13.7 m) is permitted for single-story storage warehouse facilities.

[j] Steel ordinary concentrically braced frames are permitted in single-story buildings up to a structural height, h_n, of 60 ft (18.3 m) where the dead load of the roof does not exceed 20 lb/ft² (0.96 kN/m²) and in penthouse structures.

[k] See Section 12.2.5.7 for limitations in structures assigned to Seismic Design Categories D, E, or F.

[l] See Section 12.2.5.6 for limitations in structures assigned to Seismic Design Categories D, E, or F.

[m] In Section 2.3 of ACI 318. The definition of "special moment frame" includes precast and cast-in-place construction.

[n] Cold-formed steel—special bolted moment frames shall be limited to one story in height in accordance with ANSI/AISI S400.

[o] Alternately, the seismic load effect including overstrength, E_{mh}, is permitted to be based on the expected strength determined in accordance with ANSI/AISI S400.

[p] Ordinary moment frame is permitted to be used in lieu of intermediate moment frame for Seismic Design Categories B or C.

ASCE Table 12.14-1 Design Coefficients for SFRSs for the Simplified Method

Seismic Force-Resisting System	ASCE 7 Where Detailing is Specified	Response Modification Coefficient, R^a	Limitations[b] Seismic Design Category		
			B	C	D, E
A. BEARING WALL SYSTEMS					
1. Special reinforced concrete shear walls	14.2	5	P	P	P
2. Ordinary reinforced concrete shear walls	14.2	4	P	P	NP
3. Detailed plain concrete shear walls	14.2	2	P	NP	NP
4. Ordinary plain concrete shear walls	14.2	1½	P	NP	NP
5. Intermediate precast shear walls	14.2	4	P	P	40[c]
6. Ordinary precast shear walls	14.2	3	P	NP	NP
7. Special reinforced masonry shear walls	14.4	5	P	P	P
8. Intermediate reinforced masonry shear walls	14.4	3½	P	P	NP
9. Ordinary reinforced masonry shear walls	14.4	2	P	NP	NP
10. Detailed plain masonry shear walls	14.4	2	P	NP	NP
11. Ordinary plain masonry shear walls	14.4	1½	P	NP	NP
12. Prestressed masonry shear walls	14.4	1½	P	NP	NP
13. Light-frame (wood) walls sheathed with wood structural panels rated for shear resistance	14.5	6½	P	P	P
14. Light-frame (cold-formed steel) walls sheathed with wood structural panels rated for shear resistance or steel sheets	14.1	6½	P	P	P
15. Light-frame walls with shear panels of all other materials	14.1 & 14.5	2	P	P	NP[d]
16. Light-frame (cold-formed steel) wall systems using flat strap bracing	14.1	4	P	P	P
B. BUILDING FRAME SYSTEMS					
1. Steel eccentrically braced frames	14.1	8	P	P	P
2. Steel special concentrically braced frames	14.1	6	P	P	P
3. Steel ordinary concentrically braced frames	14.1	3¼	P	P	P
4. Special reinforced concrete shear walls	14.2	6	P	P	P
5. Ordinary reinforced concrete shear walls	14.2	5	P	P	NP
6. Detailed plain concrete shear walls	14.2 & 14.2.2.7	2	P	NP	NP
7. Ordinary plain concrete shear walls	14.2	1½	P	NP	NP
8. Intermediate precast shear walls	14.2	5	P	P	40[c]
9. Ordinary precast shear walls	14.2	4	P	NP	NP
10. Steel and concrete composite eccentrically braced frames	14.3	8	P	P	P
11. Steel and concrete composite special concentrically braced frames	14.3	5	P	P	P
12. Steel and concrete composite ordinary braced frames	14.3	3	P	P	NP
13. Steel and concrete composite plate shear walls	14.3	6½	P	P	P
14. Steel and concrete composite special shear walls	14.3	6	P	P	P
15. Steel and concrete composite ordinary shear walls	14.3	5	P	P	NP
16. Special reinforced masonry shear walls	14.4	5½	P	P	P
17. Intermediate reinforced masonry shear walls	14.4	4	P	P	NP
18. Ordinary reinforced masonry shear walls	14.4	2	P	NP	NP
19. Detailed plain masonry shear walls	14.4	2	P	NP	NP
20. Ordinary plain masonry shear walls	14.4	1½	P	NP	NP
21. Prestressed masonry shear walls	14.4	1½	P	NP	NP
22. Light-frame (wood) walls sheathed with wood structural panels rated for shear resistance	14.5	7	P	P	P
23. Light-frame (cold-formed steel) walls sheathed with wood structural panels rated for shear resistance or steel sheets	14.1	7	P	P	P
24. Light-frame walls with shear panels of all other materials	14.1 & 14.5	2½	P	P	NP[d]
25. Steel buckling-restrained braced frames	14.1	8	P	P	P
26. Steel special plate shear walls	14.1	7	P	P	P

[a] Response modification coefficient, R, for use throughout the standard.

[b] P = permitted; NP = not permitted.

[c] Light-frame walls with shear panels of all other materials are not permitted in Seismic Design Category E.

[d] Light-frame walls with shear panels of all other materials are permitted up to 35 ft (10.6 m) in structural height, h_n, in Seismic Design Category D and are not permitted in Seismic Design Category E.

ASCE Table C12.2-1 Conditions for OMFs and IMFs in Structures at SDC D, E or F

Section	Frame	SDC	Max. Number Stories	Light-Frame Construction	Max. h_n ft	Max. roof/floor D_L (lb/ft²)	Exterior Wall DL Max. (lb/ft²)	Wall[a] Height (ft)
12.2.5.6.1(a)	OMF	D, E	1	NA	65	20	20	35
12.2.5.6.1(a)-Exc	OMF	D, E	1	NA	NL	20	20	35
12.2.5.6.1(b)	OMF	D, E	NL	Required	35	35	20	0
12.2.5.6.2	OMF	F	1	NA	65	20	20	0
12.2.5.7.1(a)	IMF	D	1	NA	65	20	20	35
12.2.5.7.1(a)-Exc	IMF	D	1	NA	NL	20	20	35
12.2.5.7.1(b)	IMF	D	NL	NA	35	NL	NL	NA
12.2.5.7.2(a)	IMF	E	1	NA	65	20	20	35
12.2.5.7.2(a)-Exc	IMF	E	1	NA	NL	20	20	35
12.2.5.7.2(b)	IMF	E	NL	NA	35	35	20	0
12.2.5.7.3(a)	IMF	F	1	NA	65	20	20	0
12.2.5.7.3(b)	IMF	F	NL	Required	35	35	20	0

NL means No Limit, NA means Not Applicable. For metric units, use 20 m for 65 ft and use 10.6 m for 35 ft. For 20 lb lb/ft², use 0.96 kN/m² and for 30 lb/ft², use 1.68 kN/m².

[a] Applies to portion of wall above listed wall height.

ASCE Table 12.12-1 Allowable Story Drift, Δ_a [a, b]

Structure	Risk Category I or II	III	IV
Structures, other than masonry shear wall structures, four stories or less above the base as defined in Section 11.2, with interior walls, partitions, ceilings, and exterior wall systems that have been designed to accommodate the story drifts	$0.025h_{sx}$[c]	$0.020h_{sx}$	$0.015h_{sx}$
Masonry cantilever shear wall structures[d]	$0.010h_{sx}$	$0.010h_{sx}$	$0.010h_{sx}$
Other masonry shear wall structures	$0.007h_{sx}$	$0.007h_{sx}$	$0.007h_{sx}$
All other structures	$0.020h_{sx}$	$0.015h_{sx}$	$0.010h_{sx}$

[a] h_{sx} is the story height below level x.

[b] For seismic force-resisting systems solely comprising moment frames in Seismic Design Categories D, E, and F, the allowable story drift shall comply with the requirements of Section 12.12.1.1.

[c] There shall be no drift limit for single-story structures with interior walls, partitions, ceilings, and exterior wall systems that have been designed to accommodate the story drifts. The structure separation requirement of Section 12.12.3 is not waived.

[d] Structures in which the basic structural system consists of masonry shear walls designed as vertical elements cantilevered from their base or foundation support that are so constructed that moment transfer between shear walls (coupling) is negligible.

Seismic Characteristics of Engineered System

This page is intentionally left blank

Problems & Solutions

The following questions have been carefully designed to provide comprehensive coverage of the material presented in this book, while also being educational in nature, as you will notice. Additionally, you may come across a few questions not directly covered in this chapter; these have been intentionally included to ensure broader coverage of the content.

➤ Problem 2.1: System Connection Capacity
➤ Problem 2.2: Dual SFRS System Seismic Force Share
➤ Problem 2.3: Height Limitation for an Eccentrically Braced System
➤ Problem 2.4: Column Capacity
➤ Problem 2.5: SFRS Selection
➤ Problem 2.6: Shear Wall Bays
➤ Problem 2.7: Accidental Torsional Moment
➤ Problem 2.8: Story Drift (1)
➤ Problem 2.9: Story Drift (2)
➤ Problem 2.10: Out of Plane Irregularity Design Treatment
➤ Problem 2.11: Stability Method for Addressing Story Drift
➤ Problem 2.12: Minimum Distance Between Two Buildings
➤ Problem 2.13: Building Drift
➤ Problem 2.14: Stiffness vs. Ductility
➤ Problem 2.15: Foundation Isolators

PROBLEM 2.1 *System Connection Capacity*

In the below sketch, the smaller system to the right is attached to the rest of the building's shear wall SFRS system to the left via connections as shown.

Assume each slab associated with the smaller system to the right weighs a total of 9 *kip*, and that there will be two equidistant connections to the shear wall system at each level (the above sketch shows one only). The following design parameters apply to this building location:

$$S_S = 1.5$$
$$S_1 = 0.6$$
$$F_a = 1.2$$
$$F_v = 1.4$$

The design capacity of each connection should be:

(A) 0.25 *kip*
(B) 0.75 *kip*
(C) 1.25 *kip*
(D) 1.75 *kip*

PROBLEM 2.2 *Dual SFRS System Seismic Force Share*

Moment resisting frames are sometimes designed as part of the seismic force resisting system to support a braced frame system. What should be the portion of the lateral seismic force the braced frames should be designed to resist?

(A) 25%
(B) 50%
(C) 75%
(D) 100%

PROBLEM 2.3 *Height Limitation for an Eccentrically Braced System*

The below floor plan is 60 *ft* × 60 *ft* and is constructed using steel W sections for columns. Columns are eccentrically braced and designed not to take more than 60% of the total seismic load per plane in a way to create more resiliency.

Floor Plan

Elevation

If this building is in a SDC E, what would be its height limitation?

(A) 65 *ft*
(B) 100 *ft*
(C) 160 *ft*
(D) 240 *ft*

PROBLEM 2.4 *Column Capacity*

The below is a short residential complex supported on columns as depicted in the sketch below to create parking spaces for residents.

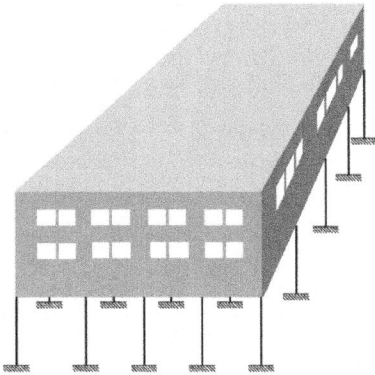

Using the strength design method, the axial load combination 6 and 7 as defined in the ASCE 7 Standard for any internal first floor column were as follows:

(6) $1.2D + E_v + E_h + L + 0.2S = 4.1 \ kip$

(7) $0.9D - E_v + E_h = 2.9 \ kip$

Assuming this building belongs to SDC E, what is the minimum axial capacity those internal columns should be designed for?

(A) 2.9 *kip*

(B) 4.1 *kip*

(C) 18.0 *kip*

(D) 27.3 *kip*

PROBLEM 2.5 SFRS Selection

A developer is considering the addition of a high-rise building exceeding 1,000 *ft* in height to one of his projects that will be used as residential. The seismic parameters for this location are as follows:

$S_S = 1.5$

$S_1 = 0.9$

$F_a = 1.2$

$F_v = 1.4$

What is the best Seismic Force Resisting System that shall be used to build this structure?

(A) Special reinforced concrete shear walls in a bearing wall system.

(B) Steel and concrete composite ordinary shear walls in a dual system with special moment frames that can resist 25% of seismic loads.

(C) Steel eccentrically braced frames in a building frame system.

(D) Steel special moment frames in a moment resisting frames system.

PROBLEM 2.6 *Shear Wall Bays*

The plan below is for a bearing wall system for an SFRS system. Shear walls are placed on axis 1, A and D as shown below.

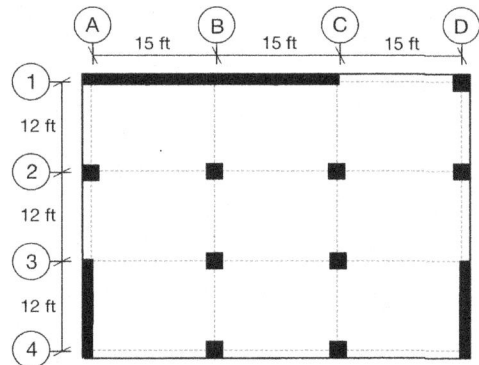

Knowing that the story height for this building is 10 *ft* and is made of reinforced concrete, how many shear wall bays are placed at axis 1?

(A) 1

(B) 2

(C) 3

(D) 4

PROBLEM 2.7 *Accidental Torsional Moment*

The story plan below illustrates the deflected shape, shown in light grey, resulting from seismic loading of 1,000 lb/ft on a floor that belongs to a reinforced concrete building assigned to SDC D.

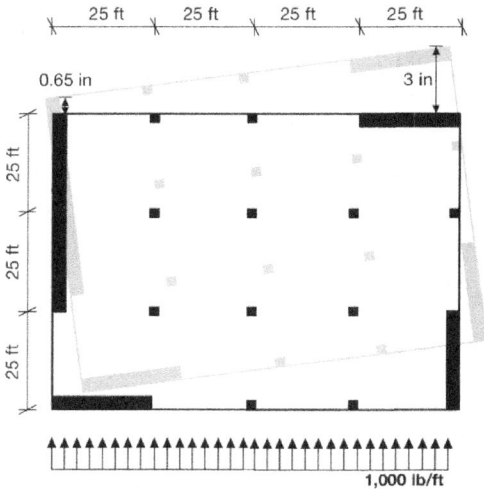

Based on the deflected floor shape and drift shown above, what is the approximate effect of accidental torsional moment for this floor?

(A) 500 $kip.ft$

(B) 1,000 $kip.ft$

(C) 1,500 $kip.ft$

(D) 2,000 $kip.ft$

PROBLEM 2.8 *Story Drift (1)*

What is the allowable story drift for a story with a height of 14 ft for a moment frame six-story building assigned to an SDC D in a risk category III having a torsional irregularity type 1a as defined per ASCE 7 Standard?

(A) 1.6 in

(B) 2.6 in

(C) 3.4 in

(D) 4.3 in

PROBLEM 2.9 *Story Drift (2)*

What is the allowable story drift for a story with a height of 14 ft for a solely moment frame six-story building assigned to an SDC D in a risk category III having a torsional irregularity type 1b as defined per ASCE 7 Standard?

(A) 1.6 in

(B) 2.6 in

(C) 3.4 in

(D) 4.3 in

PROBLEM 2.10 *Out of Plane Irregularity Design Treatment*

What is the proper design treatment for a building with a floor supported from below by a seismic force-resisting system that is not continuous along the same vertical line, with the building classified under SDC F?

(A) This design is prohibited in Seismic Design Category F.

(B) The seismic loads used for design shall include the overstrength factor.

(C) Allowable story drift shall be divided by the redundancy factor of this building.

(D) Nonlinear Response History Analysis shall be used to design this building.

PROBLEM 2.11 *Stability Method for Addressing Story Drift*

What is the ratio of a building's base shear to one of its story's vertical loads for a Risk Category II building in a Seismic Design Category B that eliminates the requirements for considering the P-Delta effect for an ordinary plain masonry shear wall system while ensuring an allowable story drift?

(A) 6%

(B) 11%

(C) 16%

(D) 21%

PROBLEM 2.12 *Minimum Distance Between Two Buildings*

What should be the minimum separation distance between the following two buildings when building No. 1 is made of intermediate reinforced concrete moment frames, while the taller building's SFRS system is a special steel moment frame.

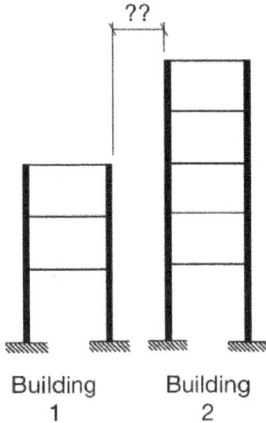

Building 1   Building 2

The two buildings are a risk category III, and both are assigned to SDC C. The following table represents the maximum elastic deflection per floor for each building:

Floor No	Building 1	Building 2
1	1.3 *in*	1.5 *in*
2	2.3 *in*	2.5 *in*
3	3.1 *in*	3.4 *in*
4	–	4.2 *in*
5	–	5.1 *in*

(A) 26.1 *in*

(B) 18.7 *in*

(C) 6.50 *in*

(D) 4.60 *in*

PROBLEM 2.13 *Building Drift*

A 40 *ft* tall, three-story building, located on soil type B and classified under Risk Category II with an occupant load $OL = 200$, requires a calculation of its lateral drift to assess the appropriate cladding installation. The actual elastic drift calculation for this building were found to be 0.2 *in*. The building has no irregularities or overhang.

What is the building structural drift that should be used for this purpose?

(A) *zero*

(B) 0.2 *in*

(C) 0.5 *in*

(D) 0.8 *in*

PROBLEM 2.14 *Stiffness vs. Ductility*

Which of the following statements best describes the difference between stiffness and ductility in seismic design:

(A) Stiffness refers to a structure's ability to deform plastically during an earthquake, while ductility measures its resistance to deformation under seismic forces.

(B) Stiffness measures a structure's resistance to lateral displacement during an earthquake, while ductility measures its ability to absorb and dissipate seismic energy through inelastic deformation.

(C) Stiffness is only relevant to the structural elements of a building, while ductility applies only to non-structural components.

(D) Stiffness and ductility are essentially the same in seismic design, as both describe how a structure resists seismic loads.

PROBLEM 2.15 *Foundation Isolators*

Which of the following is the primary benefit of using elastomeric or bearing-based foundation isolators in seismic design:

(A) They increase the overall lateral stiffness of the structure, reducing the displacement of the building during an earthquake.

(B) They decouple the building from ground motion, allowing for significant reductions in seismic forces transmitted to the structure while maintaining the building's stability.

(C) They absorb seismic energy through plastic deformation, effectively preventing damage to the foundation but potentially causing structural failure at higher seismic intensities.

(D) They rely on dynamic amplification of seismic waves to strengthen the building's resistance to earthquakes by increasing base shear forces.

SOLUTION 2.1

Reference in this question is made to ASCE 7 Standard Section 12.1.3 which refers to load path.

The smaller portion of the structure in this case should be tied to the rest of the structure and capable of transmitting the following force (divided by two because the question says that there are two connections per level):

$$F = \frac{0.133 \times S_{DS} \times Weight}{2} \ (*) \ or \ \frac{0.05 \times Weight}{2}$$

$$S_{DS} = \frac{2}{3} S_{MS}$$
$$= \frac{2}{3} F_a S_S$$
$$= \frac{2}{3} \times 1.2 \times 1.5$$
$$= 1.2$$

$$F = \frac{0.133 \times 1.2 \times 9}{2} \ or \ \frac{0.05 \times 9}{2}$$
$$= \frac{0.133 \times 1.2 \times 9}{2} \ or \ \frac{0.05 \times 9}{2}$$
$$= 0.72 \ kip \ or \ 0.23 \ kip$$

The larger value governs which is **0.72 kip**.

Correct Answer is (B)

(*) Worth noting that the first equation is multiplied by 0.2 for the "Simplified Method" as stated in ASCE 7 Section 12.14.7.1 in place of the 0.133 shown here.

SOLUTION 2.2

Reference in this question is made to ASCE 7 Standard Section 12.2.5.1 which states that the moment resisting frames system should be able to resist 25% of the overall seismic load, and that the seismic load should be distributed to the two systems based on each system's rigidity. Based on this, the lateral force that the braced system should be able to resist is the balance of that which is 75%.

Correct Answer is (C)

SOLUTION 2.3

The height limitation for an eccentrically braced frame per Table 12.2-1 of ASCE 7 Standard item B.1 is $160 \, ft$ for an SDC E. However, Section 12.2.5.4 states that for eccentrically braced frame, the height limitation can be revised for SDC E to $240 \, ft$ provided that:
- No extreme torsional eccentricity, which is obvious from the floor plan provided.
- The braced frames in any plane shall resist no more than 60% of the total seismic force in each direction.

Based on the above, the new height limitation for this building can be revised to **240 ft**.

Correct Answer is (D)

SOLUTION 2.4

Reference in this question is made to ASCE 7 Standard Section 12.2.5.2 which provides the required (i.e., applied) axial capacity for columns in a cantilever system. It states that they should not exceed 15% of their capacity (i.e., available axial strength) for the load combinations that include seismic effect, which are load combination 6 and 7 of the strength design method presented in Section 2.3.6 of the ASCE 7 Standard.

In this case, the column's capacity should not be less than the maximum of:

$$4.1/15\% = 27.3 \ kip$$

or

$$2.7/15\% = 18.0 \ kip$$

Correct Answer is (D)

SOLUTION 2.5

Reference in this question is made to ASCE 7 Standard Section 12.2. Also, with reference to Section 11.6, while $S_1 > 0.75$ with a risk category III (high occupancy residential), the Seismic Design Category is SDC E.

Table 12.2-1 in Section 12.2 lists all SFRS systems along with their height limitations in accordance with the Seismic Design Categories. We will be looking in this table to find a Not Limited (NL) height in this case.

For SDC E this applies to items C.1 in the table: **Steel special moment frames in a moment resisting frames system.**

Correct Answer is (D)

SOLUTION 2.6
Reference in this question is made to ASCE 7 Standard Section 12.3.4.2.(b). The number of shear wall bays in case of heavy building frames equals to the length of shear wall(s), divided by the story height as follows:

$$= \frac{30\,ft}{10\,ft} = 3\,bays$$

Correct Answer is (C)

SOLUTION 2.7
Reference in this question is made to ASCE 7 Standard Sections 12.8.4.2 & 3 which calculates the accidental torsional moment (M_{ta}).

Prior to proceeding with (M_{ta}), we need to assess whether the floor diaphragm has any irregularities.

Referring to Figure 2-16 here (or Table 12.3-1 of the ASCE 7 Standard), it appears that this floor has a Type 1b horizontal irregularity, as follows:

$$\frac{\delta_{max}}{\delta_{ave}} = \frac{3}{0.5 \times (3 + 0.65)} = 1.64 > 1.4$$

Based on this irregularity, and in reference to Table 2-1 here, accidental torsional moment effect shall be magnified by (A_x) as

discussed in Section 12.8.4.3 of the ASCE 7 Standard for not flexible diaphragms assigned to SDC D as follows:

$$A_x = \left(\frac{\delta_{max}}{1.2\delta_{avg}}\right)^2$$
$$= \left(\frac{3}{1.2 \times 1.825}\right)^2$$
$$= 1.9 < 3$$

Per Section 12.8.4.2, and for not flexible diaphragms, the accidental torsional moment (M_{ta}) is caused by the assumed displacement of center of mass by a distance that equals to 5% of the dimension perpendicular to the direction of the applied force. See below:

$$M_{ta} = 1.9 \times \left(1,000\tfrac{lb}{ft} \times 100\,ft\right) \times 5\% \times 100\,ft$$
$$= 950,000\,lb.ft\,(950\,kip.ft)$$

Correct Answer is (B)

SOLUTION 2.8
Reference in this question is made to Table 2-1 here, or Table 12.3-1 of the ASCE 7 Standard along with Table 12.12-1.

$$\Delta_a = 0.020h_{sx}$$
$$= 0.020 \times \left(14\,ft \times 12\tfrac{in}{ft}\right)$$
$$= 3.36\,in$$

Correct Answer is (C)

SOLUTION 2.9
Reference in this question is made to Table 2-1 here, or Table 12.3-1 of the ASCE 7 Standard along with Table 12.12-1.

$$\Delta_a = 0.020h_{sx}$$
$$= 0.020 \times \left(14\,ft \times 12\tfrac{in}{ft}\right)$$
$$= 3.36\,in$$

Since this is a torsional irregularity Type 1b assigned to SDC D, Section 12.3.4.2 of the ASCE 7 Standard assigns the building a redundancy factor $\rho = 1.3$, in which case Section 12.12.1.1 requires the allowable story drift (Δ_a) for moment frames assigned to SDC D to be divided by (ρ).

$$= \frac{3.36}{1.3} = 2.58 \; in$$

Correct Answer is (B)

SOLUTION 2.10
The description provided in this question corresponds to an out-of-plan offset irregularity, also known as Type 4 horizontal irregularity, as illustrated in Figure 2-16 here and described in Table 12.3-1 of the ASCE 7 Standard.

According to Section 12.3.3.3 of the ASCE 7 Standard (see Table 2-1 here), addressing this irregularity requires that seismic loads account for the overstrength factor.

Correct Answer is (B)

SOLUTION 2.11
Reference in this question is made to ASCE 7 Standard Section 12.8.7 (Section 2.4.2 here) along with Table 12.2-1 and Table 12.12-1.

Start with determining the allowable drift that the question requests to comply with using Table 12.12-1 for masonry shear wall structures as follows:

$$\Delta_a = 0.007h_{sx}$$

Use the stability equation presented in ASCE 7 Section 12.8.7 and discussed in Section 2.4.2 of this book. Set the stability coefficient θ to 0.1, which is the threshold below which P-Delta effect can be disregarded. Also, use Table 12.2-1, line-item A.11, to extract the deflection amplification factor $C_d = 1\frac{1}{4}$ for the SFRS system described in this question, along with an importance factor of $I_e = 1.0$ for risk category II per Table 1.5-2 of the ASCE 7 Standard.

$$\theta = \frac{P_x \Delta I_e}{V_x h_{sx} C_d} \le 0.1$$

$$\frac{V_x}{P_x} \ge \frac{\Delta I_e}{\theta h_{sx} C_d}$$

$$\ge \frac{0.007 h_{sx} \times 1.0}{0.1 \times h_{sx} \times 1.25}$$

$$\ge 0.056 \; (5.6\%)$$

Correct Answer is (A)

SOLUTION 2.12
Reference in this question is made to ASCE 7 Standard Section 12.12.3 along with Table 12.12-1.

Start with determining the deflection amplification factor (C_d) for each building, after which, determine the maximum inelastic response at floor 3 only for the two buildings as this is the floor where pounding is most likely to occur.

Building 1 SFRS is line C.6 of Table 12.12-1:

$$C_d = 4.5$$
$$I_e = 1.25$$
$$\delta_{M1} = \frac{C_d \delta_{max}}{I_e}$$
$$= \frac{4.5 \times 3.1 \; in}{1.25}$$
$$= 11.16 \; in$$

Building 2 SFRS is line C.1 of Table 12.12-1:

$$C_d = 5.5$$
$$I_e = 1.25$$
$$\delta_{M2} = \frac{C_d \delta_{max}}{I_e}$$
$$= \frac{5.5 \times 3.4 \; in}{1.25}$$
$$= 14.96 \; in$$

$$\delta_{MT} = \sqrt{(\delta_{M1})^2 + (\delta_{M2})^2}$$
$$= \sqrt{(11.16)^2 + (14.96)^2}$$
$$= 18.66 \; in$$

Correct Answer is (B)

SOLUTION 2.13

Reference in this question is made to ASCE 7 Standard Section 12.14.8.5.

The building described in this question is qualified to use the simplified method. Refer to section 1.5.3 of this book for more details or Section 12.14 of the ASCE 7 Standard.

Using the simplified method, 1% of the building height can be used for structural drift (i.e., 0.4 *in*), **unless computed drift is less (i.e., 0.2 *in*).**

Correct Answer is (B)

SOLUTION 2.14

Stiffness measures the structure's resistance to lateral displacement during an earthquake, while ductility measures its ability to absorb and dissipate seismic energy through inelastic deformation.

Correct Answer is (B)

SOLUTION 2.15

Foundation isolators are primarily designed to decouple the building from the ground motion during an earthquake, reducing the seismic forces that are transmitted to the structure. This allows for less displacement and potential damage, improving the building's overall performance during a seismic event.

Correct Answer is (B)

This page is intentionally left blank

CHAPTER
3

Seismic Forces: Building Structures

3.1. Purpose of this Chapter

Building on the concepts presented in the previous two chapters, this chapter consolidates that information and puts it to use in this chapter using the Equivalent Lateral Force (ELF) method.

We will examine the roles of mass and stiffness, methods for determining a structure's fundamental period, and the integration of various seismic force-resisting systems. Key topics include calculating seismic base shear, vertical force distribution, and the design forces on diaphragms and structural elements, including out-of-plane and anchorage forces on walls.

3.2. Mass, Lateral Stiffness and Buildings Fundamental Period
3.2.1. The Single/Multi Degree of Freedom

Consider the one-story building shown in Figure 1-8 of Chapter 1 and let's assume that the floor of this buildings rests on columns that have certain lateral flexibility. During an earthquake, these columns will deform laterally. This building can be modeled as a single degree of freedom (SDOF) system, where a mass on top of a flexible stick represents the building as shown in Figure 3-1 here.

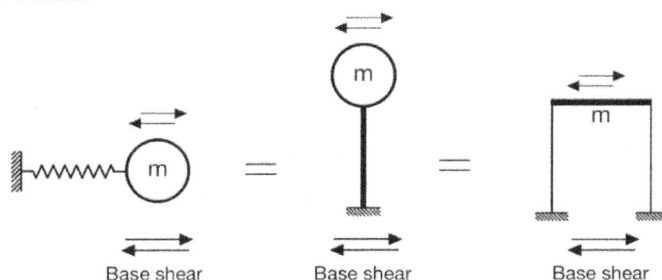

Figure 3-1 Single Degree of Freedom System

In this example, when the ground shakes with a certain degree of acceleration, the mass vibrates while the stick (the building columns) deflects. This deflection acts like a spring, as the lateral stiffness of the columns returns the mass to its original position, assuming there is no inelastic deformation or failure.

Theoretically speaking, the lateral stiffness of the columns, combined with the floor mass above them, can be used to calculate the fundamental period (T) of the structure, which represents its frequency of vibration. This period can then be used with the design response spectrum discussed in Chapter 1 to derive the spectral acceleration (S_a), allowing us to determine the lateral force acting on each floor using the Newton's second law of motion $F = m.a$ – Note that mass (m) in this case should be divided by the ground acceleration (g) when using US unit to convert the mass force (*lbm*) into weight force (*lbf*) – it is important to note that the method used to determine lateral seismic force is more complex than this simplified approach.

Example 3.1 – Simple Lateral Force Calculation for an SDOF System

Using the SDOF provided in Figure 3-1, assuming this model is located at the Los Angeles Airport with the same parameters provided in Example 1.3 of Chapter 1. While the mass attached to this SDOF is 500 *lbm*, and the natural frequency is $T = 1.5\ sec$. What is the horizonal force acting on this SDOF? **(ignore other seismic coefficients that are part of the spectral acceleration as those will be discussed later in Section 3.3).**

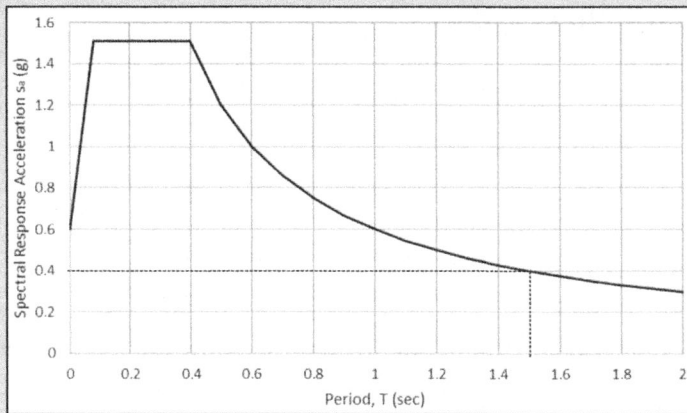

With a $T = 1.5\ sec$ → $S_a = 0.4g$, which leads to the following:

$$F = \frac{m.S_a}{g} = \frac{500 \times 0.4g}{g} = 200\ lb$$

However, the situation can be more complicated, especially when dealing with multiple stories (Multi Degree Of Freedom systems – or MDOF). Additionally, inelastic conditions may complicate the analysis further. The ASCE 7 Standard addresses these complexities by providing various methods and analytical procedures to calculate the appropriate coefficients for multiplying by story weights (or the effective seismic weight) in order to determine the base shear, which is then distributed vertically and horizontally across each floor.

In this section we only discussed the Single Degree Of Freedom (SDOF). You can refer to the appendix of this chapter for an overview of the Multi Degree Of Freedom (MDOF).

3.2.2.   The Effective Seismic Weight

The previous section highlighted that the building floor masses, and any of their permanent attachments, contribute to the lateral forces experienced by a building during an earthquake. To quantify this, the ASCE 7 Standard defines the effective seismic weight as the total weight of the floor and its permanent fixtures, which can swing and create inertia in response to seismic activity. For this reason, you can observe that in the following paragraph, only a small percentage of the live load is considered in this calculation, and there are several exclusions that pertain to non-inertia-generating masses.

According to ASCE 7 Standard Section 12.7.2, the effective seismic weight of a structure includes the dead load above the base – i.e., the foundation or upper basement level – along with additional specified loads. For buildings with storage areas, at least 25% of the floor live load must be factored in, unless storage load < 5% of the effective seismic weight, or if the structure is a public garage or open parking area. The effective seismic weight also encompasses partition weights, requiring either the actual weight or a minimum of 10 psf of floor area, as well as the total weight of permanently attached operating equipment and the weight of landscape or roof landscape if existing. Additionally, if the snow load exceeds 30 psf, only 20% of that load is included, reflecting the low likelihood of a major earthquake coinciding with a heavy snowstorm.

3.2.3. Lateral Stiffness

To clarify the concept of lateral stiffness, we can compare it to the stiffness of a spring, as illustrated in Figure 3-1. The stiffness of a spring (k) is defined as its resistance to deformation, calculated by dividing the force (P) applied to the spring by the resulting deformation (Δ).

$$k = \frac{P}{\Delta}$$

When we apply this concept to buildings, the stiffness of a Structural Frame Resisting System (SFRS) refers to its ability to resist lateral deformation caused by earthquakes or other lateral forces. In this context, the determination of stiffness depends on several factors: the material used (such as concrete, wood, steel, or masonry), characterized by its modulus of elasticity (E); the shape of the cross-sections, represented by their moment of inertias (I); and the height of the structural element (h). It is normally expressed in the following format:

$$k = coefficient \times \frac{EI}{h^3}$$

Lateral deflection of a wall or a column depends on the shape of the applied force (point or distributed) and the end conditions (fixed, pinned or free). Consequently, the $coefficient$ in the above equation varies. Several end and load conditions related to stiffness (k) are detailed in the appendix at the end of this chapter that could apply for both walls and columns.

When it comes to concrete walls, it is important to differentiate between two types of deflection, which correspond to two types of stiffnesses. Flexural deflection occurs in relatively tall shear walls, when height-to-length ratio $h/l > 3.0$ (4.0 for masonry walls) (see Figure 3-2). Shear deflection occurs when $h/l < 0.3$ (0.25 for masonry walls). Stiffness equations for these cases can be found in the chapter's appendix along with some tables to simplify their calculation.

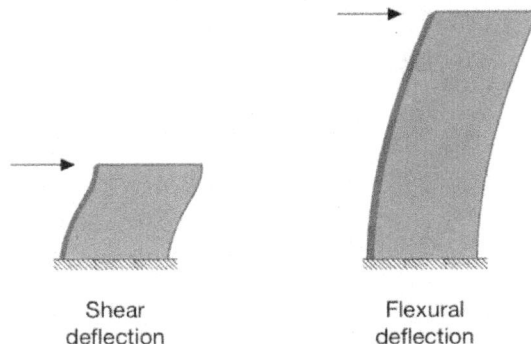

Shear deflection

Flexural deflection

Figure 3-2 Shear Wall Deflection Types

When height-to-length falls between 0.3 and 3.0, which could be the case of most of the buildings, both stiffnesses should be added together using the following equation:

$$\frac{1}{\sum k_i} = \frac{1}{\sum k_{si}} + \frac{1}{\sum k_{fi}}$$

Where:

$\sum k_i$ = Summation of total stiffness at level i.

$\sum k_{si}$ = Summation of shear stiffness at level i.

$\sum k_{fi}$ = Summation of flexural stiffness at level i.

A simplified method for combining the above is to use an effective moment of inertia (I_{eff}). I_{eff} approximates the moment of inertia that would result from a flexural deflection that equals to the combined flexural and shear deflection of the wall.

Example 3.2 – Single Story Shear Wall Stiffness Calculation

The plan cross-sectional area for a cantilever concrete shear wall is 12 ft long and 8 in thick. The wall height is 10 ft and is a single story. The diaphragm lateral load is applied to the wall similar to the first case presented in Table 3-5 in the chapter's Appendix. Concrete modulus of elasticity is $E = 4.2 \times 10^6\ psi$. What is the lateral stiffness for this wall?

The height-to-length for this wall $0.3 < (h/l = 10/12) < 3$, in which case, flexural and shear stiffness shall be calculated and added together as discussed in this section:

$$I = \frac{tl^3}{12} = \frac{(8/12)\times 12^3}{12} = 96\ ft^4 \quad \&\quad A = 12 \times \left(\frac{8}{12}\right) = 8\ ft^2.$$

$$k_s = \frac{AE}{3h} = \frac{8\ ft^2 \times 4.2 \times 10^6\ \frac{lb}{in^2} \times \left(\frac{144\ in^2}{ft^2}\right)}{3 \times 10\ ft} = 161.3 \times 10^6\ lb/ft$$

$$k_f = \frac{3EI}{h^3} = \frac{3 \times 4.2 \times 10^6\ \frac{lb}{in^2} \times \left(\frac{144\ in^2}{ft^2}\right) \times 96\ ft^4}{(10\ ft)^3} = 174.2 \times 10^6\ lb/ft$$

$$\frac{1}{k} = \frac{1}{k_s} + \frac{1}{k_f} = \frac{1}{161.3\times 10^6} + \frac{1}{174.2\times 10^6} \rightarrow k = 83.8 \times 10^6\ lb/ft$$

Alternatively, let's try use the (I_{eff}) method and apply it to the flexural stiffness equation alone to verify if we get the same result:

$$I_{eff} = \frac{I}{1 + \frac{9.0I}{Ah^2}} = \frac{96\ ft^4}{1 + \frac{9.0 \times 96\ ft^4}{8\ ft^2 \times (10\ ft)^2}} = 46.2\ ft^4$$

$$k = \frac{3EI_{eff}}{h^3} = \frac{3 \times 4.2 \times 10^6\ \frac{lb}{in^2} \times \left(\frac{144\ in^2}{ft^2}\right) \times 46.2\ ft^4}{(10\ ft)^3} = 83.8 \times 10^6\ lb/ft\ (*)$$

(*) Refer to the Appendix of this chapter and Table 3-6 for a quick method to calculate (k) with the use of (I_{eff}). **The quick method is shown in Example 3.8.**

Rigidity (R) and stiffness (k) are often used interchangeably, but they have subtle differences. Rigidity refers to the ability to resist deformation, while stiffness is the quantitative measure of a wall's resistance to elastic deformation. These terms are used throughout this book interchangeably.

3.2.4.   Buildings Fundamental Period

Calculating the Fundamental Period

Following the section on calculating the stiffness of walls and columns, we can now use this information to estimate the fundamental period of a structure using a simple method. This approach is particularly useful for basic structures like single columns or low-rise buildings.

The fundamental period (T) can be calculated using the following equation:

$$T = 2\pi \sqrt{\frac{W}{kg}}$$

Where (W) is the entire mass of the building in (lb) or (kip), (k) in this case is the total stiffness of contributing elements in the desired direction, and $g = 32.174\ ft/sec^2$.

The simplified method provides a quick way to estimate the fundamental period of a structure. For more accurate results, especially for complex structures, detailed analysis using structural software is recommended.

Example 3.3 – Simple Building Fundamental Period Calculation

This symmetrical simple building has four square concrete columns each with a moment of inertia $I = 1,700\ in^4$. The concrete modulus of elasticity is $E = 4.2 \times 10^3\ ksi$. The building height is $12\ ft$. Assume that columns are pinned at the slab level and fixed to their footings, and the total building weight of $300\ kip$. Calculate the building fundamental period in any direction.

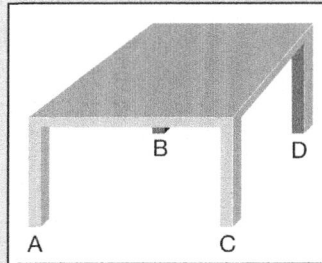

Per Table 3-5 (Case 1): $k = 4 \times \left(\frac{3EI}{h^3}\right) = 4 \times \left(\frac{3 \times 4.2 \times 10^3 \frac{kip}{in^2} \times 1,700\ in^4 \times \frac{ft^2}{144\ in^2}}{(12\ ft)^3}\right) = 344.3\ \frac{kip}{ft}$

$$T = 2\pi \sqrt{\frac{W}{kg}} = 2\pi \sqrt{\frac{300\ kip}{344.3\ kip/ft \times 32.174\ ft/sec^2}} = 1.03\ sec\ (*)$$

(*) The same example is repeated in Problem 3.23 with a fixed both end columns to show how the period is affected (i.e., reduced due to increased stiffness).

Approximate Fundamental Period

In this section, we will explain the approximate fundamental period (T_a), as defined by the ASCE 7 Standard Section 12.8.2. This standard provides three equations for calculating (T_a) for buildings, which do not apply to non-building structures, and those are summarized in Table 3-1 below:

Table 3-1 Approximate Fundamental Period Equations

Equation	Conditions for use
$T_a = C_t h_n^x$	This is the **general equation** and is a product of empirical studies carried out in California, hence, only applicable in California. In this equation, values for (C_t) and (x) are taken from Table 12.8-2 of the ASCE 7 Standard, while (h_n) is the story height. **Table 12.8-2 is copied in the chapter's appendix with permission from ASCE 7 for ease of reference.**
$T_a = 0.1N$	This equation can be used when there are less than 12 stories and only used for either **Concrete or Steel Moment Frames**, with an average story height of at least $10\ ft$. Where (N) in this case is the number of stories above the base.
$T_a = \dfrac{C_q}{\sqrt{C_w}}\, h_n$	This is used for **Masonry or Concrete Shear Walls only** with $120\ ft$ maximum height. Where: $C_w = \dfrac{100}{A_B} \sum_{i=1}^{x} \dfrac{A_i}{\left[1+0.83\left(\frac{h_n}{D_i}\right)^2\right]}$ $C_q = 0.0019\ ft$ A_B is the area of base of structure. A_i is the web area of shear wall (i) in (ft^2). D_i is the length of shear wall (i) in (ft). x is the number of shear walls resisting lateral forces in the direction under consideration.

If the building is complex, it may be necessary to calculate the fundamental period rather than using the approximate method explained above. This calculation can be performed using computer software. However, ASCE 7 Table 12.8-1 sets an upper limit for the calculated fundamental period using the coefficient (C_u). This means the calculated fundamental period cannot exceed $(T_a \times C_u)$.

$$T_{calculated} < T_a C_u$$

The reason for this upper limit is that lateral forces decrease as the fundamental period of the structure increases. To better understand this concept, refer to the response spectrum in Section 1.5 of this book. The spectrum shows that spectral acceleration (S_a) decreases as the fundamental period (T) increases, which means more flexible buildings experience reduced lateral forces. If there was an error in calculating the fundamental period, this upper limit will help catch this error.

For further clarity, see Table 3-2 and Figure 3-3 below. Table 3-2 demonstrates that as the acceleration (S_{D1}) decreases (due to an increase in T), the upper limit coefficient (C_u) increases slightly. This table is reproduced here with permission from ASCE.

Table 3-2 Coefficient for Upper Limit on Calculated Period

S_{D1}	C_u
≥ 0.4	1.4
0.3	1.4
0.2	1.5
0.15	1.6
≤ 0.1	1.7

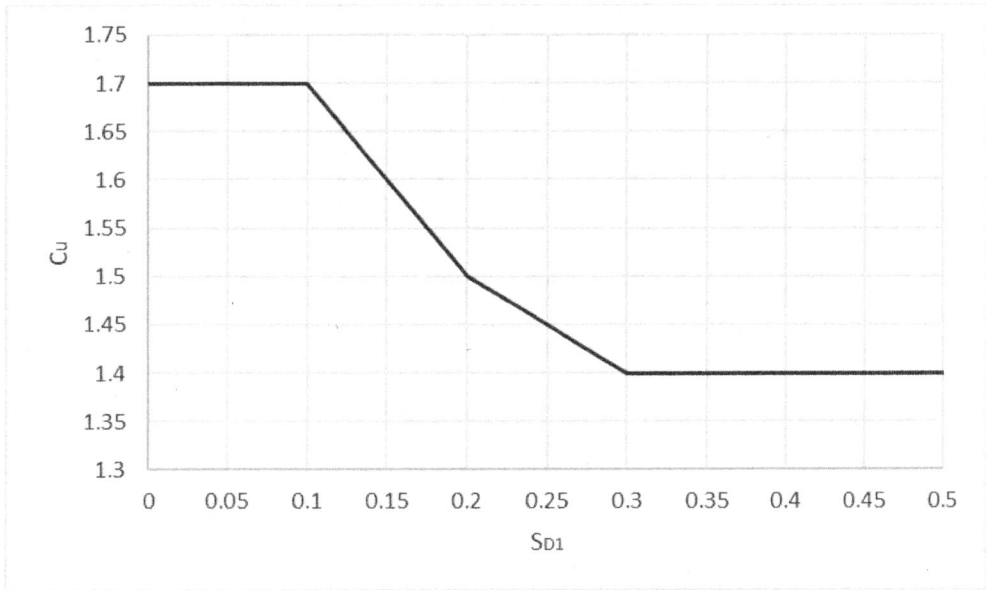

Figure 3-3 Coefficient for Upper Limit for Calculated Period

Additionally[8], ASCE 7 Standard allows the use of the approximate fundamental period (T_a) for base shear calculations in place of the calculated one ($T_{calculated}$), which offers conservative limits. Refer to Table 3-3 below for further details on use.

Table 3-3 Summary of Period Values to be used in Calculations

Situation	Period (T) to be used in <u>base shear</u> calculations	Period (T) to be used in <u>drift</u> calculations
$T_{calculated} \leq T_a$	T_a	T_a
$T_a < T_{calculated} < C_u T_a$	$T_{calculated}$	$T_{calculated}$
$T_{calculated} \geq C_u T_a$	$C_u T_a$	$T_{calculated}$

For elastic drift calculations (δ_{xe}) as outlined in ASCE 7 Standard Section 12.8.6.2, the code permits the use of the calculated fundamental period, as it tends to yield conservative results for drift calculations.

---

[8] (T) is also used to calculate the (k) factor used in the vertical load distribution. This is discussed later in Section 3.3.3 of this chapter.

PE Essential Guides | California Seismic Principles

Example 3.4 – Building Approximate Fundamental Period Calculation

The five stories above the basement are made of concrete moment-resisting frames while the basement is made of rigid shear walls imbedded into stiff soil.

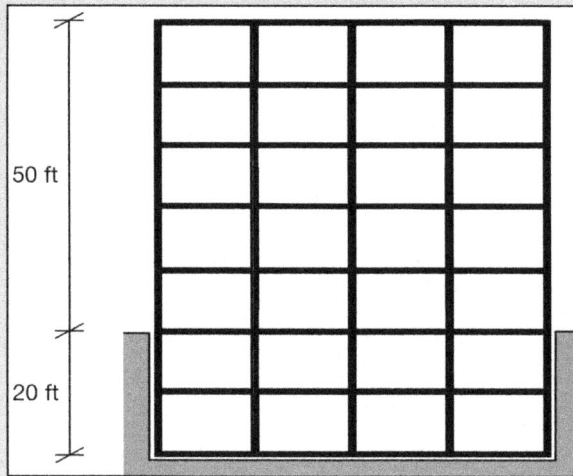

What is the approximate fundamental period for this building?

Using ASCE 7 Section 12.8.2.1 (see Table 3-1 here), the following general equation can be used while consulting with Table 12.8-2 of ASCE. Also, consider that the building height is 50 ft since it is built on a rigid system and on stiff soil per ASCE 7 Section C11.2.

$$T_a = C_t h_n^x = 0.016 \times 50^{0.9} = 0.54 \ sec$$

Equation 12.8-8 can also be used since this is a concrete moment frame, with no other system combined with it, and less than 12 stories:

$$T_a = 0.1N = 0.1 \times 5 = 0.5 \ sec$$

Example 3.5 – Applications for the Fundamental Period

A building is located in a region of moderate seismicity with $S_{D1} = 0.25g$. The calculated building's fundamental period is $T_{calculated} = 0.75 \ sec$, while the approximate fundamental period of the building using ASCE 7 relevant equations is $T_a = 0.45 \ sec$.

Which fundamental period shall be used for elastic drift calculations, and which one shall be used for base shear calculations?

Start with determining the upper limit coefficient from ASCE 7 Table 12.8-1 (Table 3.2 here) as $C_u = 1.45$, in which case, the upper limit for the fundamental period is:

$$T_{upper} = C_u \times T_a = 1.45 \times 0.45 = 0.65 \ sec < T_{calculated}$$

In this case, and in reference to Table 3-3 here, $T = 0.65 \ sec$ shall be used for base shear calculations and $T = 0.75 \ sec$ shall be used for elastic drift calculations.

3.3. Seismic Design Loads
3.3.1.   The Equivalent Lateral Force Procedure

Section 2.3.5 of this book provided a high-level definition of the Equivalent Lateral Force (ELF) procedure. Additionally, ASCE 7 Code Table 12.6-1 outlines two more complex analytical procedures alongside the ELF, detailing the conditions for their application. Generally, the ELF method is appropriate for structures without significant irregularities and when the risk categories are low.

The Equivalent Lateral Force (ELF) procedure treats the structure as a Single Degree Of Freedom (SDOF) system with mass participation in the first mode (see Appendix of this chapter for more information about this). This method is beneficial for preliminary design and is often permitted for final design in most structures. The ELF procedure is applicable to structures with consistent mass and stiffness along their height, primarily responding to horizontal ground motion without significant torsion.

The ELF procedure consists of three main steps: (1) it starts with calculating the base shear, (2) it then distributes it vertically, and (3) it distributes shear horizontally across the diaphragm width. These steps are discussed in the following sections except for the horizontal distribution of floor shear, this is covered separately in Chapter 4.

3.3.2.   Base Shear

To determine the seismic base shear (V) using the ELF procedure, it is expressed as the product of the effective seismic weight (W) discussed earlier and a period-dependent response coefficient (C_s), which is modified to account for inelastic behavior and enhanced performance for critical structures.

$$V = C_s W$$

C_s is calculated per Table 3-4, where (R) in this case is the response modification factor provided by Table 12.2-1 of ASCE 7 and (I_e) is the importance factor discussed earlier in Chapter 1, Table 1-5, or Table 1.5-2 of the ASCE 7 Standard.

Table 3-4 Seismic Response Coefficient (C_s)

Equation	Condition for Use
$C_s = \dfrac{S_{DS}}{\left(\frac{R}{I_e}\right)}$	Used when $T \leq T_s$ *C_s calculated here cannot exceed C_s calculated below when $T_s < T \leq T_L$*
$C_s = \dfrac{S_{D1}}{T\left(\frac{R}{I_e}\right)}$	Used when $T_s < T \leq T_L$ .... Where $T_s = S_{D1}/S_{DS}$
$C_s = \dfrac{S_{D1}T_L}{T^2\left(\frac{R}{I_e}\right)}$	Used when $T > T_L$
$C_{s,min} = 0.044 S_{DS} I_e \geq 0.01$	General use (*)
$C_{s,min} = \dfrac{0.5 S_1}{\left(\frac{R}{I_e}\right)}$	Used when $S_1 \geq 0.6g$

(*) S_{DS} here can be taken as $1.0 \geq 0.7 \times S_{DS,actual}$ provided: no irregularities, stories ≤ 5 with mezzanine considered as a story, $T \leq 0.5\ sec$, $\rho = 1$, not Site Class E or F, or Risk Category III or IV. For more details refer to ASCE 12.8.1.3.

3.3.3.   Vertical Seismic Force Distribution

Now that we have determined the base shear, we will discuss how it is distributed vertically across all floors. This analysis uses the Equivalent Lateral Force (ELF) method, which assumes the first mode of vibration (see the chapter appendix for more details on higher modes). To simplify the integration of higher vibration modes into the ELF method, ASCE introduces the (k) coefficient. This coefficient is proportional to the fundamental period of the building – i.e., a higher fundamental period indicates a more flexible structure, which tends to experience higher modes of vibration, resulting in an increase in the (k) coefficient.

The following equations are taken from ASCE 7 Standard Section 12.8.3:

$$F_x = C_{vx}V$$

$$C_{vx} = \frac{w_x h_x^k}{\sum_{i=1}^{n} w_i h_i^k}$$

Where:

C_{vx} is the vertical distribution factor.
$w_{x\ or\ i}$ is the portion of the effective seismic weight at level (x) or (i).
$h_{x\ or\ i}$ is the height from base to either level (x) or (i).
k coefficient, as discussed above, is directly proportional to period (T) as follows:
$k = 1\ for\ T \leq 0.5\ \text{sec}$

$k = 2\ for\ T \geq 2.5\ sec$

$k = 2\ for\ 0.5 < T < 2.5$ – or interpolate using $k = 0.75 + 0.5T$

Example 3.6 – Seismic Load Calculation and Vertical Distribution

The four-story building below has a total effective seismic weight of 1,050 *kip* while each floor weighs 300 *kip* and the roof weighs 150 *kip* only.

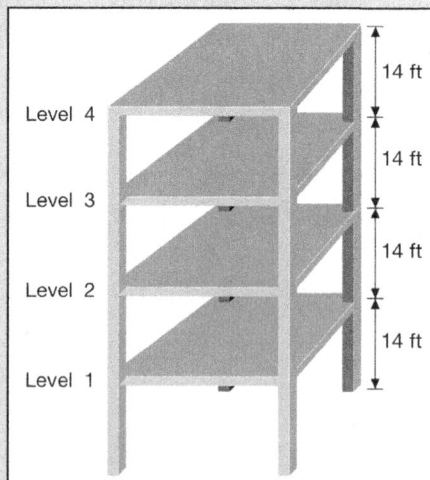

The following seismic coefficients apply to this building:
$R = 3$
$S_1 = 0.65, S_{D1} = 0.60, S_{DS} = 0.75$
$T = 1.0\ sec, T_L = 3.0\ sec, I_e = 1.0$

What is the lateral seismic force applied at each floor?

Using ASCE 7 Section 12.8.1 to calculate the base shear as follows while complying with the limits presented in the same section:

$$T_s = \frac{S_{D1}}{S_{Ds}} = \frac{0.6}{0.75} = 0.8$$

Use the following upper limit for (C_s) since $T > T_s$:

$$C_s = \frac{S_{D1}}{T\left(\frac{R}{I_e}\right)} = \frac{0.6}{1.0\left(\frac{3.0}{1.0}\right)} = 0.2$$

Check against the following lower limit for (C_s) since $S_1 > 0.6$:

$$C_s = \frac{0.5S_1}{\left(\frac{R}{I_e}\right)} = \frac{0.5 \times 0.65}{\left(\frac{3.0}{1.0}\right)} = 0.11$$

Use $C_s = 0.2$ to calculate the base shear as follows:

$$V = C_s W = 0.2 \times 1,050 = 210 \ kip$$

Moving to the vertical seismic load distribution per Section 3.3.3 of this book. The following table is designed to calculate the coefficient (C_{vx}) for each floor. It is important to note that the height of each floor is measured from the base up to the floor in question. Additionally, the (k) coefficient is interpolated to $k = 1.25$ to account for a higher vibration mode since $T = 1.0 \ sec$ – note that the ASCE permits (k) to be taken as 2.0 in this case (see 12.8.3 of the ASCE 7 Standard).

Floor	h_x	h_x^k	w_x	$w_x h_x^k$	C_{vx}	F_x
	(ft)		(kip)			(kip)
4	56	153.19	150	22,978.7	0.28	58.5
3	42	106.92	300	32,076.2	0.39	81.6
2	28	64.41	300	19,322.7	0.23	49.2
1	14	27.08	300	8,124.2	0.10	20.7
			1,050	82,501.9	1.0	210.0

The below is an example for demonstration purposes for the calculation for the lateral seismic force for the third floor:

$$C_{v,3} = \frac{w_3 h_3^k}{\sum_{i=1}^{n} w_i h_i^k} = \frac{300 \times 42^{1.25}}{\left[\begin{array}{c} 300 \times (14^{1.25} + 28^{1.25} + 42^{1.25}) \\ + \\ (150 \times 56^{1.25}) \end{array}\right]} = 0.39$$

$$F_3 = 0.39 \times 210 \approx 82 \ kip$$

3.3.4. Seismic Forces on Diaphragms

Seismic forces from vertically distributed seismic loads cannot be used for the design of diaphragms as those are typically used for the design of the lateral SFRS elements. Lateral forces used in diaphragm design on the other hand are larger because those can be amplified when the building vibrates at higher modes. ASCE 7 Standard Section 12.10.1.1 outlines the method of how hose forces are calculated as follow:

$$F_{px} = \frac{\sum_{i=x}^{n} F_i}{\sum_{i=x}^{n} w_i} w_{px}$$

Where:

F_{px} is the diaphragm design force at floor level (x).
F_i is the design force applied at floor level (i).
w_i is the weight tributary to level (i).
w_{px} is the weight tributary to the diaphragm at level (x).

It is also important to note the following regarding the above equation:

- The summation shown in the equation is from level (x), which is the level under consideration, to the very top of the building ($x \rightarrow n$). This is better shown in Example 3.7 below.

- You must comply with the minimums and maximums for this force as follows:

  $F_{px,min} = 0.2 S_{DS} I_e w_{px}$

  $F_{px,max} = 0.4 S_{DS} I_e w_{px}$

- For structures with horizontal irregularity Type 4, the transfer from the vertical seismic force resisting elements above the diaphragm to other vertical seismic force resisting elements below the diaphragm shall be increased by the overstrength factor (Ω_o).

Example 3.7 – Diaphragm Design Forces

Use the information in Example 3.6 to determine the diaphragm design forces per floor.

Floor	w_i	F_i	$\sum_{i=x}^{n} F_i$	$\sum_{i=x}^{n} w_i$	F_{px}	$F_{px,min}$	$F_{px,max}$	$F_{px,final}$
	(kip)	(kip)			$\frac{\sum_{i=x}^{n} F_i}{\sum_{i=x}^{n} w_i} w_{px}$	(kip)	(kip)	(kip)
4	150	58.5	58.5	150	58.5	22.5	45.0	45.0
3	300	81.6	140.1	450	93.4	45.0	90.0	90.0
2	300	49.2	189.3	750	75.7	45.0	90.0	75.7
1	300	20.7	210.0	1,050	60.0	45.0	90.0	60.0

If all story weights were equal, the roof diaphragm design force will match its base shear portion $F_{p4,roof} = F_{roof}$, while design forces for other diaphragms will be higher than the distributed story forces. Additionally, you can observe from this table that you can determine the first floor F_{p1} with only using base shear (210 kip) divided by full seismic weight (1,050 kip) multiplied by weight of the first floor.

3.3.5.   Seismic Forces on Structural Elements

Out-of-Plane Forces

It is important to understand that out-of-plane forces can push elements[9], such as walls, away from the supported diaphragms causing possible failures and catastrophes. These out-of-plane forces arise from the inertia of the walls acting outside their line of action during an earthquake.

To clarify, out-of-plane forces applied to walls are those forces that are not resisted by the line of action of the wall. Instead, these forces exert pressure on the wall's surface, similar to how wind loads affect a structure – See Figure 3-4 for an example of out-of-plane inertia forces acting on a masonry wall.

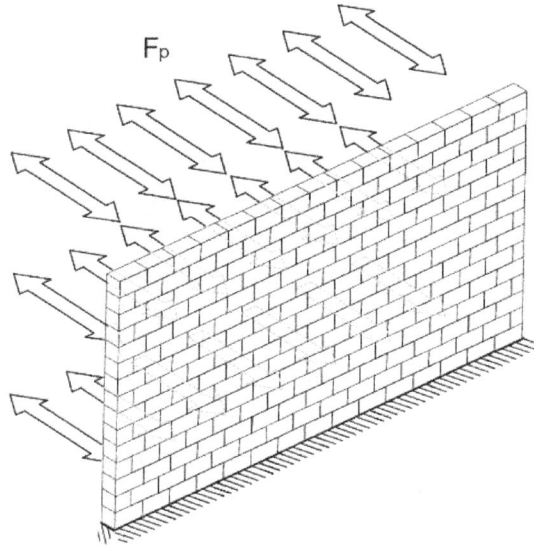

Figure 3-4 Out-of-Plane Force on a Masonry Wall

ASCE 7 Standard Section 12.11.1 specifies that this load should equal to the following:

$$F_p = 0.4 S_{DS} I_e \times (Weight\ of\ the\ Wall)$$

$$F_{p,min} = 10\% \times (Weight\ of\ the\ Wall)$$

Wall Anchorage Forces

Numerous instances during earthquakes have demonstrated that heavy walls can detach from their supporting roofs, leading to the collapse of both the wall and the roof. To prevent such displacement caused by out-of-plane forces, diaphragms are anchored to the walls. The calculation for the anchor force is similar to that presented above, but it also considers the diaphragm's flexibility and the tributary area of the wall relative to the anchor. For more details, see Figure 3-5.

---

[9] This chapter focuses on structural elements only, particularly walls, as outlined in ASCE 7 Standard Section 12.11. ASCE also addresses non-structural elements, such as parapets in Section 13.3, this is discussed in Chapter 5 of this book.

$$F_p = 0.4S_{DS}k_a I_e W_p$$

$$F_{p,min} = 0.2k_a I_e W_p$$

Where:

W_p is the weight of the tributary area explained with the use of Figure 3-5.

k_a is an amplification factor and is an indication of the diaphragm flexibility. If the diaphragm was rigid or not flexible this factor is taken as 1.0, while, if the diaphragm was flexible, it should be calculated as follows:

$$k_a = 1.0 + \frac{L_f \ (in \ ft)}{100} \leq 2.0$$

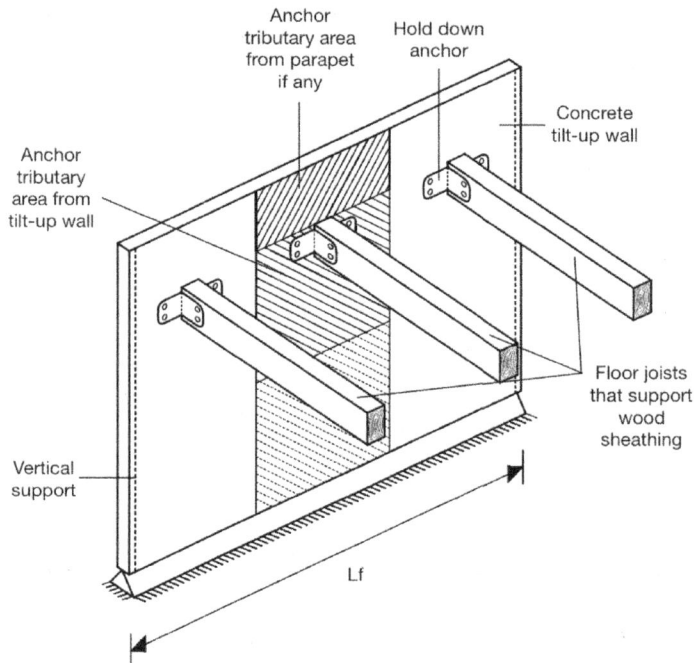

Figure 3-5 Tributary Area for Wall Anchorage

If you have multiple floors (i.e., multiple layers of anchorages), and the anchorage in question is located in the middle of the building elevation (h) at a height (z) from base rather than on the roof, and the floors are not flexible, you can adjust the initial force calculation (F_p) by multiplying it by a factor of $(1 + 2z/h)/3$. Also, in this case $F_{p,min} = 0.2W_p$

Sub-Diaphragms

Wooden diaphragms providing out-of-plane support for concrete or masonry walls must adhere to additional wall anchorage requirements outlined in ASCE 7 Standard Section 12.11.2.2, especially for structures assigned to Seismic Design Category C through F. Continuous ties in this case must connect to the wall across the entire diaphragm length, ensuring that all sections effectively resist wall anchorage forces. This requirement is informed by past earthquake incidents where concrete tilt-up[10] walls detached from wooden diaphragms, sometimes resulting

---

[10] A concrete tilt-up wall is a construction method where large concrete panels are poured on-site, typically on the building's foundation slab, and then "tilted up" into a vertical position to form walls.

in roof collapses. However, in some cases this may not be achievable with the main diaphragm. To address this, the concept of the sub-diaphragm was developed.

A sub-diaphragm is a smaller diaphragm within the main diaphragm, designed to more efficiently distribute wall anchor forces better than the main diaphragm (see Figure 3-6). These forces are then transferred into the main diaphragm, with continuous ties placed at each end rather than at every wall anchor. The diaphragm sheathing (even metal deck diaphragms – see ASCE 12.11.2.2.4) shall not be considered effective in place of ties for tension connections.

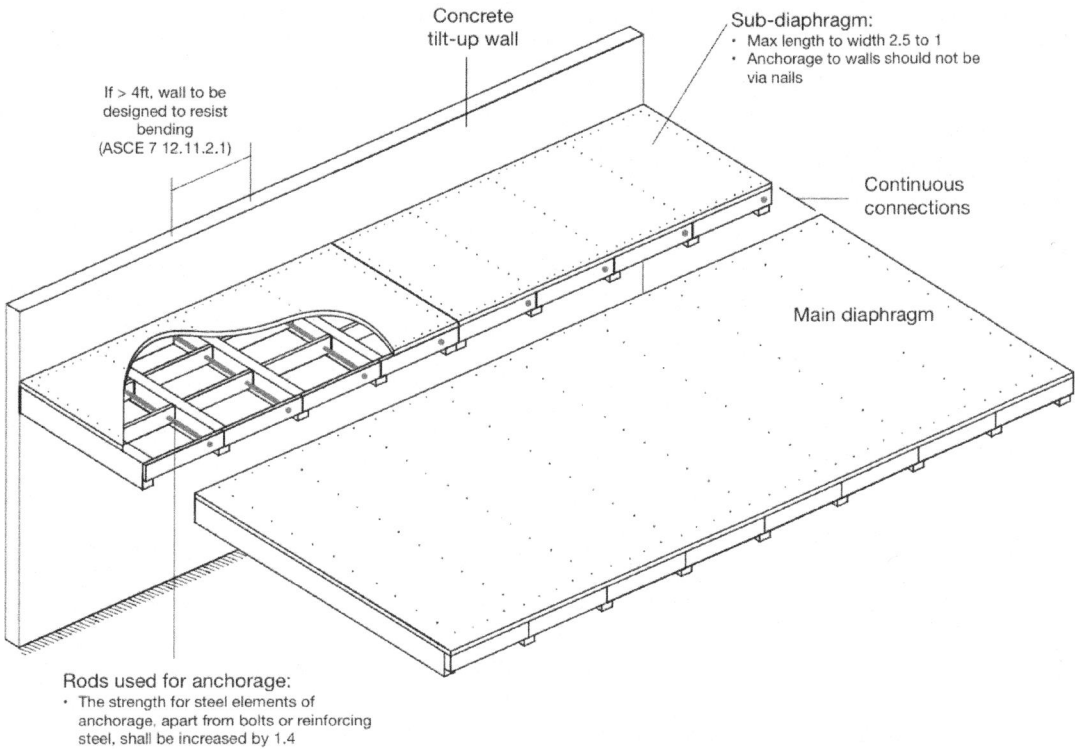

Figure 3-6 Diaphragm, Sub-Diaphragm, and Concrete Wall Anchorage

Figure 3-7 here offers details on diaphragm or sub-diaphragm anchorage to concrete or masonry walls, the section to the right corresponds to the schematic provided in Figure 3-6.

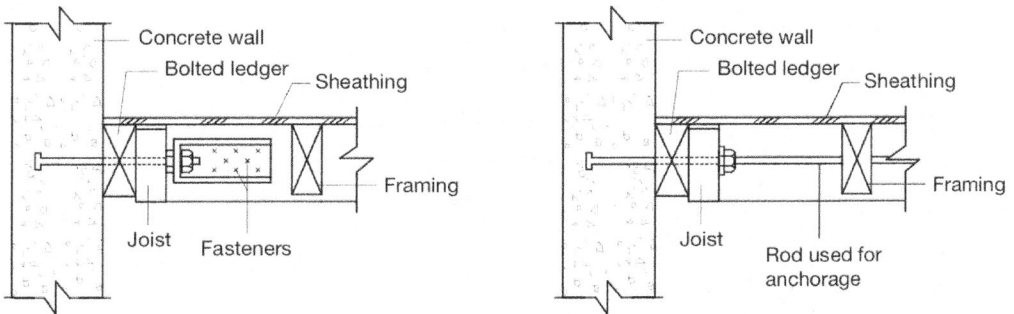

Figure 3-7 Concrete Wall Anchorage to Diaphragm or Sub-Diaphragm

3.3.6.   The Simplified Method

The method used to calculate seismic forces in this chapter follows the standard approach. However, the "Simplified Method" described in Section 1.5.3 of this book and Section 12.14 of the ASCE 7 Standard, is largely similar but includes key variations that simplify the process. These variations are discussed here.

Base Shear Calculation

In base shear calculation, the equation simplifies the determination of two coefficients: the fundamental period (T) and the site class coefficient (F_a). Additionally, it no longer uses the importance factor (I_e). The relevant response modification factor (R) is collected from ASCE 7 Table 12.14-1 – copied in this chapter's appendix with permission from ASCE.

$$V = \frac{F S_{DS}}{R} W$$

Where:

$S_{DS} = \frac{2}{3} F_a S_S$ in this case (F_a) is taken as follows as discussed in section 1.5.3 here:

$F_a = 1.0$ for rock or when soils $< 10\,ft$ between footing and rock.
$F_a = 1.4$ for soil.

$F = 1.0$ for buildings with one story.

$F = 1.1$ for two story buildings.

$F = 1.2$ for three story buildings.

Vertical Seismic Force Distribution

In vertical forces distribution, the simplified method does not rely on the vertical distribution factor (C_{vx}), instead, it uses the portion of effective weight (w_x) for each floor.

$$F_x = \frac{w_x}{W} V$$

Out of Plane and Wall Anchorage Forces

The Simplified Method uses the same equations and methods presented in Section 3.3.6 here, with the key difference being the exclusion of the importance factor (I_e) from the said equations.

3.4. Combination of Seismic Force Resisting Systems

3.4.1.   Horizontal Combination

Horizontal combination of two different Seismic Force Resisting Systems (SFRS) is permitted, subject to the following conditions:

- If the systems are combined in different directions: Each system must be analyzed in its respective direction, using the appropriate response modification factor (R), deflection amplification factor (C_d), and overstrength factor (Ω_0) as specified in ASCE 7, Table 12.2-1.

- If the systems are combined in the same direction other than the dual system: Use the smaller – i.e., most conservative – (R) from the two systems, and the most stringent for the other coefficients of SFRS.

3.4.2.   Vertical Combination

Vertical Combination of Different Seismic Force Resisting Systems (SFRS) is permitted, subject to the following conditions:

- If the lower system is more rigid/less ductile (i.e., lower R value): The upper system uses its own $(R, C_d \& \Omega_0)$ values, and the lower system uses its own as well. The forces transferred from the upper system to the lower system must be increased by a factor $= R_{upper}/R_{lower}$.

- If the upper system is more rigid/less ductile (i.e., lower R value): The $(R, C_d \& \Omega_0)$ values of the upper system apply to the entire design, including both the upper and lower systems.

The above first two bullet points are modified in the CBC 2022 for Community Colleges (CBC Section 1617), Hospitals and Correctional Treatment Centers (CBC Section 1617A.1.5.2) as follows:

- *The response modification factor (R), used for design at any story shall not exceed the lowest value of (R) at any story above it.*

- *Likewise, the deflection amplification factor and the overstrength factor $(C_d \& \Omega_0)$ for any story shall not be less than the largest value at any story above that story.*

Exception to this rule are rooftops less than two stories in height with a weight less than 10% of the overall weight of the structure and detached one- and two-family dwellings of light-frame construction.

3.4.3.   Vertical Combination Two Stage Analysis

You can perform a two-stage analysis, treating the upper and lower portions of the structure separately using the Equivalent Lateral Force (ELF) method for each portion. The upper portion may also be eligible for a different analysis method. This approach is only applicable under the following conditions – also summarized in Figure 3-8 below:

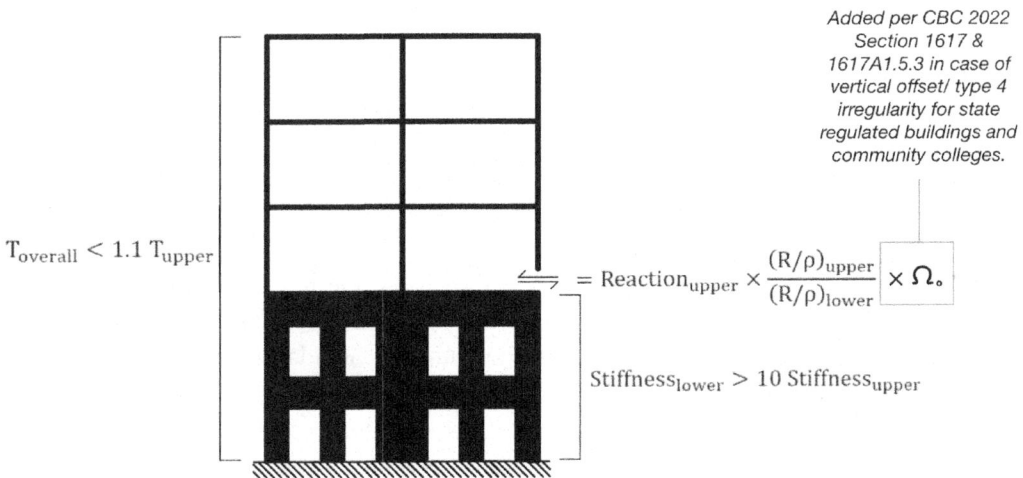

$T_{overall} < 1.1\, T_{upper}$

Added per CBC 2022 Section 1617 & 1617A1.5.3 in case of vertical offset/ type 4 irregularity for state regulated buildings and community colleges.

$$\Longleftrightarrow = \text{Reaction}_{upper} \times \frac{(R/\rho)_{upper}}{(R/\rho)_{lower}} \times \Omega_0$$

$\text{Stiffness}_{lower} > 10\, \text{Stiffness}_{upper}$

Figure 3-8 Vertical Combination Two Stage Analysis

- The stiffness of the lower portion must be at least ten times that of the upper portion (refer to Figure 3-8).

- The period of the entire structure should not exceed 1.1 times the period of the upper portion when considered as a separate structure.

- The reactions from the upper portion should be determined by amplifying them using the following ratio $= \frac{(R/\rho)_{upper}}{(R/\rho)_{lower}}$

- *In case of a vertical offset irregularity for <u>Community Colleges, Hospitals and Correctional Treatment Centers (See CBC 1617.11.4.3 and 1617A.1.5.3)</u> at the level of interaction between the two vertical systems (i.e., horizontal irregularity type 4), the reaction calculated for the upper portion as indicated in the previous bullet point should be increased by the overstrength factor (Ω_o).*

3.5. Chapter References

The following references were used in this chapter. The reference(s) in italics is/are required for the exam, as noted in the preface of this book.

[1] *American Society of Civil Engineers, ASCE/SEI 7-16 Minimum Design Loads for Buildings and Other Structures.*

[2] Armouti, N. (2015). Earthquake Engineering: Theory and Implementation with the 2015 International Building Code, Third Edition. McGraw Hill.

[3] *California Building Code (CBC 2022) California Code of Regulations, Title 24, Part 2, Volumes 1 and 2. Accessed through https://codes.iccsafe.org in January 2025*

[4] Cobeen, J.E., Dolan, J. & Lindt, J. (2017). Seismic Design of Wood Light-Frame Structural Diaphragm Systems – A Guide for Practicing Engineers. NEHRP Seismic Design Technical Brief No. 10.

[5] Faherty, K. & Williamson, T. (1998). Wood Engineering and Construction Handbook 3rd Edition. McGraw Hill.

[6] Gosh, S.K., Cleland, N. & Naito, C. (2017). Seismic Design of Precast Concrete Diaphragms. NEHRP Seismic Design Technical Brief No. 13.

[7] Moehle, J., Hooper, J. & Meyer, T. R. (2010). Seismic Design of Cast-in-Place Concrete Diaphragms, Chords, and Collectors: A Guide for Practicing Engineers. NEHRP Seismic Design Technical Brief No.3. National Institute of Standards and Technology. U.S. Department of Commerce.

[8] PCI Design Handbook: Precast and Prestressed Concrete 7th edition.

[9] *TMS 402/602-16 Building Code Requirements and Specifications for Masonry Structures.*

Chapter Appendix

This appendix contains the following information:

➤ Multi Degree Of Freedom (MDOF)

➤ Stiffness Calculations for Shear Walls *(and Columns)*

➤ Relevant Rigidity for Walls Simplified
- Table 3-5 Stiffness Calculation for Concrete and Masonry Shear Walls (and columns)
- Table 3-6 Rigidity Concrete & Masonry Cantilever Walls 12 *in* & $0.25 \leq h/l \leq 4.0$
- Table 3-7 Rigidity Concrete & Masonry Walls 12 *in* & $0.25 \leq h/l \leq 4.0$ Fixed Ends
- Example 3.8 – Single Story Shear Wall Stiffness Calculation Simplified
- Example 3.9 – Single Story Stiffness for a Cantilever Fixed Ends Wall

➤ Table 12.8-2 of ASCE 7 Standard
Values of Approximate Period Parameters

Multi Degree Of Freedom (MDOF)

In Section 3.2.1, we discussed the Single Degree Of Freedom (SDOF) system, which provides a basic representation of how single-story structures oscillate in one direction. However, in seismic design, this simplification is inadequate, especially for buildings with two or more stories. In such cases, multiple modes of lateral movement can occur, all of which must be considered during analysis.

The concept of Multi Degree Of Freedom (MDOF) systems is crucial for understanding how structures respond to ground motion. In MDOF systems, buildings can exhibit higher modes of vibration that involve more complex movements. These higher-frequency modes may cause different parts of the structure to move out of phase from one another, as illustrated in Figure 3-9 below. The available modes can include the following:

- First Mode: Building floors oscillate in one direction without any out of phase movements between each other.
- Second Mode: Often involves the structure bending or twisting at higher frequencies.
- Third Mode and Beyond: Higher modes may exhibit more intricate patterns of movement, which can be critical for understanding the overall seismic response.

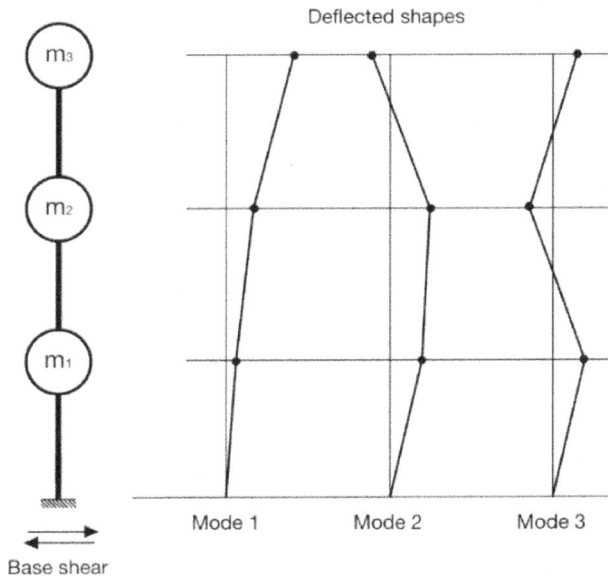

Figure 3-9 Example of Different Modes for an MDOF System

MDOF systems require analysis of how each mode contributes to the overall response during an earthquake. Modal analysis is typically used to determine the natural frequencies and mode shapes of a structure, which helps in predicting how it will respond to seismic loading.

Stiffness Calculation for Shear Walls (and Columns)

As discussed in Section 3.2.3, Table 3-5 provides equations to calculate the flexural stiffness for cantilever walls and fixed-fixed walls (this also applies to columns), along with the shear stiffness calculations. As mentioned in Section 3.2.3, when the height-to-length ratio falls between 0.3 and 3.0, both the flexural and shear stiffnesses should be combined as outlined below, or alternatively, the effective moment of inertia (I_{eff}) can be used in the relevant flexural stiffness equation.

Table 3-5 Stiffness Calculation for Concrete and Masonry Shear Walls (and Columns)

Shear Wall Load Type and Support Condition	Flexural Stiffness (k_{fi})	Shear Stiffness[11] (k_{si})	I_{eff} for single-story
Case 1	$\dfrac{3EI}{h^3}$ Where $I = \dfrac{l^3 t}{12}$	$\dfrac{AE}{3.0h}$ A is the wall cross sectional area where (t) is wall thickness $A = lt$	$\dfrac{I}{1 + \dfrac{9.0I}{Ah^2}}$
Case 2	$\dfrac{8EI}{h^3}$ Where $I = \dfrac{l^3 t}{12}$	$\dfrac{AE}{1.5h}$ Where $A = lt$	NA
Case 3	$\dfrac{12EI}{h^3}$ Where $I = \dfrac{l^3 t}{12}$	$\dfrac{AE}{3.0h}$ Where $A = lt$	$\dfrac{I}{1 + \dfrac{36.0I}{Ah^2}}$

Relevant Rigidity for Walls Simplified

The following two equations are used to calculate the relevant rigidity for **1 ft thick walls**. We named it "relevant" here as we removed the modulus of elasticity (E) from the equations. Tables 3-6 and 3-7 provide a quick solution for these equations. For walls with varying thicknesses, multiply the result by the new thickness in (ft), and to account for material changes, multiply by the relevant rigidity by modulus of elasticity (E).

$$R_{rel} = \frac{3 l_{eff}}{h^3} = \frac{1}{4\left(\frac{h}{l}\right)^3 + 3\left(\frac{h}{l}\right)}$$ Case 1 of Table 3-5. Solution to this equation is in Table 3-6.

$$R_{rel} = \frac{12 l_{eff}}{h^3} = \frac{1}{\left(\frac{h}{l}\right)^3 + 3\left(\frac{h}{l}\right)}$$ Case 3 of Table 3-5. Solution to this equation is in Table 3-7.

In some references, the values mentioned above (and in both Tables 3-6 and 3-7) are multiplied by **10** to provide additional weight. This does not change the method as long as we are consistent throughout the solution.

---

[11] The shear modulus of elasticity (E_v) for concrete or masonry shear walls is typically assumed to be 40% of the normal modulus of elasticity (E), which is accounted for in the table. Only (E) is used in Shear Stiffness (k_{si}) calculations. Any change to $E_v = 40\% E$ assumption due to material change could slightly affect the coefficients in the Shear Stiffness (k_{si}) calculations above, but this is outside the scope of the exam.

Table 3-6 Rigidity for Concrete & Masonry Cantilever Walls 12 in & $0.25 \leq h/l \leq 4.0$

12 in		Height (ft)											
Cantilever	8	9	10	11	12	13	14	15	16	17	18	19	20
3	0.0119	0.0085	0.0063	0.0048	0.0037								
4	0.0263	0.0191	0.0143	0.0109	0.0085	0.0068	0.0055	0.0045	0.0037				
5	0.0472	0.0348	0.0263	0.0203	0.0160	0.0128	0.0104	0.0085	0.0071	0.0060	0.0051	0.0043	0.0037
6	0.0742	0.0556	0.0425	0.0332	0.0263	0.0212	0.0173	0.0143	0.0119	0.0101	0.0085	0.0073	0.0063
7	0.1064	0.0809	0.0627	0.0494	0.0395	0.0321	0.0263	0.0218	0.0183	0.0155	0.0132	0.0113	0.0098
8	0.1429	0.1102	0.0865	0.0689	0.0556	0.0454	0.0375	0.0313	0.0263	0.0223	0.0191	0.0165	0.0143
9	0.1826	0.1429	0.1134	0.0912	0.0742	0.0610	0.0507	0.0425	0.0360	0.0307	0.0263	0.0227	0.0198
10	0.2248	0.1781	0.1429	0.1160	0.0951	0.0788	0.0659	0.0556	0.0472	0.0404	0.0348	0.0302	0.0263
11	0.2688	0.2153	0.1744	0.1429	0.1181	0.0985	0.0829	0.0703	0.0600	0.0515	0.0446	0.0388	0.0339
12	0.3140	0.2540	0.2077	0.1715	0.1429	0.1200	0.1015	0.0865	0.0742	0.0640	0.0556	0.0485	0.0425
13	0.3599	0.2938	0.2422	0.2015	0.1691	0.1429	0.1216	0.1041	0.0897	0.0777	0.0677	0.0593	0.0521
14	0.4064	0.3343	0.2777	0.2327	0.1964	0.1670	0.1429	0.1229	0.1064	0.0926	0.0809	0.0711	0.0627
15	0.4531	0.3754	0.3140	0.2647	0.2248	0.1922	0.1652	0.1429	0.1242	0.1084	0.0951	0.0838	0.0742
16	0.5000	0.4168	0.3507	0.2974	0.2540	0.2182	0.1885	0.1637	0.1429	0.1252	0.1102	0.0975	0.0865
17	0.5469	0.4583	0.3878	0.3306	0.2837	0.2449	0.2126	0.1854	0.1624	0.1429	0.1262	0.1119	0.0996
18	0.5936	0.5000	0.4251	0.3641	0.3140	0.2722	0.2372	0.2077	0.1826	0.1612	0.1429	0.1270	0.1134
19	0.6403	0.5417	0.4625	0.3979	0.3445	0.3000	0.2624	0.2306	0.2035	0.1802	0.1602	0.1429	0.1278
20	0.6868	0.5833	0.5000	0.4319	0.3754	0.3280	0.2880	0.2540	0.2248	0.1997	0.1781	0.1592	0.1429
21	0.7331	0.6248	0.5375	0.4659	0.4064	0.3564	0.3140	0.2777	0.2466	0.2198	0.1964	0.1762	0.1584
22	0.7793	0.6662	0.5749	0.5000	0.4375	0.3849	0.3401	0.3018	0.2688	0.2402	0.2153	0.1935	0.1744
23	0.8252	0.7074	0.6123	0.5341	0.4688	0.4136	0.3665	0.3261	0.2912	0.2609	0.2345	0.2113	0.1909
24	0.8710	0.7485	0.6496	0.5681	0.5000	0.4423	0.3931	0.3507	0.3140	0.2820	0.2540	0.2294	0.2077
25	0.9165	0.7895	0.6868	0.6021	0.5312	0.4712	0.4197	0.3754	0.3369	0.3032	0.2737	0.2478	0.2248
26	0.9619	0.8303	0.7239	0.6361	0.5625	0.5000	0.4465	0.4002	0.3599	0.3247	0.2938	0.2664	0.2422
27	1.0071	0.8710	0.7608	0.6699	0.5936	0.5288	0.4732	0.4251	0.3831	0.3463	0.3140	0.2853	0.2599
28	1.0521	0.9115	0.7977	0.7037	0.6248	0.5577	0.5000	0.4500	0.4064	0.3681	0.3343	0.3044	0.2777
29	1.0970	0.9518	0.8344	0.7373	0.6558	0.5865	0.5268	0.4750	0.4297	0.3900	0.3548	0.3236	0.2958
30	1.1417	0.9921	0.8710	0.7709	0.6868	0.6152	0.5536	0.5000	0.4531	0.4119	0.3754	0.3429	0.3140
31	1.1863	1.0322	0.9074	0.8044	0.7177	0.6439	0.5803	0.5250	0.4766	0.4339	0.3960	0.3624	0.3323
32	1.2308	1.0721	0.9438	0.8377	0.7485	0.6725	0.6070	0.5500	0.5000	0.4559	0.4168	0.3819	0.3507
33		1.1119	0.9800	0.8710	0.7793	0.7011	0.6337	0.5749	0.5234	0.4779	0.4375	0.4015	0.3692
34		1.1517	1.0161	0.9041	0.8099	0.7296	0.6603	0.5999	0.5469	0.5000	0.4583	0.4211	0.3878
35		1.1913	1.0521	0.9372	0.8405	0.7580	0.6868	0.6248	0.5703	0.5221	0.4792	0.4408	0.4064
36		1.2308	1.0881	0.9701	0.8710	0.7864	0.7133	0.6496	0.5936	0.5441	0.5000	0.4605	0.4251
37			1.1239	1.0030	0.9014	0.8146	0.7397	0.6744	0.6170	0.5661	0.5208	0.4803	0.4438
38			1.1596	1.0358	0.9317	0.8428	0.7661	0.6992	0.6403	0.5881	0.5417	0.5000	0.4625
39			1.1952	1.0685	0.9619	0.8710	0.7924	0.7239	0.6636	0.6101	0.5625	0.5197	0.4813
40			1.2308	1.1011	0.9921	0.8990	0.8187	0.7485	0.6868	0.6321	0.5833	0.5395	0.5000

(*) Concrete codes specify an aspect ratio range of 0.3 to 3.0 for medium-height walls, while masonry codes extend this to 0.25 to 4.0. The additional range for masonry walls is highlighted in grey.

Table 3-7 Rigidity for Concrete & Masonry Walls 12 in & $0.25 \leq h/l \leq 4.0$ Fixed Ends

12 in		Height (ft)											
Fixed Ends	8	9	10	11	12	13	14	15	16	17	18	19	20
3	0.0371	0.0278	0.0213	0.0166	0.0132								
4	0.0714	0.0551	0.0432	0.0344	0.0278	0.0227	0.0187	0.0156	0.0132				
5	0.1124	0.0890	0.0714	0.0580	0.0476	0.0394	0.0329	0.0278	0.0236	0.0202	0.0174	0.0151	0.0132
6	0.1570	0.1270	0.1038	0.0857	0.0714	0.0600	0.0508	0.0432	0.0371	0.0320	0.0278	0.0242	0.0213
7	0.2032	0.1672	0.1389	0.1164	0.0982	0.0835	0.0714	0.0615	0.0532	0.0463	0.0405	0.0355	0.0314
8	0.2500	0.2084	0.1753	0.1487	0.1270	0.1091	0.0943	0.0819	0.0714	0.0626	0.0551	0.0487	0.0432
9	0.2968	0.2500	0.2125	0.1821	0.1570	0.1361	0.1186	0.1038	0.0913	0.0806	0.0714	0.0635	0.0567
10	0.3434	0.2916	0.2500	0.2159	0.1877	0.1640	0.1440	0.1270	0.1124	0.0999	0.0890	0.0796	0.0714
11	0.3896	0.3331	0.2875	0.2500	0.2188	0.1925	0.1701	0.1509	0.1344	0.1201	0.1076	0.0968	0.0872
12	0.4355	0.3743	0.3248	0.2841	0.2500	0.2212	0.1965	0.1753	0.1570	0.1410	0.1270	0.1147	0.1038
13	0.4810	0.4152	0.3619	0.3180	0.2812	0.2500	0.2232	0.2001	0.1800	0.1624	0.1469	0.1332	0.1211
14	0.5261	0.4557	0.3988	0.3518	0.3124	0.2788	0.2500	0.2250	0.2032	0.1840	0.1672	0.1522	0.1389
15	0.5709	0.4960	0.4355	0.3854	0.3434	0.3076	0.2768	0.2500	0.2266	0.2059	0.1877	0.1715	0.1570
16	0.6154	0.5361	0.4719	0.4189	0.3743	0.3363	0.3035	0.2750	0.2500	0.2279	0.2084	0.1909	0.1753
17	0.6596	0.5758	0.5081	0.4521	0.4050	0.3648	0.3301	0.2999	0.2734	0.2500	0.2292	0.2106	0.1939
18	0.7037	0.6154	0.5440	0.4851	0.4355	0.3932	0.3567	0.3248	0.2968	0.2721	0.2500	0.2303	0.2125
19	0.7475	0.6547	0.5798	0.5179	0.4658	0.4214	0.3831	0.3496	0.3202	0.2941	0.2708	0.2500	0.2313
20	0.7911	0.6939	0.6154	0.5505	0.4960	0.4495	0.4093	0.3743	0.3434	0.3160	0.2916	0.2697	0.2500
21	0.8346	0.7329	0.6508	0.5830	0.5261	0.4775	0.4355	0.3988	0.3666	0.3379	0.3124	0.2894	0.2687
22	0.8780	0.7718	0.6861	0.6154	0.5560	0.5053	0.4615	0.4233	0.3896	0.3598	0.3331	0.3091	0.2875
23	0.9212	0.8105	0.7212	0.6476	0.5857	0.5330	0.4874	0.4476	0.4126	0.3815	0.3537	0.3287	0.3062
24	0.9643	0.8491	0.7562	0.6797	0.6154	0.5606	0.5132	0.4719	0.4355	0.4032	0.3743	0.3483	0.3248
25	1.0073	0.8876	0.7911	0.7117	0.6449	0.5880	0.5389	0.4960	0.4583	0.4247	0.3948	0.3678	0.3434
26	1.0502	0.9260	0.8259	0.7435	0.6743	0.6154	0.5645	0.5201	0.4810	0.4462	0.4152	0.3872	0.3619
27	1.0930	0.9643	0.8606	0.7753	0.7037	0.6426	0.5900	0.5440	0.5036	0.4676	0.4355	0.4066	0.3804
28	1.1358	1.0025	0.8953	0.8070	0.7329	0.6698	0.6154	0.5679	0.5261	0.4889	0.4557	0.4259	0.3988
29	1.1784	1.0407	0.9298	0.8386	0.7621	0.6969	0.6407	0.5917	0.5485	0.5102	0.4759	0.4451	0.4172
30	1.2211	1.0787	0.9643	0.8701	0.7911	0.7239	0.6659	0.6154	0.5709	0.5314	0.4960	0.4642	0.4355
31	1.2636	1.1168	0.9987	0.9016	0.8201	0.7509	0.6911	0.6390	0.5932	0.5525	0.5161	0.4833	0.4537
32	1.3061	1.1547	1.0330	0.9329	0.8491	0.7777	0.7162	0.6626	0.6154	0.5735	0.5361	0.5024	0.4719
33		1.1927	1.0673	0.9643	0.8780	0.8045	0.7412	0.6861	0.6375	0.5945	0.5560	0.5213	0.4900
34		1.2305	1.1016	0.9956	0.9068	0.8313	0.7662	0.7095	0.6596	0.6154	0.5758	0.5403	0.5081
35		1.2683	1.1358	1.0268	0.9356	0.8580	0.7911	0.7329	0.6817	0.6362	0.5956	0.5591	0.5261
36		1.3061	1.1699	1.0580	0.9643	0.8846	0.8160	0.7562	0.7037	0.6570	0.6154	0.5779	0.5440
37			1.2040	1.0891	0.9930	0.9112	0.8408	0.7795	0.7256	0.6778	0.6351	0.5967	0.5619
38			1.2381	1.1202	1.0216	0.9378	0.8656	0.8028	0.7475	0.6985	0.6547	0.6154	0.5798
39			1.2721	1.1513	1.0502	0.9643	0.8903	0.8259	0.7693	0.7192	0.6743	0.6340	0.5976
40			1.3061	1.1823	1.0787	0.9908	0.9150	0.8491	0.7911	0.7398	0.6939	0.6527	0.6154

(Length (ft) is labeled along the left vertical axis.)

(*) Concrete codes specify an aspect ratio range of 0.3 to 3.0 for medium-height walls, while masonry codes extend this to 0.25 to 4.0. The additional range for masonry walls is highlighted in grey.

Example 3.8 – Single Story Shear Wall Stiffness Calculation Simplified

This is a simplified solution for Example 3.2 with the use of Table 3.6 of this appendix.

The plan cross-sectional area for a cantilever concrete shear wall is 12 *ft* long and 8 *in* thick. The wall height is 10 *ft* and is single-story. Concrete modulus of elasticity is $E = 4.2 \times 10^6$ *psi*. What is the lateral stiffness of this wall?

The height-to-length for this wall $0.3 \le (h/l = 10/12) \le 3.0$, in which case, flexural and shear stiffness shall be calculated.

We can collect the coefficient $\left(\frac{3l_{eff}}{h^3}\right)$ from Table 3-6 in this appendix and multiply it with the modulus of elasticity (E) and thickness in (ft).

From Table 3-6 → $\frac{3l_{eff}}{h^3} = 0.208\ ft$

Remember that $k = \frac{3EI_{eff}}{h^3}$ (Table 3-5 for Case 1)

$$k = 0.208E \times \left(\frac{8}{12}\right)ft = 0.208 \times \left(\frac{8}{12}\right) \times \left[4.2 \times 10^6 \frac{lb}{in^2} \times \left(\frac{144\ in^2}{ft^2}\right)\right] = 83.8 \times 10^6\ lb/ft$$

In some other resources, you may find that the rigidity in Tables 3-6 and 3-7 multiplied by 10. This only adds more weight to the number. As long as consistency is maintained throughout the solution, both approaches are acceptable.

Example 3.9 – Single Story Stiffness for Two Fixed Ends Wall

The plan cross-sectional area for a two fixed end concrete shear wall is 12 *ft* long and 8 *in* thick. The wall height is 10 *ft* and is a single story. Concrete modulus of elasticity is $E = 4.2 \times 10^6$ *psi*. What is the lateral stiffness of this wall?

The height-to-length for this wall $0.3 \le (h/l = 10/12) \le 3.0$, in which case, flexural and shear stiffness shall be calculated.

We can collect the coefficient $\left(\frac{12l_{eff}}{h^3}\right)$ from Table 3-7 in this appendix and multiply it with the modulus of elasticity (E) and thickness in (ft).

From Table 3-7 → $\frac{12l_{eff}}{h^3} = 0.325\ ft$

Remember that $k = \frac{12EI_{eff}}{h^3}$ (Table 3-5 for Case 3)

$$k = 0.325E \times \left(\frac{8}{12}\right)ft = 0.325 \times \left(\frac{8}{12}\right)ft \times \left[4.2 \times 10^6 \frac{lb}{in^2} \times \left(\frac{144\ in^2}{ft^2}\right)\right] = 131.0 \times 10^6\ lb/ft$$

Information in this appendix, and hence Tables 3-6 and 3-7, can be used to calculate the center of rigidity, which is covered in Chapter 4 of this book. The center of rigidity is used to determine the horizonal distribution of seismic forces for the various SFRS elements within the diaphragm.

ASCE Table 12.8-2 Values of Approximate Period Parameters

Structure Type	C_t	x
Moment-resisting frame systems in which the frames resist 100% of the required seismic force and are not enclosed or adjoined by components that are more rigid and will prevent the frames from deflecting where subjected to seismic forces:		
Steel moment-resisting frames	0.028 (0.0724)[a]	0.8
Concrete moment-resisting frames	0.016 (0.0466)[a]	0.9
Steel eccentrically braced frames in accordance with Table 12.2-1 lines B1 or D1	0.03 (0.0731)[a]	0.75
Steel buckling-restrained braced frames	0.03 (0.0731)[a]	0.75
All other structural systems	0.02 (0.0488)[a]	0.75

[a] Metric equivalents are shown in parentheses.

Problems & Solutions

The following questions have been carefully designed to provide comprehensive coverage of the material presented in this book, while also being educational in nature, as you will notice. Additionally, you may come across a few questions not directly covered in this chapter; these have been intentionally included to ensure broader coverage of the content.

- ➤ Problem 3.1: Base Shear for a Tall Building
- ➤ Problem 3.2: Adjusted Base Shear
- ➤ Problem 3.3: Vertical Seismic Load Distribution (1)
- ➤ Problem 3.4: Vertical Seismic Load Distribution (2)
- ➤ Problem 3.5: Shear Reactions at a Particular Story Level
- ➤ Problem 3.6: Shear Reactions in a Vertically Combined System
- ➤ Problem 3.7: Wall Anchorage Force
- ➤ Problem 3.8: Sub-diaphragms
- ➤ Problem 3.9: Anchorage Hold-Down Design
- ➤ Problem 3.10: Horizontal SFRS Combination
- ➤ Problem 3.11: Wall Anchorage Force for a Non-Flexible Diaphragm
- ➤ Problem 3.12: Vertical SFRS Combination
- ➤ Problem 3.13: Foundation Load
- ➤ Problem 3.14: Diaphragm Seismic Design Force
- ➤ Problem 3.15: Stiffness and the Fundamental Period
- ➤ Problem 3.16: Resonance
- ➤ Problem 3.17: Overturning Moment
- ➤ Problem 3.18: Applications for the Fundamental Period
- ➤ Problem 3.19: Effective Seismic Weight
- ➤ Problem 3.20: Building Rigidity
- ➤ Problem 3.21: Stiffness for Various Seismic Force Resisting Systems
- ➤ Problem 3.22: Sub-diaphragm Sizing
- ➤ Problem 3.23: Fundamental Period Calculation
- ➤ Problem 3.24: Shear Walls Stiffness
- ➤ Problem 3.25: Anchorage Multiplier

PROBLEM 3.1 *Base Shear for a Tall Building*

A six-story risk category II retail building located in California (*) with equal story heights of 9 ft each. The building is founded on soil type C and is built using steel and concrete composite ordinary braced frames. All floors are equal in mass and each floor weighs 520 kip.

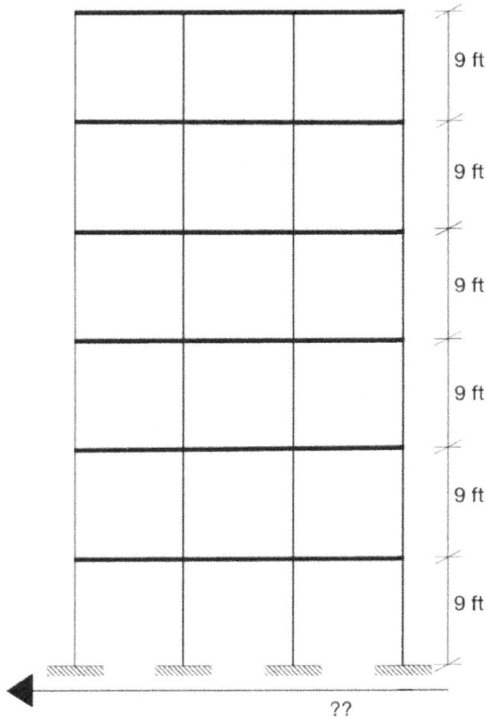

Assume that the effective seismic weight equals to the weight of all floors plus an extra 25% to account for other code requirements.

What is the base shear for this building?

(A) 3,430 kip

(B) 1,130 kip

(C) 905 kip

(D) 150 kip

(*) Use the following spectral parameters:

$S_s = 1.1$

$S_1 = 0.35$

$T_L = 4\ Sec$

PROBLEM 3.2 *Adjusted Base Shear*

A five-story building constructed on a soil type C has a base shear of 400 kip. The building is made of steel and concrete composite ordinary braced frame.

Considering the effects of Soil-Structure Interaction (SSI), what is the minimum limit below which base shear cannot be reduced?

(A) 280 kip

(B) 320 kip

(C) 340 kip

(D) 360 kip

PROBLEM 3.3 *Vertical Seismic Load Distribution (1)*

A four-story hospital located in Miami (*) with equal story heights of 14 ft each. The building is founded on soil type B and is built using reinforced concrete ordinary moment resisting frames. All floors are equal in mass and the portion of dead weight per story is 900 kip.

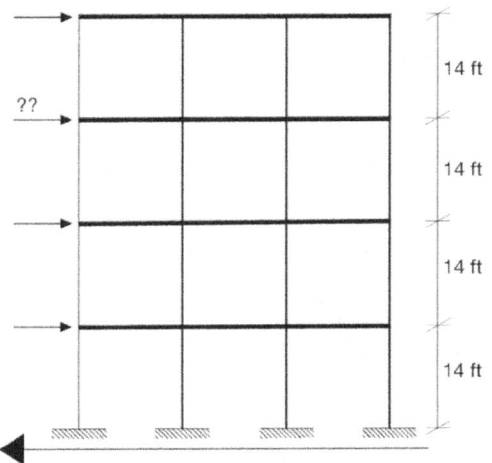

What is the lateral seismic force on the third floor?

(A) 2 kip

(B) 9 kip

(C) 11 kip

(D) 45 kip

(*) Use the following spectral parameters for Miami.

$$S_s = 0.048$$
$$S_1 = 0.021$$
$$T_L = 8\,Sec$$

PROBLEM 3.4 *Vertical Seismic Load Distribution (2)*

The three-story building below has a total effective seismic weight of 750 *kip* while each floor's effective weight is 300 *kip* and the roof 150 *kip*. The building is sitting on soil and is made of intermediate concrete moment resisting frame.

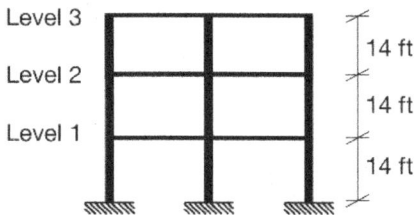

The following seismic coefficients apply:

$R = 3$
$S_1 = 0.30$
$S_S = 0.45$
$I_e = 1.0$

Using the simplified method, what is the lateral seismic force applied at the second floor?

(A) 50 *kip*

(B) 90 *kip*

(C) 125 *kip*

(D) 165 *kip*

PROBLEM 3.5 *Shear Reactions at a Particular Story Level*

The four-story building shown below is supported by an SFRS system of ordinary steel moment resisting frames. A seismic analysis for the building, located in a region with specified earthquake conditions, has determined the base shear as 300 *kip*, and

each story has an effective weight of 225 *kip* with the roof alone of 175 *kip*.

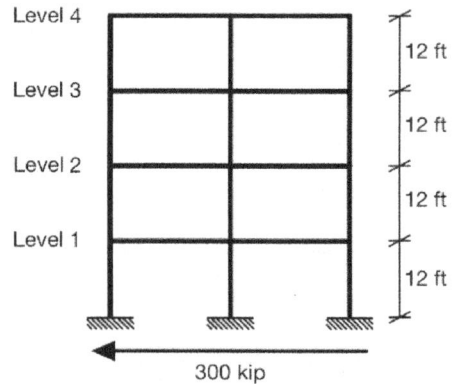

Based on this information, what is the total story reaction (or floor shear) at the third story of the building?

(A) 99 *kip*

(B) 204 *kip*

(C) 270 *kip*

(D) 300 *kip*

PROBLEM 3.6 *Shear Reactions in a Vertically Combined System*

The building below incorporates two Seismic Force-Resisting Systems. The lower portion consists of a podium with a shear wall system, which has a response modification factor of 4.0 and a redundancy factor of 1.0. Above the podium, is a special moment-resisting frames with a response modification factor of 8.0 and a redundancy factor of 1.3 due an irregularity type 1b in its floors.

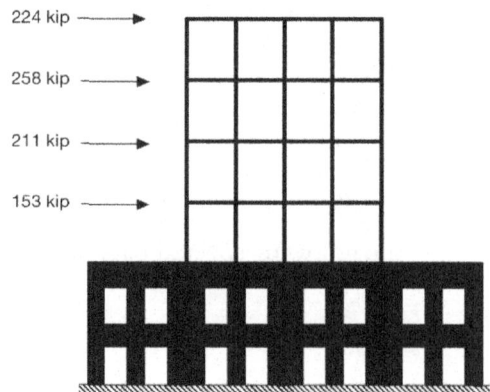

Due to the higher rigidity of the lower portion compared to the upper portion, ASCE allows separate analysis of the two systems. The lateral forces on the upper portion, as determined from this analysis, are shown in the provided sketch.

Based on this, what are the total reactions from the upper portion that should be transferred to the podium?

(A) 549 *kip*

(B) 846 *kip*

(C) 1,303 *kip*

(D) 1,692 *kip*

PROBLEM 3.7 *Wall Anchorage Force*
The below is a one-story building and its roof is made of a flexible diaphragm attached to 20 *ft* high concrete tilt-up-walls. The tilt-up-walls are made of 8 *in* thick 150 *pcf* concrete. The diaphragms are attached to the wall with the use of anchors at 3 *ft* intervals.

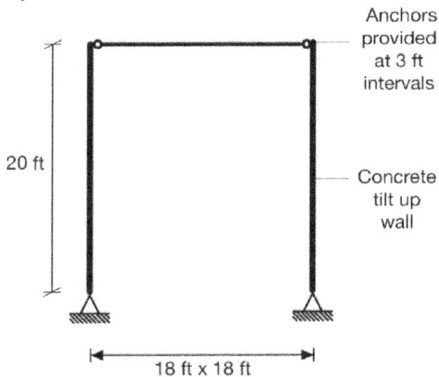

The building is a risk category II with an importance factor of 1.0, and is located at an $S_{DS} = 0.45$.

Based on this information, what is the wall anchorage force for any internal anchor at the floor level?

(A) 0.71 *kip*

(B) 1.20 *kip*

(C) 1.28 *kip*

(D) 1.42 *kip*

PROBLEM 3.8 *Sub-diaphragms*
Why are sub-diaphragms added to floors in building structures?

(A) Breaking the diaphragm into smaller pieces to enhance the flexibility of the diaphragm, allowing it to better adapt to seismic forces.

(B) To break the diaphragm into smaller sections, reducing its flexibility and improving force distribution.

(C) To reduce the amount of reinforcement required in the diaphragm.

(D) To improve the diaphragm's resistance to local deformations caused by concentrated loads.

PROBLEM 3.9 *Anchorage Hold-Down Design*
The steel hold-down anchor for the diaphragm to the concrete tilt-up wall is to be designed using the strength design method. What multiplier should be applied to the design force for this steel hold down anchor?

(A) 1.0

(B) 1.2

(C) 1.4

(D) 2.0

PROBLEM 3.10 *Horizontal SFRS Combination*
A building structure in SDC B has an SFRS that combines two systems: an ordinary plain concrete shear wall system, and, an ordinary reinforced concrete moment frame system, both used to resist seismic forces in the two horizontal directions.

What response modification factor should be used for seismic design in either direction?

(A) 1.5

(B) 2.25

(C) 3.0

(D) 4.5

PROBLEM 3.11 *Wall Anchorage Force for a Non-Flexible Diaphragm*

The below is a two-story building made of precast concrete panels as diaphragms attached to the 20 ft high concrete tilt-up-walls. The tilt-up-walls are made of 8 in thick 150 pcf concrete. The diaphragms are attached to the walls at 3 ft intervals.

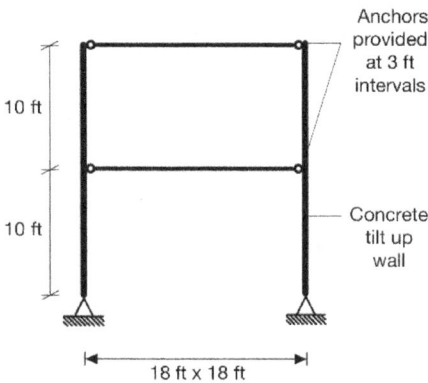

The building is a risk category II with an importance factor of 1.0, and is located at an $S_{DS} = 0.95$.

Based on this information, what is the wall anchorage force for any internal anchor on level 1?

(A) 0.76 kip

(B) 1.20 kip

(C) 1.52 kip

(D) 1.83 kip

PROBLEM 3.12 *Vertical SFRS Combination*

A five-story building in Seismic Design Category B has the following seismic force-resisting systems: The first two stories are constructed using ordinary reinforced concrete shear walls, while the upper three stories are built with steel eccentrically

braced frames. Both systems are used to resist seismic forces in the two horizontal directions.

What response modification factor (R) and overstrength factor (Ω_o) should be used for seismic design for each system in either direction?

(A) Lower system: $R = 4$ & $\Omega_o = 2.5$
Upper system: $R = 8$ & $\Omega_o = 2.0$

(B) Lower system: $R = 4$ & $\Omega_o = 2.5$
Upper system: $R = 4$ & $\Omega_o = 2.5$

(C) Lower system: $R = 4$ & $\Omega_o = 2.5$
Upper system: $R = 4$ & $\Omega_o = 2.0$

(D) Lower system: $R = 8$ & $\Omega_o = 2.0$
Upper system: $R = 8$ & $\Omega_o = 2.0$

PROBLEM 3.13 *Foundation Load*

The three below sketches are for a simple four footing two story building:

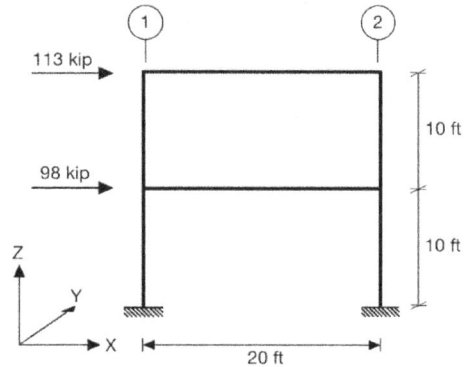

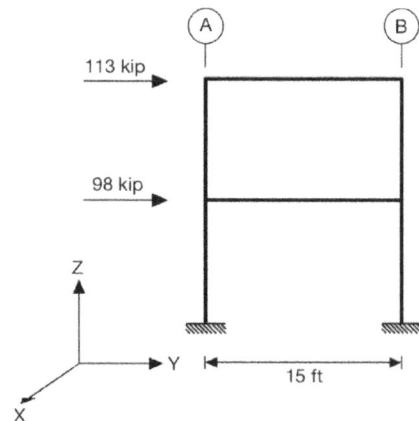

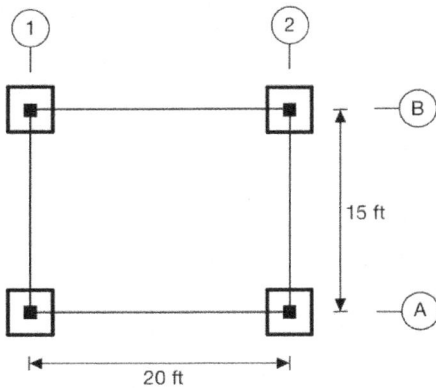

The building has been analyzed for seismic design, and the lateral forces in both directions are shown on the respective building elevations.

Based on this information, what is the maximum additional downward vertical load that could be imposed on footing at B.2 as a result of these seismic forces?

(A) 27 *kip*

(B) 108 *kip*

(C) 189 *kip*

(D) 162 *kip*

PROBLEM 3.14 *Diaphragm Seismic Design Force*

Below is base shear vertical distribution. The weight of each diaphragm is 350 *kip*. Importance factor is 1.0 and $S_{DS} = 1.1$.

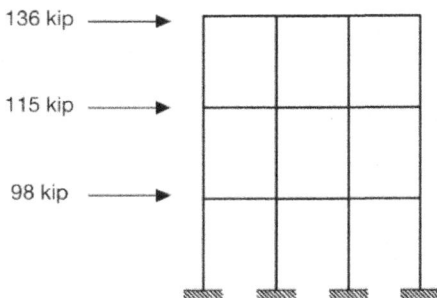

What is the diaphragm design force for the diaphragm at level 2?

(A) 50 *kip*

(B) 99 *kip*

(C) 115 *kip*

(D) 126 *kip*

PROBLEM 3.15 *Stiffness and the Fundamental Period*

The fundamental period of a building is influenced by its stiffness.

Which of the following statements is correct regarding the relationship between stiffness, the fundamental period of a structure and its seismic response?

(A) As stiffness increases, fundamental period increases, leading to higher seismic loads.

(B) As stiffness increases, fundamental period decreases, resulting in lower seismic loads.

(C) As stiffness increases, fundamental period increases, leading to lower seismic loads.

(D) As stiffness increases, fundamental period decreases, resulting in higher seismic loads.

PROBLEM 3.16 *Resonance*

What is the primary consequence of resonance for a building during an earthquake?

(A) The building will experience reduced ground motion and less damage.

(B) The building will vibrate with greater amplitude, potentially leading to increased forces and damage.

(C) The building will remain completely unaffected by the earthquake.

(D) The building will absorb all seismic energy, preventing any structural movement.

PROBLEM 3.17 *Overturning Moment*

The below diagram represents the story shear for the building next to it.

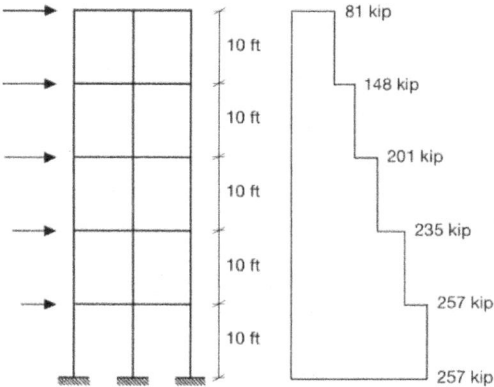

What is the overturning moment for this building around its bases?

(A) 23,270 $kip \cdot ft$

(B) 9,220 $kip \cdot ft$

(C) 4,050 $kip \cdot ft$

(D) 3,320 $kip \cdot ft$

PROBLEM 3.18 *Applications for the Fundamental Period*

A 14-story tall building with each story is 12 ft high is made of steel buckling-restrained braced frames. This building is located in a region with $S_{D1} = 0.2g$. The structural engineer calculated the fundamental period of this building as $T = 1.2\ sec$.

Which fundamental period shall be used for elastic drift calculations and for base shear calculations?

(A) For elastic drift use $T = 1.4\ sec$
  For base shear use $T = 1.4\ sec$

(B) For elastic drift use $T = 1.2\ sec$
  For base shear use $T = 1.2\ sec$

(C) For elastic drift use $T = 1.4\ sec$
  For base shear use $T = 1.2\ sec$

(D) For elastic drift use $T = 1.2\ sec$
  For base shear use $T = 1.4\ sec$

PROBLEM 3.19 *Effective Seismic Weight*

The below building is a three-story structure with concrete floors that are 9 in thick, with a unit weight of 150 pcf. The external walls have a unit weight of 15 psf. Permanent equipment on the roof has a total weight of 50 kip. The building serves as an office building with office furniture that is not attached to floors. The total area of interior partitions for stories 2 and 3 has not yet been finalized, while the first story has no partitions.

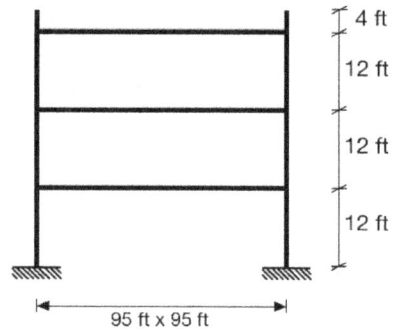

The weight of all concrete floors should be increased by 5% to account for tiles and other fixtures.

What is the estimated total effective seismic weight of the building?

(A) 3,030 kip

(B) 3,410 kip

(C) 3,620 kip

(D) 4,040 kip

PROBLEM 3.20 *Building Rigidity*

Using a computer model, a force of 100 kip is applied at the top of a 160 ft tall building, resulting in a displacement of 2 in at the top. What is the building's rigidity?

(A) 50 kip/in

(B) 7.5 kip/in

(C) 0.625 kip/in

(D) 0.03 kip/in

PROBLEM 3.21 *Stiffness for Various Seismic Force Resisting Systems*

Consider three buildings of equal height, each with a different Seismic Force Resisting System: (1) Ordinary plain concrete shear walls, (2) Ordinary reinforced concrete moment frames, and (3) Steel eccentrically braced frames.

All buildings are subjected to the same lateral load.

Which of the following statements best describes the relationship between the stiffness of these systems?

(A) The shear wall building will have the highest stiffness, followed by the steel braced frames, and finally the moment frames.

(B) Moment frames will have the highest stiffness, followed by steel braced frames, then shear walls.

(C) The steel braced frames will have the highest stiffness, followed by shear walls, and then moment frames.

(D) The shear wall building will have the highest stiffness, followed by the moment frames, and finally the steel braced frames.

PROBLEM 3.22 *Sub-diaphragm Sizing*

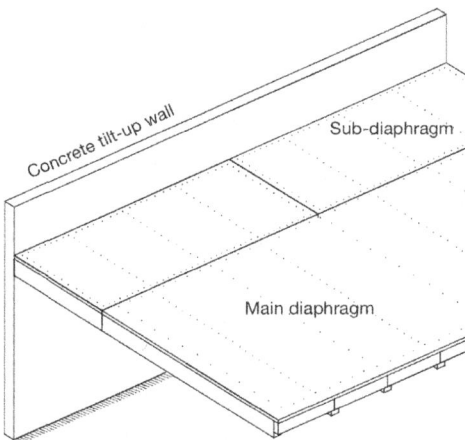

To improve force transmission between the concrete tilt-up wall and the main diaphragm, sub-diaphragms have been added to anchor the wall more effectively. These sub-diaphragms will then be connected to the main diaphragm, as shown in the figure.

The best sub-diaphragm (*Length x Width*) dimensions that they should be designed for are:

(A) $7\,ft \times 2.5\,ft$

(B) $12\,ft \times 4\,ft$

(C) $12.5\,ft \times 5\,ft$

(D) $15\,ft \times 5.5\,ft$

PROBLEM 3.23 *Fundamental Period Calculation*

This $12\,ft$ tall building has four square concrete columns each with a moment of inertia $I = 1,700\,in^4$. Concrete modulus of elasticity is $E = 4.2 \times 10^3\,ksi$.

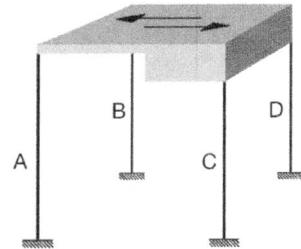

Assuming columns A & B are pinned at the slab level and fixed at the ground. While columns C & D are fixed at the ground and at the slab level. The total building weight is $300\,kip$. Calculate the building fundamental period in the shown direction.

(A) $0.35\,sec$

(B) $0.65\,sec$

(C) $0.85\,sec$

(D) $1.05\,sec$

PROBLEM 3.24 *Shear Walls Stiffness*
The below is an elevation view for two concrete shear walls located in the same single-story building, where shear wall 1 is fixed from the bottom, while you can assume it is free from the top due to the thickness of the slab it supports. Shear wall 2 is fixed from both ends.

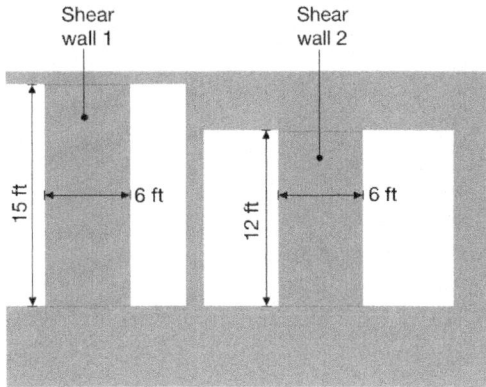

Shear wall 1   Shear wall 2

15 ft   ⊢6 ft   12 ft   ⊢6 ft

Assuming the thickness of the two shear walls is **10 in**, the relationship between the stiffness of shear wall 1 (k_1) and shear wall 2 (k_2), is as follows:

(A) $k_2 = 5.0\, k_1$

(B) $k_2 = 0.2\, k_1$

(C) $k_2 = 2.0\, k_1$

(D) $k_2 = 0.5\, k_1$

PROBLEM 3.25 *Anchorage Multiplier*
Which of the following items does not require a multiplier of 1.4 for the design of a diaphragm wall anchorage?

(A) Steel wall connectors

(B) Sub-diaphragm strapping

(C) Steel wall connectors

(D) Reinforcing steel

SOLUTION 3.1

Using the ASCE 7-16 code, base shear is obtained from the following equation:

$$V = C_s W$$

W is the effective seismic weight and is defined in ASCE 12.7.2, and C_s is the seismic response coefficient and is calculated as follows using Table 3-4 here:

$C_s = \frac{S_{DS}}{(R/I_e)}$ when $T \le T_s$ (verified later)

$$S_{DS} = \frac{2}{3} S_{MS} = \frac{2}{3} F_a S_s$$

Per Table 11.4-1 of ASCE 7 Standard (Table 1-2 here), a Site Class C and $S_s = 1.1$ → $F_a = 1.2$

$$S_{DS} = \frac{2}{3} \times 1.2 \times 1.1 = 0.88$$

S_{D1} is calculated in a similar fashion where $F_v = 1.5$ (from Table 1-3) as follows:

$$S_{D1} = \frac{2}{3} S_{M1} = \frac{2}{3} F_v S_1$$

$$= \frac{2}{3} \times 1.5 \times 0.35 = 0.35$$

$$T_s = \frac{S_{D1}}{S_{Ds}} = \frac{0.35}{0.88} = 0.4 \; sec$$

Determine the (approximate) fundamental period for the structure as follows:

$$T_a = C_t h_n^x$$

$$= 0.02 \times (6 \times 9)^{0.75}$$

$$= 0.4 \; sec = T_s$$

Equation $T_a = 0.1N$ (ASCE Eq.12.8-8) can only be used when the system is entirely concrete or steel moment resisting frames.

Based on the above and checking with Table 3-4, we can use $C_s = \frac{S_{DS}}{(R/I_e)} < \frac{S_{D1}}{T(R/I_e)}$

R is the response modification factor and can be looked up from Table 12.2-1. For steel and concrete composite ordinary braced frames → $R = 3.0$.

I_e is the importance factor and is determined from Section 11.5.1 of the ASCE 7 Standard (Table 1-5 here) = 1.0

$$C_s = \frac{S_{DS}}{(R/I_e)} = \frac{0.88}{(3/1.0)} = 0.29$$

$$C_{s,max} = \frac{S_{D1}}{T(R/I_e)} = \frac{0.35}{0.4 \times (3/1)} = 0.29$$

→ OK

Minimum value for C_s should be checked as well as follows (see Table 3-4 here):

$$C_s = 0.044 S_{DS} \, I_e \ge 0.01$$

$$= 0.044 \times 0.88 \times 1.0$$

$$= 0.039 < 0.29 \rightarrow ok$$

At this stage, we can confirm that the Seismic Design Category from Table 11.6 (or Table 1-7 here) is SDC D.

Base shear can now be calculated as follows:

$$V = C_s W$$

$$= 0.29 \times (6 \times 520 \times 1.25)$$

$$= 1,131 \; kip$$

Correct Answer is (B)

SOLUTION 3.2

Reference is made to the ASCE 7 Standard, Section 19.2 SSI Adjusted Structural Demands, also to Table 12.2-1 of the same code to choose the Response Modification Factor R where $R = 3$ for the type of frame provided in this question – i.e., steel and concrete composite ordinary braced frame.

$$V' \ge \propto V$$

Where $\propto = 0.7$ for $R \le 3$ per Equation 19.2-3.

$$V' \ge 0.7 \times 400 = 280 \; kip$$

Correct Answer is (A)

SOLUTION 3.3

ASCE 7 Standard explains in Section 11.4.2 that when $S_1 \leq 0.04$ and $S_s \leq 0.15$ (*), it is permitted to use Seismic Design Category A and compliance with Section 11.7 of the code shall suffice.

Section 11.7 of the code states that all buildings assigned to Seismic Design Category A shall comply with the requirements of Section 1.4 of the ASCE 7.

Based on this, ASCE 7 Section 1.4.2 specifies that lateral force F_x should be calculated as follows:

$$F_x = 0.01 \, W_x$$

W_x being the portion of total dead load assigned per floor = 900 kip

$$F_x = 0.01 \times 900 \, kip = 9 \, kip$$

Correct Answer is (B)

(*) We tried to design this question for a location in California, but even Sacramento – which is considered one of the safest areas in terms of earthquakes – has an $S_s = 0.5 \, sec$ and $S_1 = 0.25 \, sec$.

SOLUTION 3.4

Using the simplified method, and as per Section 12.14 of the ASCE 7 Standard (Section 3.3.6 here), the following equations apply to calculate the base shear and its vertical distribution:

$$V = \frac{FS_{DS}}{R} W$$

Where $S_{DS} = \frac{2}{3} F_a S_s$ and $F_a = 1.4$ for buildings sitting on soil while $F = 1.2$ for three story buildings.

$$V = \frac{F \times \left(\frac{2}{3} F_a S_s\right)}{R} W$$

$$= \frac{1.2 \times \left(\frac{2}{3} \times 1.4 \times 0.45\right)}{3.0} \times 750$$

$$= 126 \, kip$$

In vertical force distribution, the simplified method uses the effective weight (w_x) for each floor to determine the vertical distribution. The second-floor lateral force is therefore calculated as follows:

$$F_x = \frac{w_x}{W} V = \frac{300}{750} \times 126 = 50.4 \, kip$$

Correct Answer is (A)

SOLUTION 3.5

The location of the third story is shown below, and hence its shear (or reactions), shall include the lateral force from floors at levels 3 and 4.

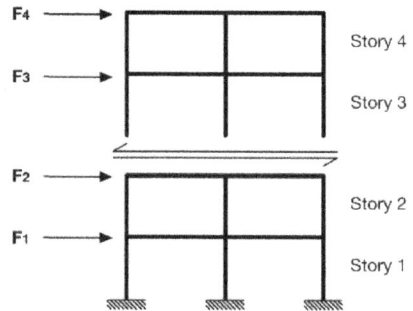

The procedure explained in Section 3.3.3 here is followed, which refers to ASCE 7 Sections 12.8.3.

Start with estimating the approximate fundamental period for the steel frame using Table 12.8-2 from ASCE as follows:

$$T_a = C_t h_n^x$$
$$= 0.028 \times (4 \times 12)^{0.8}$$
$$= 0.62 \, sec \, (*)$$

Determine the required vertical distribution coefficient (C_{vx}) as follows:

$$C_{vx} = \frac{w_x h_x^k}{\sum_{i=1}^{n} w_i h_i^k}$$

$$k = 0.75 + 0.5 \times T$$
$$= 0.75 + 0.5 \times 0.062$$
$$= 1.06$$

$$C_{v,3} = \frac{w_3 h_3^k}{\sum_{i=1}^n w_i h_i^k}$$

$$= \frac{225 \times (12 \times 3)^{1.06}}{\left[\begin{array}{c} 225 \times (12^{1.06} + (12 \times 2)^{1.06} + (12 \times 3)^{1.06}) \\ + \\ 175 \times (12 \times 4)^{1.06} \end{array}\right]}$$

$$= \frac{10,043.0}{30,307.8}$$

$$= 0.33$$

$$C_{v,4} = \frac{w_4 h_4^k}{\sum_{i=1}^n w_i h_i^k}$$

$$= \frac{175 \times (12 \times 4)^{1.06}}{\left[\begin{array}{c} 225 \times (12^{1.06} + (12 \times 2)^{1.06} + (12 \times 3)^{1.06}) \\ + \\ 175 \times (12 \times 4)^{1.06} \end{array}\right]}$$

$$= \frac{10,596.3}{30,307.8}$$

$$= 0.35$$

$$Story\ 3\ shear = (0.33 + 0.35) \times 300$$

$$= 204\ kip$$

Correct Answer is (B)

(*) If you choose $T_a = 0.1N = 0.4\ sec$, in which case $k = 1.0$. Applying this value consistently throughout the entire problem gets you the same results.

SOLUTION 3.6

ASCE 7 explains in Section 12.2.3.2 that in such case, the upper reactions shall be multiplied by the ratio (see Figure 3-8 here):

$$\frac{(R/\rho)_{upper}}{(R/\rho)_{lower}} = \frac{8/1.3}{4/1} = 1.54$$

In which case the requested sum of reactions is calculated as follows:

$$(224 + 258 + 211 + 153) \times 1.54 \cong 1,303\ kip$$

Correct Answer is (C)

SOLUTION 3.7

ASCE Standard explains how to derive these forces in Section 12.11.2.1, which is explained in this book in Section 3.3.5.

Initially, and given that this is a flexible diaphragm, these forces should be amplified as explained in the above sections by the following factor:

$$k_a = 1.0 + \frac{L_f\ (in\ ft)}{100}$$

$$= 1.0 + \frac{18}{100}$$

$$= 1.18 < 2.0 \rightarrow OK$$

The force is calculated using half the area shown below – only the upper half of this area is carried by the anchor (*):

$$F_p = 0.4 S_{DS} k_a I_e W_p$$

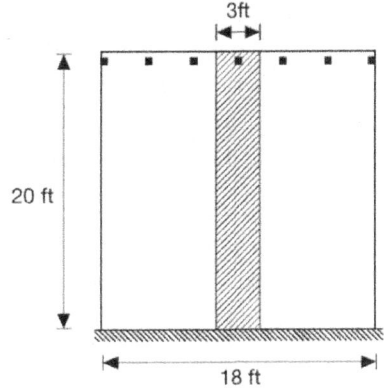

$$Where\ W_p = \left(20/2 \times 3 \times \frac{8}{12}\right) \times 150 \frac{lb}{ft^3}$$

$$= 3,000\ lb\ (3\ kip)$$

$$F_p = 0.4 \times 0.45 \times 1.18 \times 1.0 \times 3 = 0.64\ kip$$

This force should not be less than:

$$F_{p,min} = 0.2 k_a I_e W_p$$

$$= 0.2 \times 1.18 \times 1 \times 3$$

$$= 0.71\ kip$$

$$Use\ F_p = 0.71\ kip$$

Correct Answer is (A)

(*) Area above anchorage – such as a parapet, if given, should be added to the relevant tributary area to calculate w_p.

SOLUTION 3.8

Sub-diaphragms (*) divide large, flexible floor diaphragms into smaller, more rigid sections, which helps improve the overall stiffness and efficiency of the diaphragm. By reducing flexibility, they prevent excessive deformation under lateral loads (such as wind or seismic forces), ensuring more effective load transfer to vertical elements like shear walls or braces. This increased stiffness allows the diaphragm to better resist torsional effects and reduces the risk of localized failures, such as tearing of the diaphragm

Additionally, sub-diaphragms enhance force distribution by creating more direct and efficient load paths, allowing lateral forces to be spread evenly across the diaphragm. This helps distribute the forces more uniformly to the shear walls or bracing systems, improving the building's resistance to lateral forces during seismic or wind events.

Correct Answer is (B)
(*) It is crucial to remember that sub-diaphragm dimensions are 1 width to a maximum of 2.5 length as stated in ASCE 7 Section 12.11.2.2.1.

SOLUTION 3.9
A steel hold-down anchor is shown in Figure 3-5 here, also in Figure C12.11-2 of the ASCE 7 Standard.

Per ASCE 7 Standard Section 12.11.2.2.2, the steel elements for anchors – apart from the anchor bolt itself or steel reinforcements – strength design force should be multiplied by 1.4.

Correct Answer is (C)

SOLUTION 3.10
As explained in Section 3.4.1 in this book, also in ASCE 7 Standard Section 12.2.3, both sections explain that if the SFRS systems are horizontally combined in the same direction, the smallest, which is the most conservative, response modification factor (R) from the two systems shall be used for the analysis.

Ordinary plain concrete shear walls system is item No. A.4 in ASCE Table 12.2-1 with $R = 1.5$.

Ordinary reinforced concrete moment frames system is item No. C.7 in ASCE Table 12.2-1 with $R = 3.0$.

The most stringent modification factor, which is the lowest, should be used in this case with $R = 1.5$.

Correct Answer is (A)

SOLUTION 3.11
ASCE Standard explains how to derive these forces in Section 12.11.2.1, which is explained in this book in Section 3.3.5.

In reference to solution 3.7 here and given that this is a non-flexible diaphragm (*), the amplification factor (k_a) should be taken as 1.0.

The force is calculated as follows using the tributary area enclosed as shown below:

$$F_p = 0.4 S_{DS} k_a I_e W_p$$

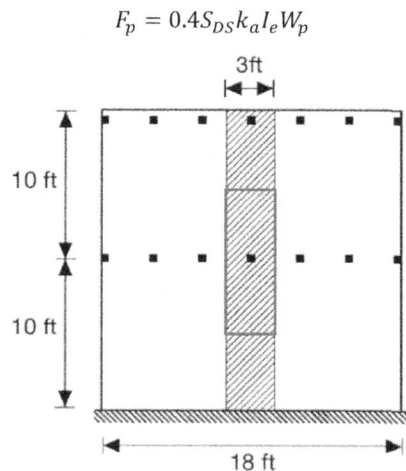

Where $W_p = \left(10\,ft \times 3\,ft \times \frac{8}{12}\,ft\right) \times 150\frac{lb}{ft^3}$
$$= 3,000\,lb\;(3\,kip)$$

Since there are multiple floors and the anchorage in question is in the middle of the building, and the floors are not flexible, (F_p) is permitted to adjusted per ASCE 12.11.2.1 by multiplying it by the following factor:

$$\left(1 + \frac{2z}{h}\right)/3 = \left(1 + \frac{2 \times 10}{20}\right)/3 = 2/3$$

$$F_p = \frac{2}{3} \times (0.4 \times 0.95 \times 1.0 \times 1.0 \times 3)$$

$$= 0.76 \; kip$$

Also, in this case $F_{p,min} = 0.2W_p = 0.6 \; kip$

Use $F_P = 0.76 \; kip$

Correct Answer is (A)
(*) Precast concrete diaphragms are much less flexible compared to wood or metal decks due to their substantial mass and stiffness. This means that they experience lesser bending or deflection, hence they are identified as not flexible.

SOLUTION 3.12
As explained in Section 3.4.2 in this book, also in ASCE 7 Standard Section 12.2.3.1.

Both sections explain that if the SFRS systems are vertically combined in the same direction, and if the lower system is more rigid with lower R value compared to the upper system, the upper system uses its own $(R, C_d \; \& \; \Omega_0)$ values, and the lower system uses its own as well.

The lower system, which is made with ordinary reinforced concrete shear walls is item No. A.2 in ASCE Table 12.2-1 with $R = 4.0 \; \& \; \Omega_o = 2.5$.

The upper system, which is made using steel eccentrically braced frames is item No. B.1 in ASCE Table 12.2-1 with $R = 8.0 \; \& \; \Omega_o = 2.0$.

Since the lower system is less ductile with lower R value, each system uses their own coefficients as follows:

Lower system: $R = 4 \; \& \; \Omega_o = 2.5$

Upper system: $R = 8 \; \& \; \Omega_o = 2.0$

Correct Answer is (A)

SOLUTION 3.13
First, let's evaluate the moment generated by the vertical forces using the X-Z elevation plane at the footings along Axis 2. Based on this, we can determine the couple (or forces) developed at the footings along Axis 1 and Axis 2 to counteract the moment caused by the seismic forces.

$$M_{Y@axis\,2\,footing} = 98 \times 10 + 113 \times 20$$

$$= 3,240 \; kip.ft$$

$$F_{axis\,2} = \frac{3,240 \; kip.ft}{20 \; ft} = 162 \; kip$$

This force is applied on two footings, and hence each footing takes 81 kip.

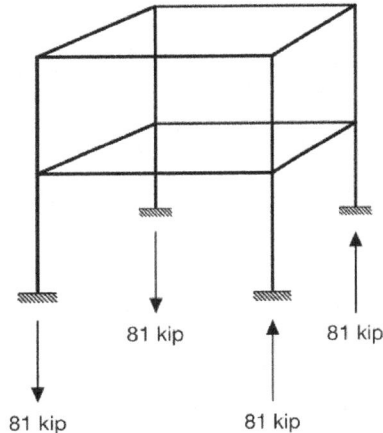

Repeat the same process for Y-Z elevation plane.

$$M_{X@axis\,B\,footing} = 98 \times 10 + 113 \times 20$$

$$= 3,240 \; kip.ft$$

$$F_{axis\,B} = \frac{3,240 \; kip.ft}{15 \; ft} = 216 \; kip$$

This force is applied on two footings, and hence each footing takes 108 kip.

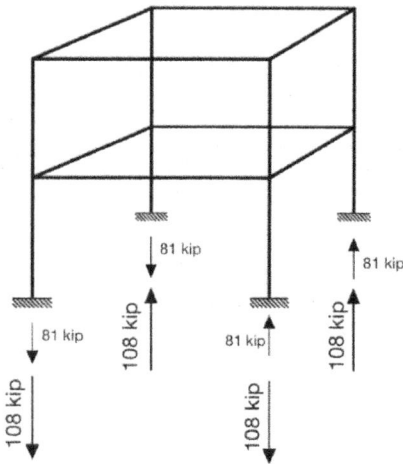

The maximum extra load on footing B.2 is $108 + 81 = 189\ kip$.

Correct Answer is (C)

SOLUTION 3.14
As explained in Section 3.3.4 in this book, also in ASCE 7 Standard Section 12.10.1.1, the following equation shall be used with attention to the comment below it.

$$F_{px} = \frac{\sum_{i=x}^{n} F_i}{\sum_{i=x}^{n} w_i} w_{px}$$

The summation shown in the equation is from level (x), which is the level under consideration, to the very top of the building.

$$= \frac{F_2+F_3}{w_2+w_3} w_2 = \frac{115+136}{350+350} \times 350 \cong 126\ kip$$

Minimum and maximum for this force shall be checked as follows:

$$F_{p2,min} = 0.2 S_{DS} I_e w_{px} = 0.2 \times 1.1 \times 1 \times 350$$
$$= 77\ kip$$

$$F_{p2,max} = 0.4 S_{DS} I_e w_{px} = 0.4 \times 1.1 \times 1 \times 350$$
$$= 154\ kip$$

Use $F_{P2} = 126\ kip$

Correct Answer is (D)

SOLUTION 3.15
The fundamental period of a structure is inversely related to its stiffness: as stiffness increases, the fundamental period decreases (i.e., the building vibrates faster).

The relationship between fundamental period and seismic response is more nuanced. In general terms, and in reference to any response spectrum diagram presented in Section 1.5 of this book, shorter fundamental periods (which corresponds to stiffer buildings) leads to a higher (S_{DS}) which leads to higher seismic forces.

On the other hand, longer fundamental periods (which corresponds to less stiff structures) leads to lower (S_{DS}), which leads to lower seismic forces.

In conclusion, as stiffness increases, fundamental period decreases, resulting in higher seismic loads.

Correct Answer is (D)

SOLUTION 3.16
Resonance occurs when the natural frequency of a building matches the frequency of the earthquake's ground motion. This causes the building to vibrate with greater amplitude, amplifying the forces acting on it. As a result, the structure can experience excessive stresses, potentially leading to severe damage or collapse. Resonance acts like a "feedback loop", where the building's vibrations are reinforced by the incoming seismic waves, increasing the risk of failure.

In conclusion, the building will vibrate with greater amplitude leading to increased forces and damage.

Correct Answer is (B)

SOLUTION 3.17

The overturning moment is simply the area under the shear diagram as follows (*):

$$= 10 \times (81 + 148 + 201 + 235 + 257)$$

$$= 9,220 \ kip.ft$$

Correct Answer is (B)

(*) The longer method is to determine the story shear and multiply it by the story height as follows:

$$= 50 \times 81 + 40 \times (148 - 81)$$

$$+30 \times (201 - 148) + 20 \times (235 - 201)$$

$$+10 \times (257 - 235)$$

$$= 9,220 \ kip.ft$$

SOLUTION 3.18

Use ASCE 7 Standard Table 12.8-2 to determine the approximate fundamental period for this building along with Table 12.8-1 to determine the upper limit for the calculated fundamental period.

$$T_a = C_t h_n^x$$

$$= 0.03 \times (14 \times 12)^{0.75}$$

$$= 1.4 \ sec$$

The upper limit is calculated using Table 12.8-1 as follows:

$$T_{upper} = C_u \times T_a$$

$$= 1.5 \times 1.4$$

$$= 2.1 \ sec$$

With reference to Table 3-3 of this book, and since $T_{calculated} < T_a$, it is recommended to **use $T_a = 1.4$ for both calculations.**

This approach introduces conservatism in the calculation of the elastic drift. However, for base shear, the use of the approximate period is already conservative. If the engineering calculation results in a higher base shear (since a lower period T leads to a higher base shear), this suggests a potential error in the calculation. In such cases, the approximate period should still be used.

Correct Answer is (A)

SOLUTION 3.19

This solution references ASCE 7 Standard, Section 12.7.2 (also Section 3.2.2 here). Also, note that only half of the first-floor walls are considered in this calculation as the first half is supported by the ground.

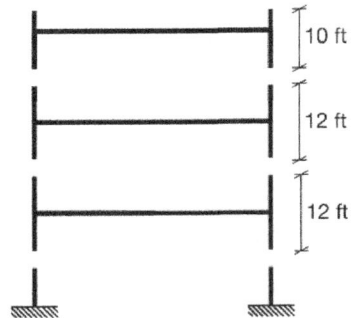

$$W_{floors} = \frac{1.05 \times 3 \times 95 \times 95 \times \left(\frac{9}{12}\right) \times 150}{1,000} \cong 3,198 \ kip$$

$$W_{external \atop walls \ 1+2} = 2 \times \frac{(4 \times 95) \times 12 \times 15}{1,000} \cong 137 \ kip$$

$$W_{external \atop walls \ 3} = 1 \times \frac{(4 \times 95) \times 10 \times 15}{1,000} = 57 \ kip$$

$$W_{interior \atop walls} = 2 \times \frac{(95 \times 95) \times 10}{1,000} \cong 181 \ kip$$

$$W_{roof \ equip.} = 50 \ kip$$

$$W_{eff. \ total} = 3,198 + 137 + 57 + 181 + 50$$

$$= 3,623 \ kip$$

Correct Answer is (C)

SOLUTION 3.20

As defined in Section 3.2.3 in this book, rigidity of a structure is its deflection per unit of displacement.

$$R = \frac{100 \ kip}{2 \ in} = 50 \ kip/in$$

Correct Answer is (A)

SOLUTION 3.21

The shear wall system offers the highest stiffness/rigidity in the described SFRSs (*).

Stiffness/rigidity is the unit force to the lateral displacement it causes, which means that the stiffer the system, the less displacement it will experience under a given lateral load.

Plain shear walls, due to their large moment of inertia and resistance to both bending and shear, provide the least displacement under lateral forces, making them the most rigid.

Moment frames, being the more flexible, experience the more displacement compared to plain shear walls for a given lateral load, making them the less rigid.

Steel braced frames offer more flexibility compared to the above two systems.

Correct Answer is (D)

(*) You can also refer to Table 12.2-1 of ASCE 7 Standard for each of the above systems. The lower the response modification factor (R) for a system means they are more rigid. See below:

o Ordinary plain concrete shear walls is item A.4 in Table 12.2-1 with $R = 1.5$.

o Ordinary reinforced concrete moment frames is item C.7 in Table 12.2-1 with $R = 3$.

o Steel eccentrically braced frames is items B.1 in Table 12.2-1 with $R = 8$.

SOLUTION 3.22

Per ASCE 7 Standard Section 12.11.2.2, the maximum length-to-width ratio for a sub-diaphragm is 2.5 to 1, which makes option C with $12.5\ ft \times 5\ ft$ the best option in this case.

Correct Answer is (C)

SOLUTION 3.23

Use Table 3-5 to calculate columns' stiffness observing the end condition for every column.

$$k_{A\ or\ B} = \frac{3EI}{h^3}$$

$$= \frac{3 \times 4.2 \times 10^3 \frac{kip}{in^2} \times 1{,}700\ in^4 \times \frac{ft^2}{144\ in^2}}{(12\ ft)^3}$$

$$= 86.1\ kip/ft$$

$$k_{A\ or\ B} = \frac{12EI}{h^3}$$

$$= \frac{12 \times 4.2 \times 10^3 \frac{kip}{in^2} \times 1{,}700\ in^4 \times \frac{ft^2}{144\ in^2}}{(12\ ft)^3}$$

$$= 344.3\ kip/ft$$

$$k_{total} = 2 \times 86.1 + 2 \times 344.3$$

$$= 860.8\ kip/ft$$

$$T = 2\pi \sqrt{\frac{W}{k_{total}\ g}}$$

$$= 2\pi \sqrt{\frac{300\ kip}{860.8\ kip/ft \times 32.174\ ft/sec^2}}$$

$$= 0.65\ sec$$

Correct Answer is (B)

SOLUTION 3.24

Checking the aspect ratio for the two concrete walls, they both falls between 0.3 and 3.0. See below:

$$0.3 \le (h_1/l_1 = 2.5) \le 3.0$$

$$0.3 \le (h_2/l_2 = 2.0) \le 3.0$$

This means that the stiffness of the two walls should include both flexural and shear stiffness per Table 3-5. For simplicity, Tables 3-6 and 3-7 (*) can be used as follows:

For shear wall No. 1, single story cantilever wall, use Table 3-6:
$$\rightarrow k_1 = 0.014 \times E \times \left(\frac{10}{12}\right)$$

For shear wall No. 2, single story two ends fixed wall, use Table 3-7:
$$\rightarrow k_2 = 0.071 \times E \times \left(\frac{10}{12}\right)$$

You can divide those two equations to derive the requested relationship as follows:

$$\frac{k_1}{k_2} = \frac{0.014}{0.071}$$

$$k_2 = 5.0\, k_1$$

Correct Answer is (A)

(*) In some other resources, you may find that the rigidity in those two tables multiplied by 10. This only adds more weight to the number. As long as consistency is maintained throughout the solution, both approaches are acceptable.

SOLUTION 3.25

The following items require 1.4 force multiplier per ASCE 7 Standard 12.11.2.2.2:

Steel wall connectors, sub-diaphragm strapping, continuous crossties, and their connections. This includes steel joists and girders part of the anchorage system, steel decking and its connections for anchoring walls, steel straps tying wood purlins together, steel wall connectors, steel ledgers, etc.

Items excluded from this force multiplier:

Reinforcing steel, anchor rods, headed-weld studs, wood bolting, and wood nailing.

Correct Answer is (D)

This page is intentionally left blank

CHAPTER
4

Seismic Analysis Procedures

4.1. Purpose of this Chapter

As we move into this chapter, you will notice how it brings together the concepts introduced in the previous chapters to draw a clear and cohesive conclusion. A key focus of this chapter is understanding the difference in analysis between a rigid and a flexible diaphragm. This distinction leads into a discussion on how seismic forces are distributed within the Seismic Force Resisting System (SFRS) within the diaphragm. The focus in this chapter is the shear wall system. To help illustrate this concept, a useful chart is provided for ease of reference in Section 4.3.1. Drawing from this understanding, this chapter covers methods for calculating the Center of Mass and Center of Rigidity of diaphragms and explains how torsional moments arise due to their interaction. These moments play a critical role in the distribution of loads to the various lateral supporting elements.

This chapter further elaborates on topics such as the forces acting on diaphragms. For example, in Chapter 2, we treated the diaphragm as a simple deep beam and demonstrated how to calculate chord forces (see Example 2.1). In this chapter, we will explore additional forces within the diaphragm, such as drag forces and those acting on other elements of the diaphragm. We will also examine how load paths are constructed to better understand the full range of forces at play.

Finally, the chapter concludes with a discussion on the application of seismic load effects and how these loads are applied within the structural system.

4.2. Seismic Force Resisting System Rigidity
4.2.1.   Overview

In the previous chapter, we discussed the stiffness and/or rigidity of the Structural Response Force Resisting System (SFRS), particularly in systems with columns and shear walls. We also explored how the height-to-length aspect ratio of a shear wall affects its deflection under lateral loads, influencing its rigidity. We also discussed how columns, acting as tall (flexural) shear walls, contribute their flexural stiffness to the overall rigidity of the structure, with end conditions at play. This concept was also used to calculate the structure's fundamental period.

In this chapter, we will apply these principles of rigidity to the analysis of diaphragms, focusing on how lateral forces are distributed to the various SFRS elements based on their relative rigidity. Simply put, the more rigid an element, the larger the portion of the lateral force it will carry proportionate to the rigidity of each element. However, factors such as flexibility of the diaphragm and the positioning of the elements also play critical roles in determining how forces are allocated.

In the following example, we demonstrate how a horizontal force acting on a rigid diaphragm supported by four columns results in the lateral force being distributed among the columns based on their relative rigidity.

Example 4.1 – Simple Horizontal Force Distribution

In the below rigid diaphragm, assume no torsional or accidental torsional moment are taking place. The moment of inertia for columns C and D are double those of columns A and B in the direction of seismic loading shown in white arrows. All columns are fixed at their both ends, and they are all made of the same material.

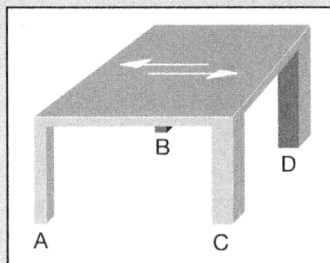

If the height of this simple structure is 14 ft and the lateral seismic loading shown is 120 kip, what is the moment experienced at column A support?

For fixed ends, and per Table 3-5, stiffness of columns is calculated as follows:

$$k_{A \text{ or } B} = \frac{12EI}{h^3} \quad \text{while} \quad k_{C \text{ or } D} = \frac{12E(2I)}{h^3}$$

Given that the following factors are common in the two equations: h, E, I and 12, they can be omitted, and we can deal with a relative stiffness as follows:

$$k_{A \text{ or } B} = K$$

$$k_{C \text{ or } D} = 2K$$

In this sense, the share of the seismic force that is applied at column A is calculated as follows:

$$F_A = \frac{k_A}{\sum k} \times F = \frac{K}{6K} \times 120 = 20 \; kip$$

$$M_A = 20 \; kip \times 14 \; ft = 280 \; kip.ft$$

The above example is a simplified case, where columns are symmetrically distributed at the edges of the diaphragm where the lateral force is applied. However, in case of any irregularities – whether horizontal irregularities within the diaphragm or a misalignment between the center of rigidity of the columns and the Center of Mass of the diaphragm – a torsional moment will be generated. This will affect the distribution of the lateral force. This concept is explored further in this chapter.

4.2.2. Relative Rigidity for the Horizontal Resisting System Elements

To analyze the lateral force distribution in diaphragms, it is not required to calculate the exact rigidity of the system or the rigidity of its horizontal resisting elements. Instead, we only need to determine the relative rigidity of the horizontal resisting structural elements for each diaphragm being analyzed. This approach simplifies the analysis and reduces the time and effort required to do so, especially since calculating exact rigidities can be quite complex – often more complex than the methods discussed in Chapter 3 of this book. As long as we apply a consistent method for determining the relative rigidity of these elements, this approach remains valid and effective.

One method is using the aspect ratio of shear walls, which is the ratio of height-to-length (h/l). For example, and as explained in Section 3.2.3 of this book, when this ratio falls below 0.3, shear walls tend to experience shear deflection, and their stiffness should be calculated using the shear stiffness equations provided in Table 3-5 in the appendix of Chapter 3. In this case, if all shear walls on a given floor share this same aspect ratio (i.e., $h/l < 0.3$), their plan cross-sectional area can be used as an indicator of relative rigidity. If these walls also have the same thickness, the length of each wall can then serve as a measure of relative rigidity. However, if those shear walls have different aspect ratios, placing them in different categories of those explained in Table 4-1 below, you can no longer rely on area or length to determine relative rigidity for all the walls on the same floor being analyzed. In this case, you would need to revert to traditional methods of calculating rigidities for all the walls.

It is important to remember that the primary purpose of calculating relative rigidity in this chapter is to determine the distribution of lateral forces and to calculate the Center of Rigidity (CR), which will be further discussed in this chapter. This method cannot be used for any other purposes. For this purpose alone, Table 4-1 below outlines the appropriate measures to use for determining relative rigidity for various cases and aspect ratios for shear walls.

Table 4-1 Walls Rigidities Similar Material Single Story

Aspect ratio (h/l)	Rigidity (R) or Measure of Relative Rigidity	
	Thickness (t) same for all walls	**Thickness (t) differs between walls**
$h/l < 0.3$	Use wall length (l)	Use wall cross-sectional area (A)
$0.3 < h/l < 3.0$ You can use 0.25 to 4.0 for masonry walls	Cantilever walls $R = \dfrac{1}{4\left(\frac{h}{l}\right)^3 + 3\left(\frac{h}{l}\right)}$ or use Table 3-6	Cantilever walls $R = \dfrac{t}{4\left(\frac{h}{l}\right)^3 + 3\left(\frac{h}{l}\right)}$ where (t) is in feet
	Two fixed end walls $R = \dfrac{1}{\left(\frac{h}{l}\right)^3 + 3\left(\frac{h}{l}\right)}$ or use Table 3-7	Two fixed end walls $R = \dfrac{t}{\left(\frac{h}{l}\right)^3 + 3\left(\frac{h}{l}\right)}$ where (t) is in feet
$h/l > 3.0$	Use moment of inertia (I) or actual stiffness if end conditions or height differ	Use moment of inertia (I) or actual stiffness if end conditions or height differ

As for columns, their rigidity is generally determined by their actual stiffness, which is typically governed by their flexural properties. Columns' stiffness should be based on those equations provided in Table 3-5 of Chapter 3 (i.e., $12EI/h^3$ for columns with both ends fixed, or $3EI/h^3$ for cantilever columns, along with other cases specified in that table). The moment of inertia can be used alone as a measure of relative rigidity, provided columns being analyzed in the same floor have consistent heights and end conditions.

Before proceeding with the determination of lateral force distribution, it is important to understand that the process differs for flexible and rigid diaphragms. Additionally, the location of the horizontal forces resisting elements plays a crucial role in these calculations. For rigid diaphragms, determining the Center of Rigidity and its distance from the Center of Mass is a critical step. In contrast, for flexible diaphragms, the Center of Rigidity is less important, as their flexibility leads to a more even distribution of forces across the diaphragm. This is explained in more detail in Section 4.3 of this Chapter.

4.2.3.   Determining the Center of Rigidity

To calculate the Center of Rigidity (CR) for shear walls for a rigid diaphragm, the relative rigidity of each wall, as explained in the previous section, is used in the following steps: (1) Identify the location of each shear wall within the diaphragm by measuring the distance from a reference point to the centerline of each wall. (2) Calculate the moment of each wall by multiplying its relative rigidity by its distance from the reference point. (3) Sum the moments of all the walls, taking into account their respective relative rigidities and distances. (4) The total moment is then divided by the sum of the relative rigidities of all the walls.

$$CR = \frac{\sum(R_{relative} \times distance)}{\sum R_{relative}}$$

Example 4.2 – Center of Rigidity for a Rigid Diaphragm

In the plan below, all shear walls have the same thickness, are made of reinforced concrete, and are part of a single-story building with a height of 18 ft.

Assuming all walls are cantilevered and ignoring the stiffness of the columns, calculate the location of the Center of Rigidity (CR) for the walls in both the South-North and East-West directions. For this calculation, use axis A as the datum for the horizontal direction and axis 1 for the vertical direction.

South-North direction using axis A as datum:

Aspect ratio for wall A (18/15 = 1.2) and wall D (18/30 = 0.6), both fall between 0.3 and 3.0, both are cantilever walls with equal thickness, use Table 3-6 to identify their relative rigidities as:

$$R_{wall\,A} = 0.095 \ \& \ R_{wall\,D} = 0.375$$

$$CR_{S-N} = \frac{\Sigma(R_{relative} \times distance)}{\Sigma R_{relative}} = \frac{0.095 \times 0 + 0.375 \times 60}{0.095 + 0.375} = 47.9\,ft$$

East-West direction using axis 1 as datum:

Aspect ratio for wall 1 (18/20 = 0.9) and wall 4 (18/20 = 0.9), both fall between 0.3 and 3.0, both are cantilever walls with equal thickness, use Table 3-6 to identify their relative rigidities as:

$$R_{wall\,1} = R_{wall\,4} = 0.178$$

$$CR_{S-N} = \frac{\Sigma(R_{relative} \times distance)}{\Sigma R_{relative}} = \frac{0.178 \times 0 + 0.178 \times 45}{0.178 + 0.178} = 22.5\,ft$$

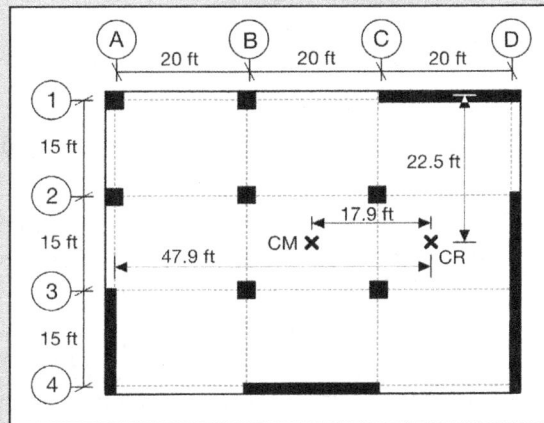

In the following diagram, Center of Rigidity (CR) is plotted against the Center of Mass (CM) showing a torsional moment would be forming in the S-N direction, but not the E-W direction (*).

(*) Accidental Torsional Moment:

It is essential to know that regardless of whether a torsional inherent moment is actually generated from the applied loads, ASCE 7 Standard Section 12.8.4.2 mandates the inclusion of an accidental torsional moment in non-flexible diaphragms. This requires displacing the CM by 5% of the dimension of the structure perpendicular to the direction of applied forces. **See Example 4.7 for more details.**

In this case, for the torsional moment in the S-N direction, the lever arm should be adjusted to 17.9 + 3 = 20.9 ft or 17.9 − 3 = 14.9 ft if loading was in that direction – Example 4.7 explains how and why – whichever generates larger results. Same applies to the W-E direction.

4.3. Seismic Force Distribution in the Shear Wall System
4.3.1.   Overview of Diaphragms

In previous chapters, we differentiated between flexible and rigid diaphragms, as defined in Section 2.3.2 of this book. For clarity, examples of rigid diaphragms include concrete floors, steel decks with concrete, or reinforced concrete slabs, while flexible diaphragms include timber floors, open-web steel joists, and metal deck roofs.

As previously discussed, the method for analyzing the distribution of lateral loads to shear walls or other lateral-resisting elements varies between these two types of diaphragms. In flexible diaphragms, the inherent flexibility allows for a more uniform distribution of lateral loads, enabling the diaphragm to function as a simple beam (in which case a tributary area method can be used), or a series of simple beams, depending on how the horizontal structural elements are arranged. This is particularly the case when horizontal resisting elements are positioned in parallel. However, when these elements are aligned in a single line of action (SLA), the relative rigidity of those elements turns into a critical factor in the analysis. This concept is explored in greater detail by way of examples in Section 4.3.2.

In case of rigid diaphragms, the Center of Rigidity (CR) must be determined and compared with the Center of Mass (CM). When the two coincide, lateral forces are distributed to elements based on their relative rigidity, as illustrated in Example 4.1. If the CR and CM are misaligned, the polar moment is used as explained in Section 4.3.3.

To simplify the analysis process, Figure 4-1 provides a flow chart that walks you through how flexible and rigid diaphragms differ in terms of lateral force distribution.

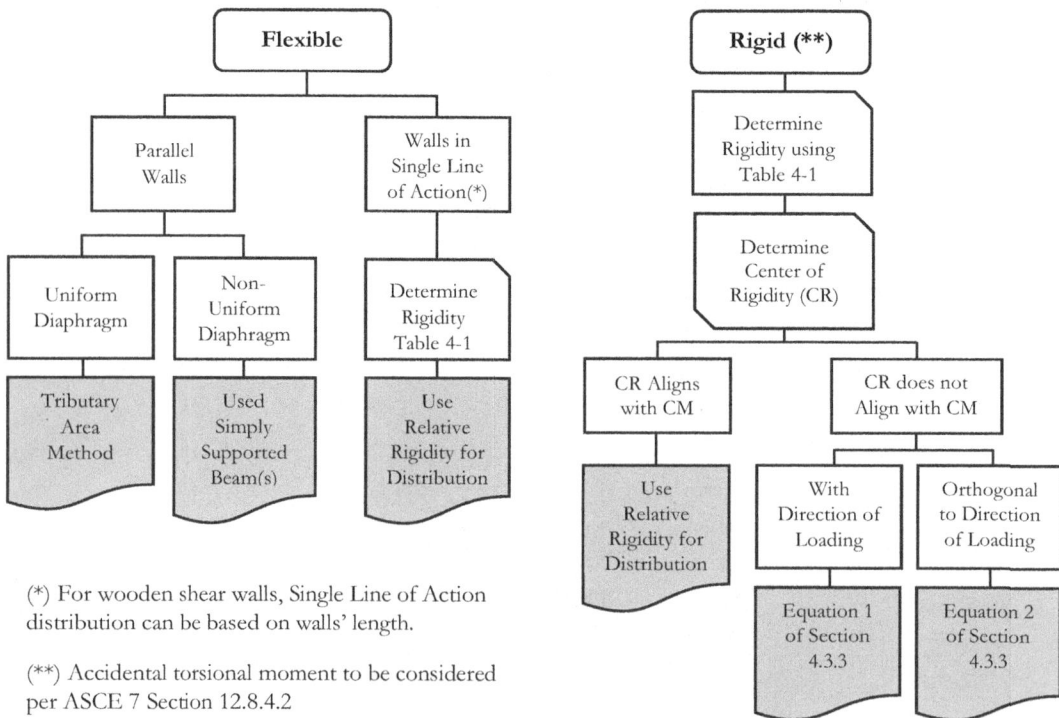

(*) For wooden shear walls, Single Line of Action distribution can be based on walls' length.

(**) Accidental torsional moment to be considered per ASCE 7 Section 12.8.4.2

Figure 4-1 Diaphragm Lateral Load Distribution Flow Chart

4.3.2. Flexible Diaphragms

The analysis of lateral loads for a flexible diaphragm is straightforward and is covered in the previous flowchart. This topic is better explained using the following two examples.

Example 4.3 – Lateral Load Distribution in a Flexible Diaphragm (1)

In the plan below, the floor is a flexible wooden diaphragm, and all shear walls have the same thickness and are made of reinforced concrete. The building is a single-story structure with a height of 18 ft.

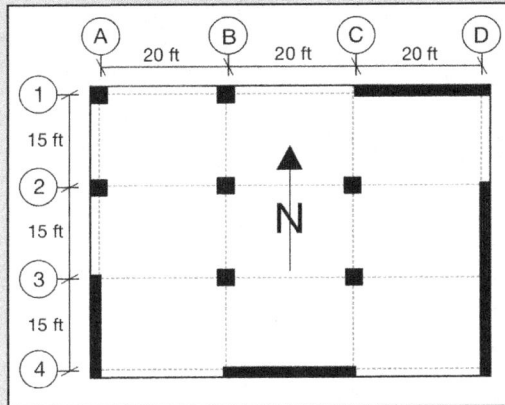

Given a lateral load of 315 kip in the South-North direction, how much of this load is transferred to walls at A and D ignoring stiffness of columns?

In reference to Figure 4-1, this is a flexible uniformly shaped diaphragm and the walls in question are parallel. Based on this, tributary area in the direction of loading can be used as follows:

$$Wall\ A = Wall\ D = 315/2 = 157.5\ kip$$

Example 4.4 – Lateral Load Distribution in a Flexible Diaphragm (2)

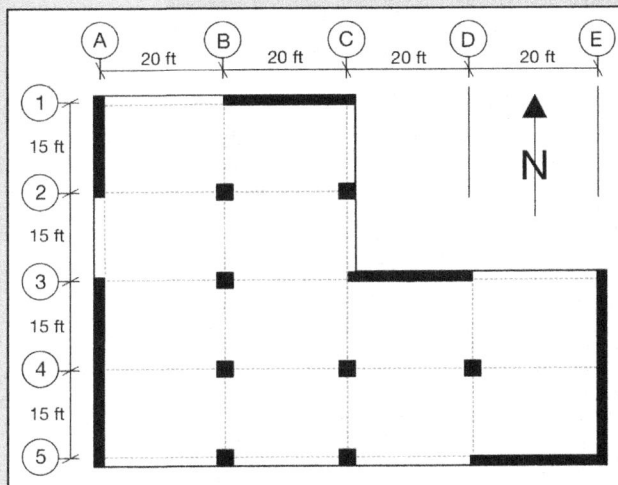

In this plan, the floor diaphragm is flexible, and the lateral load in the South-North direction is 420 kip. All shear walls are made of masonry, have the same thickness, and are 14 ft tall cantilever walls in a single-story structure assigned to SDC D. Ignore columns' stiffness.

Based on this, what is the lateral force transferred to wall A between axes 1 and 2, and what lateral connection force should be considered between the diaphragm and this wall?

In reference to Figure 4-1, this is a flexible non-uniformly shaped diaphragm and the walls in question are parallel. Based on this, the simple beam method can be used, then followed by the application of the Single Line of Action (SLA) and the relevant rigidity method.

First start with calculating the contribution of each diaphragm's area to the fictitious simple beam as follows:

$$Area\ A\ to\ C\ =\ 40 \times 60 = 2,400\ ft^2\ (66.67\%\ of\ total\ diaphragm\ area)$$

$$Area\ C\ to\ E\ =\ 40 \times 30 = 1,200\ ft^2\ (33.33\%\ of\ total\ diaphragm\ area)$$

The total load 420 kip can be distributed over a simply supported beam based on the above percentages as follows:

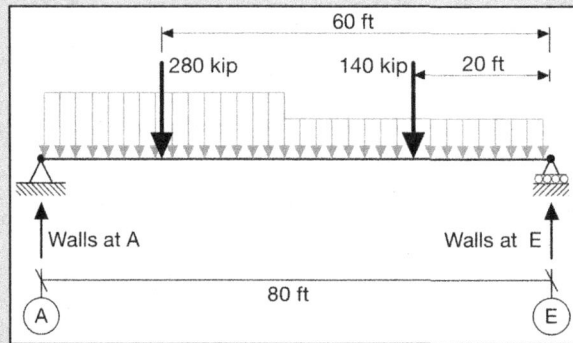

Determine the force share at walls on axis A by taking moment around E:

$$Walls\ @\ A = \frac{140 \times 20 + 280 \times 60}{80} = 245\ kip$$

This force needs to be distributed to the two walls at axis A based on their rigidities:

For the masonry wall at axes 1-2 is: $0.25 < (h/l = 14/15) \leq 4.0$, in which case, flexural and shear stiffness shall be calculated, using Table 3-6 → $R = 0.165$

For the masonry wall at axes 3-5 is: $0.25 < (h/l = 14/30) \leq 4.0$, in which case, flexural and shear stiffness shall be calculated, using Table 3-6 → $R = 0.554$

The lateral force transferred to Wall A between 1-2 $= \frac{0.165}{0.165+0.554} \times 245 = 56.22\ kip$

Finally, this diaphragm exhibits a Type 2 horizontal irregularity (reentrant corner irregularity). According to Table 2-1 of this book, the diaphragm design force for connectors to vertical elements must be increased by 25% for SDC D, as specified in ASCE 7 Section 12.3.3.4. Therefore, the adjusted design force is calculated as $1.25 \times 56.22 = 70.3\ kip$.

4.3.3.   Rigid Diaphragms

In rigid diaphragms, lateral loads are distributed to shear walls or other lateral-resisting elements, such as columns or frames, based on their relative rigidities. A basic example of this concept was introduced at the beginning of this chapter.

In typical building design, lateral loads are considered separately in two orthogonal directions. When shear walls or other structural elements are symmetrically arranged relative to the center of the applied load (i.e., when the Center of Mass coincides with the Center of Rigidity), the force resisted by each element can be calculated using the following equation:

$$F_{y,i} = \frac{R_{y,i}}{\sum R_y} \times F_y$$

Where:

$F_{y,i}$ is the force resisted by an individual element or shear wall (i) in (y) direction.
$R_{y,i}$ is the rigidity of element (i) in (y) direction.
$\sum R_y$ is the sum of all rigidities in (y) direction.
F_y is the lateral load for the diaphragm under consideration in (y) direction.

When the lateral resisting elements or shear walls are positioned asymmetrically relative to the Center of Mass (CM), the analysis must account for torsional effects, resulting in what is known as inherent torsional moment. In this case, the polar moment of rigidity is used, as explained in the equations below.

When forces are applied in the (y) direction, they are distributed to the lateral resisting elements in that direction paying close attention to the possible negative sign in Equation 1. The ($\mp$) sign used below depends on the direction of moment relative to the element being analyzed.

$$F_{y,i} = \left[\frac{R_{y,i}}{\sum R_y} \mp \frac{e_x(x)R_{y,i}}{\sum R_y(x)^2 + \sum R_x(y)^2}\right] \times F_y \dots\dots\text{Equation 1}$$

Building on the previous explanation, when forces are applied in the (y) direction, elements resisting in the (x) direction receive the distribution of those forces as follows:

$$F_{x,i} = \left[\frac{e_x(y)R_{x,i}}{\sum R_y(x)^2 + \sum R_x(y)^2}\right] \times F_y \dots\dots\text{Equation 2}$$

Where:

F_y is the applied lateral load in the (y) direction.
$R_{x,i}$ & $R_{y,i}$ are rigidities of elements resisting in the (x) and (y) directions.
$\sum R_y(x)^2 + \sum R_x(y)^2$ is the *polar moment of inertia*. This is the summation of rigidities in the (x) and (y) direction multiplied by their respective location from the Center of Rigidity (CR) in the (x) and (y) direction. This value **does not change** due to the accidental torsional moment (or accidental eccentricity).
x is the distance of the respective element from CR in (x) direction.
y is the distance of the respective element from CR in (y) direction.
e_x (when load is applied in the y direction), this is the distance between the Center of Mass (CM) to the Center of Rigidity (CR) in the (x) direction.

The above two equations can be rewritten with the use of the inherent torsional moment (M_T) which equals to $(F_y \times e_x)$:

$$F_{y,i} = \frac{R_{y,i}}{\sum R_y} \times F_y \mp \frac{M_T(x)R_{y,i}}{\sum R_y(x)^2 + \sum R_x(y)^2} \quad \text{.........................Equation 1a showing torsional moment } (M_T)$$

$$F_{x,i} = \frac{M_T(y)R_{x,i}}{\sum R_y(x)^2 + \sum R_x(y)^2} \quad \text{...............................Equation 2a showing torsional moment } (M_T)$$

Example 4.5 – Lateral Load Distribution in a Rigid Diaphragm (1)

In the plan below, the floor is a rigid diaphragm, and all shear walls have the same thickness and are made of reinforced concrete. The building is a single-story structure with a height of 18 ft.

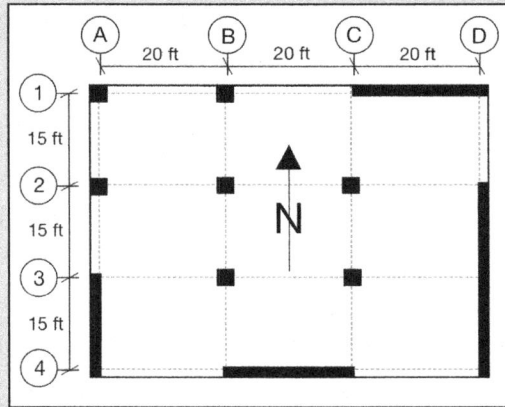

Given a lateral load of 315 kip in the South-North direction, how much of this load is transferred to walls at A and D ignoring stiffness of columns and ignoring accidental torsion?

This is the same floor plan presented in Example 4.2 where eccentricity was calculated as $e_x = 17.9\,ft$.

The north direction will be represented as $y\,axis$ while the east direction will be $x\,axis$.

In reference to the chart presented in Figure 4-1, this is a rigid diaphragm where CR does not coincide with CM, hence Equation 1 of Section 4.3.3 shall be used.

Rigidities for walls A & D were determined in Example 4.2 as follows:

$$R_{y,A} = R_{wall\,A} = 0.095$$

$$R_{y,D} = R_{wall\,D} = 0.375$$

Moreover, and per Equation 1 of Section 4.3.3, the rigidities of the walls in the x direction shall be determined as well, and they are as follows:

$$R_{x,1} = R_{wall\,1} = 0.178$$

$$R_{x,4} = R_{wall\,4} = 0.178$$

Apply Equation 1 for Wall A as follows where CR in the y direction is in the middle:

$$F_{y,A} = \left[\frac{R_{y,A}}{\sum R_y} \mp \frac{e_x(x)R_{y,A}}{\sum R_y(x)^2 + \sum R_x(y)^2}\right] \times F_y$$

$$= \left[\frac{0.095}{0.095 + 0.375} + \frac{17.9 \times (47.9) \times 0.095}{\begin{pmatrix}(0.095 \times 47.9^2 + 0.375 \times 12.1^2)\\+\\(0.178 \times 22.5^2 + 0.178 \times 22.5^2)\end{pmatrix}}\right] \times 315 \cong 120.3 \; kip$$

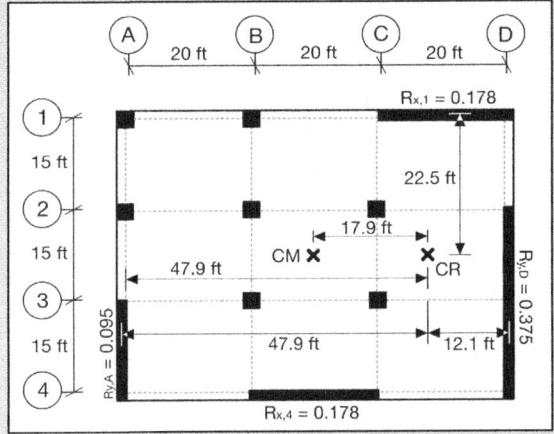

$$F_{y,D} = \left[\frac{R_{y,D}}{\sum R_y} \mp \frac{e_x(x)R_{y,D}}{\sum R_y(x)^2 + \sum R_x(y)^2}\right] \times F_y$$

$$= \left[\frac{0.375}{0.095 + 0.375} - \frac{17.9 \times (12.1) \times 0.375}{\begin{pmatrix}(0.095 \times 47.9^2 + 0.375 \times 12.1^2)\\+\\(0.178 \times 22.5^2 + 0.178 \times 22.5^2)\end{pmatrix}}\right] \times 315 \cong 194.7 \; kip$$

<u>In this equation, a negative sign was used because the torsional moment pushes wall D backwards (in the south direction) while the lateral load pushes it forward.</u>

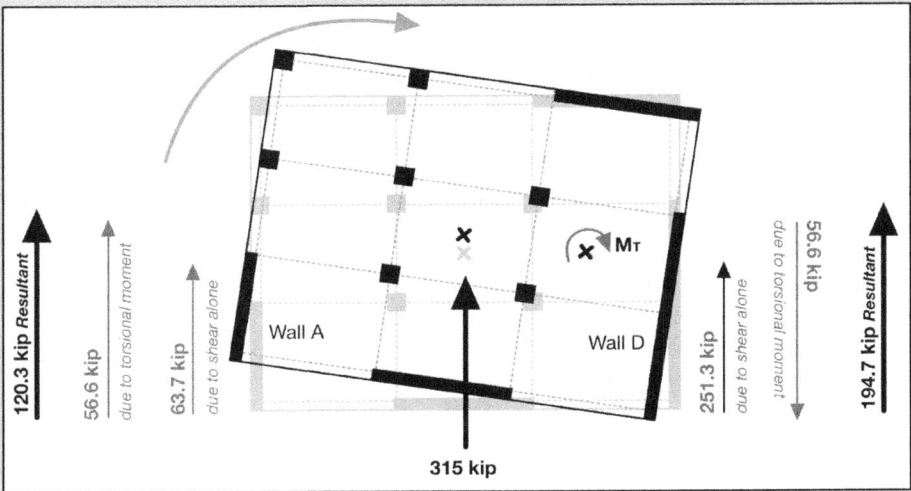

Example 4.6 – Lateral Load Distribution in a Rigid Diaphragm (2)

Solve Example 4.5 with the same lateral load direction (i.e., North direction), however, find the load transferred to walls 1 and 4, which are the walls orthogonal to the direction of loading. In this example, ignore stiffness of columns and ignore accidental torsion.

Equation 2 of Section 4.3.3 is used in this case as follows:

$$F_{x,1} = \left[\frac{e_x(y)R_{x,1}}{\sum R_y(x)^2 + \sum R_x(y)^2}\right] \times F_y = \left[\frac{17.9 \times 22.5 \times 0.178}{\begin{pmatrix}(0.095 \times 47.9^2 + 0.375 \times 12.1^2)\\ +\\ (0.178 \times 22.5^2 + 0.178 \times 22.5^2)\end{pmatrix}}\right] \times 315 = +49.8\,kip$$

$$F_{x,4} = \left[\frac{e_x(y)R_{x,4}}{\sum R_y(x)^2 + \sum R_x(y)^2}\right] \times F_y = \left[\frac{(-1) \times 17.9 \times 22.5 \times 0.178}{\begin{pmatrix}(0.095 \times 47.9^2 + 0.375 \times 12.1^2)\\ +\\ (0.178 \times 22.5^2 + 0.178 \times 22.5^2)\end{pmatrix}}\right] \times 315 = -49.8\,kip$$

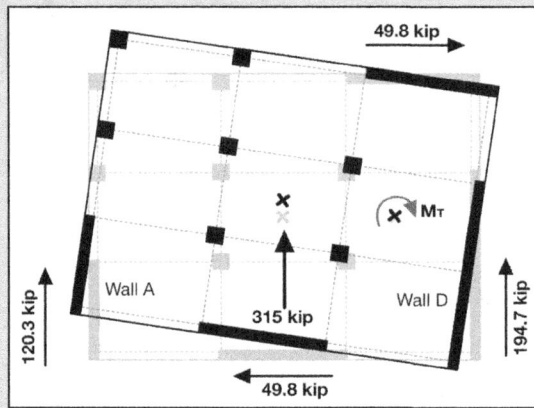

Example 4.7 – Accidental Torsional Moment

In Examples 4.5 and 4.6, recalculate the shear forces transferred to walls A, D, 1 and 4 with the inclusion of the accidental torsional moment.

Accidental displacement for the Center of Mass (CM) for the direction of loading is $5\% \times 60\,ft = 3.0\,ft$ and the accidental torsion is $M_{ta} = 3 \times 315 = 945\,kip.ft$. This value should be tested each way for each wall – i.e., add it to the inherent torsion or deducted it from it to test which value governs, where the inherent torsion is $M_T = 17.9 \times 315 = 5,638.5\,kip.ft$. You can choose to use Equation 1a and 1b for this purpose, and you can still use Equations 1 and 2 where in this case you need to adjust the eccentricity (e_x) by adding $3.0\,ft$ or deducting it.

$$F_{y,A} = \frac{R_{y,A}}{\sum R_y} \times F_y + \frac{(M_T \mp M_{ta}).(x)R_{y,A}}{\sum R_y(x)^2 + \sum R_x(y)^2}$$

$$= \frac{0.095}{0.095 + 0.375} \times 315 + \frac{(5,638.5 + 945) \times 47.9 \times 0.095}{\begin{bmatrix}(0.095 \times 47.9^2 + 0.375 \times 12.1^2)\\ +\\ (0.178 \times 22.5^2 + 0.178 \times 22.5^2)\end{bmatrix}} = 129.9\,kip.ft$$

$$F_{y,D} = \frac{R_{y,D}}{\sum R_y} \times F_y - \frac{(M_T \mp M_{ta}).(x)R_{y,D}}{\sum R_y(x)^2 + \sum R_x(y)^2}$$

$$= \frac{0.095}{0.095 + 0.375} \times 315 - \frac{(5{,}638.5 - 945) \times 12.1 \times 0.375}{\begin{bmatrix}(0.095 \times 47.9^2 + 0.375 \times 12.1^2)] \\ + \\ (0.178 \times 22.5^2 + 0.178 \times 22.5^2)]\end{bmatrix}} = 204.3 \; kip.ft$$

Observe here that we <u>deducted</u> the accidental torsion in wall D in the above equation as this generates the maximum shear transfer to this wall.

$$F_{x,1} = \frac{(M_T \mp M_{ta}).(y)R_{x,1}}{\sum R_y(x)^2 + \sum R_x(y)^2} = \frac{(5{,}638.5 + 945) \times 22.5 \times 0.178}{\begin{bmatrix}(0.095 \times 47.9^2 + 0.375 \times 12.1^2)] \\ + \\ (0.178 \times 22.5^2 + 0.178 \times 22.5^2)]\end{bmatrix}} = +58.2 \; kip.ft$$

$$F_{x,4} = \frac{(M_T \mp M_{ta}).(y)R_{x,4}}{\sum R_y(x)^2 + \sum R_x(y)^2} = \frac{(-1)(5{,}638.5 + 945) \times 22.5 \times 0.178}{\begin{bmatrix}(0.095 \times 47.9^2 + 0.375 \times 12.1^2)] \\ + \\ (0.178 \times 22.5^2 + 0.178 \times 22.5^2)]\end{bmatrix}} = -58.2 \; kip.ft$$

4.4. Diaphragm Force Behavior
4.4.1.   Seismic Force Load Path

In Section 2.3.2, we introduced the concept of diaphragms and provided a basic example of chord design in Example 2.1. In this section, we will explore diaphragm design in more depth, focusing on how diaphragms support themselves while maintaining strong connections with seismic force-resisting systems, such as shear walls. This ensures a continuous and reliable load path for seismic loads to be transferred effectively throughout the structure.

The load path in seismic design refers to the route by which seismic forces travel through a structure, from their point of application to the foundation. This path is critical for maintaining the integrity and stability of the structure during an earthquake. For diaphragms, it is essential that the load path remains continuous and uninterrupted. Any break in this path can lead to failures, such as buckling, collapse, or excessive deformation, which compromise the overall stability of the structure. Also, poor connections between the diaphragm and vertical support structures or insufficient shear capacity can lead to localized failures, disrupting the load path.

The design of diaphragms involves key components such as collectors, chords, and drag elements. Collectors, illustrated in Figure 2-15 of Chapter 2, gather forces from various parts of the diaphragm and channel them to vertical load-resisting elements. Chords serve as horizontal members that resist tension and compression forces, while drag elements facilitate the transfer of these forces to the vertical systems – refer to the following sections for more details on those elements.

This section shows how diaphragms are designed with strong connections and adequate shear capacity to withstand seismic forces and prevent rupture, maintaining a reliable load transfer throughout the structure.

4.4.2.   Chord Forces

Chords, as discussed in Chapter 2 and illustrated in Figure 2-15, can experience either tension or compression. They define the boundaries of the diaphragm, acting perpendicular to the applied load. Additionally, depending on the seismic loading direction, a chord can also function as a collector for forces in the orthogonal direction.

The calculation of forces on diaphragm chords varies slightly between rigid and flexible diaphragms, particularly when spanning multiple sections. In flexible diaphragms, we treat these sections as a series of simple beams, reflecting the diaphragm's inherent flexibility. This approach is consistent with the tributary area method. Figure 4-2 below illustrates this method and provides equations for calculating the resultant moment and shear diagrams for chords in flexible diaphragms.

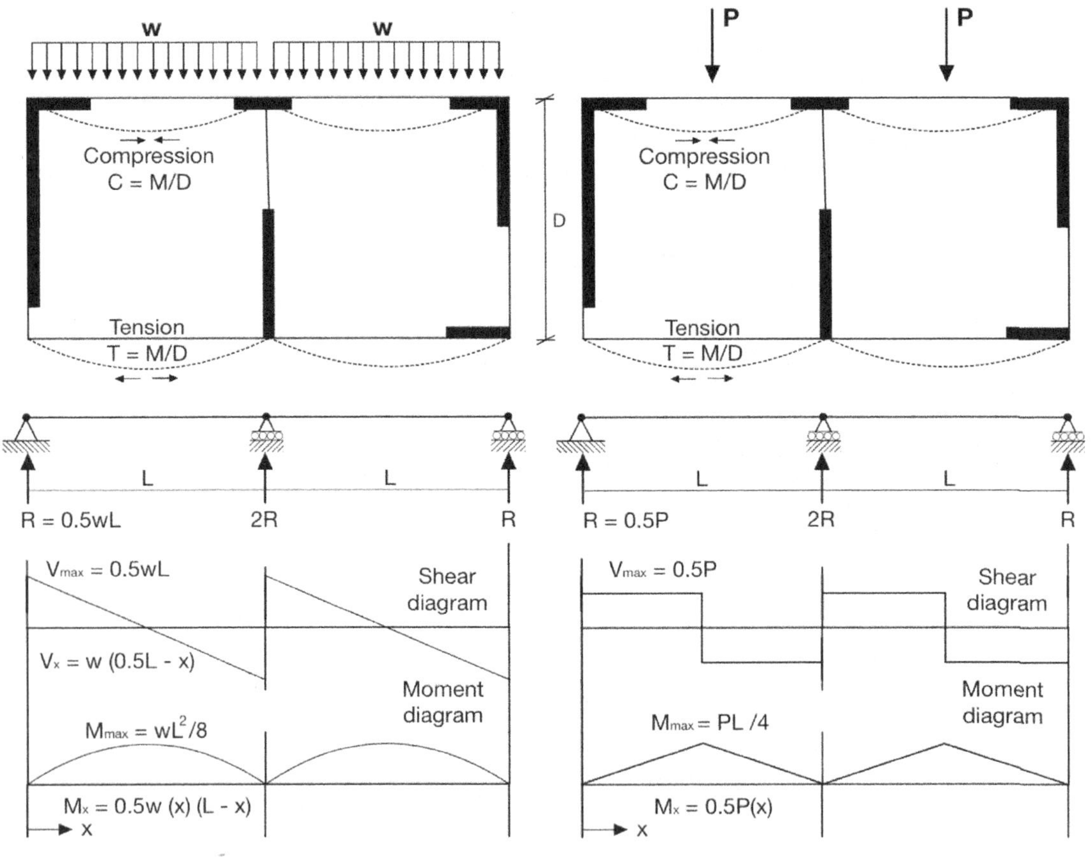

Figure 4-2 Modelling of a Series of Flexible Diaphragms

As for rigid diaphragms, if they were one span only, you can approximate it with a simple beams analysis, however, if they span multiple sections laterally, the calculation method changes. A rigid diaphragm lacks the inherent flexibility needed to act as a series of simple beams and instead functions as a continuous beam, depending on its setup. An example for that is shown in Figure 4-3.

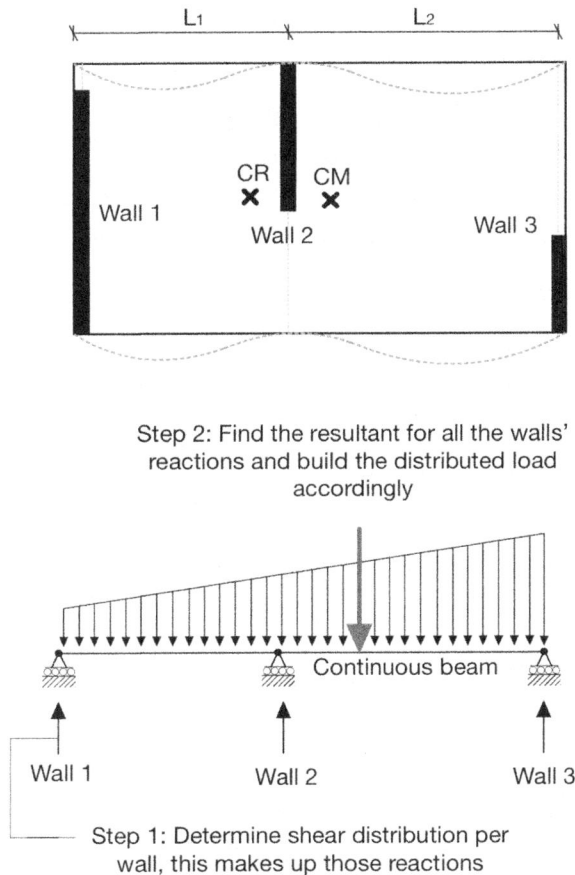

Figure 4-3 Modelling of a Series of Rigid Diaphragms

The example demonstrated in Figure 4-3 shows how forces on continuous rigid diaphragms are determined. While we will not go into detail on the calculations here as it could be quite lengthy, we will outline the process only while referring to Figure 4-3 for relevance:

- Step (1): distribute the lateral loads to walls 1, 2, and 3, this distribution takes CM and CR into consideration as discussed in Section 4.3.3.
- Step (2): once these "reactions" are determined, the lateral load on the diaphragm is reconstructed.
- Step (3): use the reconstructed load to calculate chord forces or any other forces within the rigid diaphragm.

Tension forces in diaphragms are often more critical than compression forces, as they require additional considerations. For instance, reinforced concrete rigid diaphragms may need extra tension reinforcements, while less flexible diaphragms might require elements such as ties to address these forces.

In reinforced concrete diaphragms, chord reinforcements are typically placed near the edge of the slab when no edge beams are present. If perimeter beams are included, the reinforcements are positioned just outside the edge beam, as shown in Figure 4-4. This placement prevents the introduction of additional tension or compression forces into the edge beams.

Chord reinforcements
Cannot be part of the edge beam reinforcements
Can be part of shrinkage reinforcements

Slab reinforcements

Edge beam reinforcements

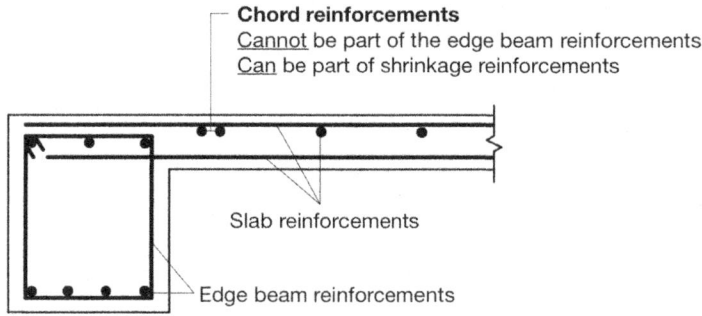

Figure 4-4 Typical Chord Reinforcement in a Slab with Edge Beam

While most diaphragms can normally withstand compression forces in the chords, the primary focus is typically on tension forces to prevent diaphragm tearing. However, large slab openings can create thinner sub-diaphragms or chords, which may be prone to buckling under compression. This introduces an additional design consideration especially in Seismic Design Categories D, E and F. Furthermore, on the tension side of these thinner sub-diaphragms, extra tension reinforcement may be required due to the reduced lever arm created by the thin diaphragm. Figure 4-5 provides an example of a large opening with a thin sub-diaphragm formed which may require some attention.

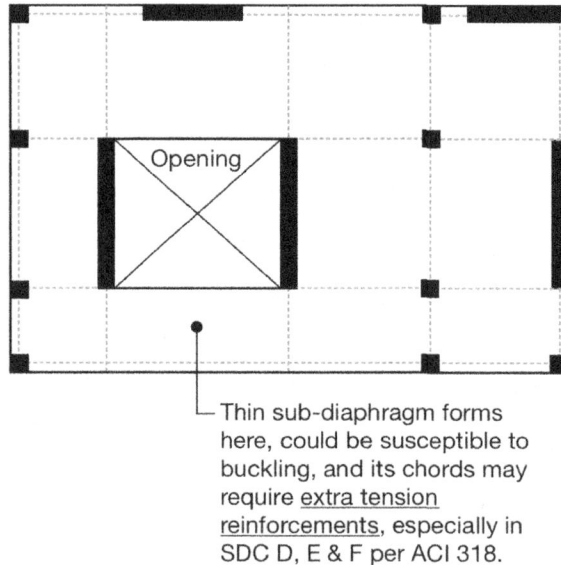

Opening

Thin sub-diaphragm forms here, could be susceptible to buckling, and its chords may require extra tension reinforcements, especially in SDC D, E & F per ACI 318.

Figure 4-5 Large Openings in Diaphragms and their Effect on Chord Design

4.4.3.   Shear Forces Delivered to Shear Walls

As the title of this section suggests, and as part of the load path discussed earlier, shear forces – typically measured in pounds per linear foot (*plf*) – are calculated at each shear wall to determine the required wall/diaphragm connection capacity. It is important to note that not all shear walls are poured monolithically with diaphragms, which is common in reinforced concrete construction. In many cases, shear walls are constructed first, and then diaphragms are added later with the use of dowels in reinforced concrete construction, or anchors in wood construction, or bolts/anchors in masonry or steel construction.

Figure 4-6 illustrates two shear walls of different lengths (L_1 & L_2). Assuming a lateral load (w) is applied to the diaphragm, the load is distributed between the two walls in a flexible diaphragm scenario, with each wall receiving ($0.5wB$). The shear force transferred to wall 1 is $v_1 = 0.5wB/L_1$, and the shear force transferred to wall 2 is $v_2 = 0.5wB/L_2$. These shear forces help determine the design requirements for connections, such as dowels or fasteners, or they help determine what shear wall or diaphragm shear capacities are required for design.

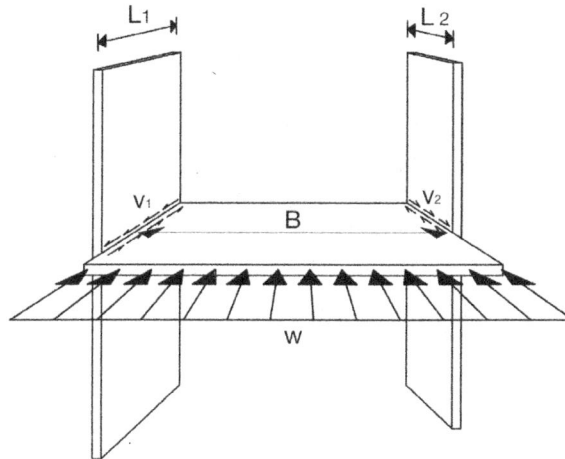

Figure 4-6 Shear Forces Transfer to Shear Walls

For example, in wood construction, Table 4.3A-D of the *AWC Special Design Provisions for Wind and Seismic (SDPWS 2021)* – a reference required for the California Seismic Principles exam – specifies the nominal shear capacity for sheathed wood frame shear walls, which can be referred to during the design process. A similar approach is outlined in the ACI 318 for concrete design, though the latter reference is not required for the California Seismic Principles exam.

When it comes to wood construction, and as discussed in Chapter 6 Section 6.3.2 in this book, and in Section 4.1.4 of the SDPWS 2021, the latter specifies that the allowable shear capacities for walls and diaphragms for seismic design in this case must be divided by a reduction factor of 2.8 for the Allowable Stress Design (ASD) method, or the nominal shear capacity must be multiplied by a resistance factor of $\emptyset_D = 0.5$ for the Strength Design method.

4.4.4.   Collectors

As discussed earlier in this chapter, diaphragm forces are distributed across the diaphragm based on the distribution of seismic weight and the forces (F_{px}) derived in Section 3.3.4 of this book. The locations of Center of Mass (CM) and Center of Rigidity (CR), along with the location and orientation of lateral-resisting elements, play a critical role in this force distribution. Therefore, we explained why it is essential to provide a continuous positive load path within the diaphragm to ensure it connects seamlessly to the vertical load-resisting system and ultimately to the ground.

With this in mind, what if openings such as doors and windows are present in a shear wall? Or, what if the available length of the shear wall is less than the depth of the diaphragm attached to it as illustrated in Figure 4-7?

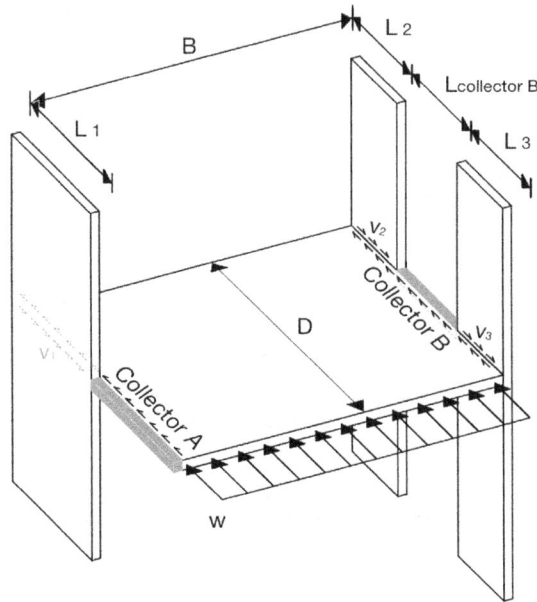

Figure 4-7 Collectors with Shear Walls

The effective length of a shear wall can be extended using a member called a collector, sometimes referred to as a drag or strut. These elements "drag" the lateral force applied to the diaphragm and transfer it to the shear wall, as shown in Figure 4-7.

Calculating the force in drag elements or collectors can be complex due to the need for detailed tracking of how the drag force accumulates and is transferred to the shear wall – similar to a bookkeeping exercise. In some cases, residual forces from the collector may be transferred to the next collector, depending on the geometry of the system. Also, in most of the cases you will need to account for collectors carrying load from its two sides and delivering it to a shear wall for a non-uniform diaphragm which increases the level of complexity.

Let's walk through a simple example using collector A of Figure 4-7. Refer to Example 4.8 below for a better understanding of this concept.

Example 4.8 – Collector Shear Force Profile (1)

In reference to Figure 4-7, assume a distributed load of $w = 10\,kip/ft$ across a flexible diaphragm with a width of $B = 20\,ft$, resulting in a total lateral load of 200 kip. Half of this load (i.e., 100 kip) is carried by collector A and shear wall 1.

Assume $D = 20\,ft$, $L_1 = 10\,ft$, and the length of collector A is also $10\,ft$. Determine the shear forces profile in collector A and shear wall 1.

The unit shear forces dragged on the left side of the diaphragm (including collector and shear wall is $100\,kip/20\,ft = 5\,kip/ft$.

Shear force carried by shear wall 1 is the $100\,kip$, which makes the unit force resisting in the opposite direction at this shear wall $v_1 = 100\,kip/10\,ft = 10\,kip/ft$. These forces are displayed in the following profile:

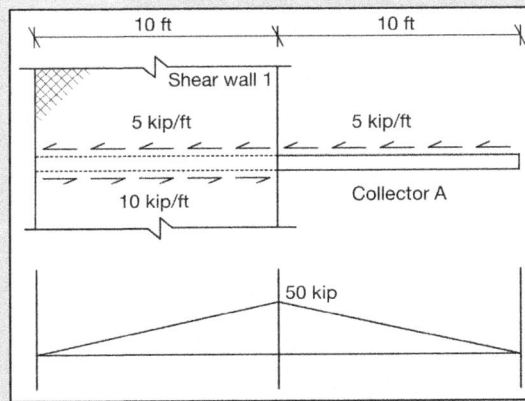

As you can see from the above shear profile, the maximum shear delivered by collector A is 50 *kip* in compression.

As shown in Example 4.8, the drag force generated in a collector is not always constant. It varies based on the location of the collector and can be either in compression or in tension, or even experience both simultaneously especially if placed between two diaphragms.

Example 4.9 – Collector Shear Force Profile (2)

In reference to Figure 4-7, assume a distributed load of $w = 10\,kip/ft$ across a flexible diaphragm with a width of $B = 20\,ft$, resulting in a total lateral load of 200 *kip*. Half of this load (i.e., 100 *kip*) is carried by collector B and shear walls 2 and 3.

Assume $D = 20\,ft$, $L_2 = L_3 = 5\,ft$, and the length of collector B is 10 *ft*. Determine the shear forces profile in collector B and the two shear walls.

The unit shear forces dragged on the right side of the diaphragm including collector B and the two shear walls is $100\,kip/20\,ft = 5\,kip/ft$.

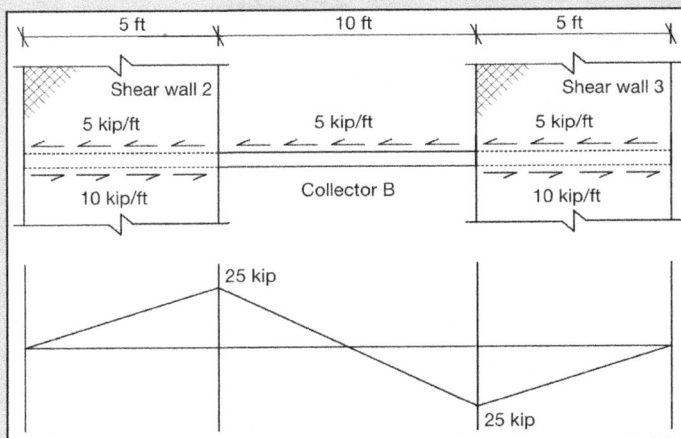

Shear force carried by shear wall 2 and 3 is the 50 *kip* each (same shear walls' length and assume similar height as well), which makes the unit force resisting in the opposite direction at any of those shear walls $v_2 = v_3 = 50\,kip/5\,ft = 10\,kip/ft$.

As observed in the shear profile above, wall 3 experiences a tension drag force distribution. Collector B pulls on this wall and then pushes wall 2, consequently, this collector carries both tension and compression stresses. Meanwhile, the diaphragm on the left is pulling the collector and walls and subjecting them to a constant force of 5 *kip/ft*.

Additionally, as shown in the profiles of Examples 4.8 and 4.9, these profiles should always close at zero, which confirms that the calculation is accurate.

4.4.5.   Drag Forces

As discussed in the previous section, the terms "collectors" and "drags" are used interchangeably to describe the same concept. This section does not introduce significant new information beyond what was already covered regarding collectors, with the main addition being the introduction of the concept of drags within the diaphragm itself.

One of the initial assumptions for diaphragms, as illustrated in Examples 4.8 and 4.9, is that shear is uniformly distributed across the diaphragm's depth. This assumption implies that loads are applied to the diaphragm throughout the full depth and width of the "web" (web – as defined in the context of deep beams in earlier discussions). However, in practice, buildings often have irregularities, such as offset shear walls, or diaphragms with no shear walls, or any other Seismic Force Resisting System (SFRS) attached to them.

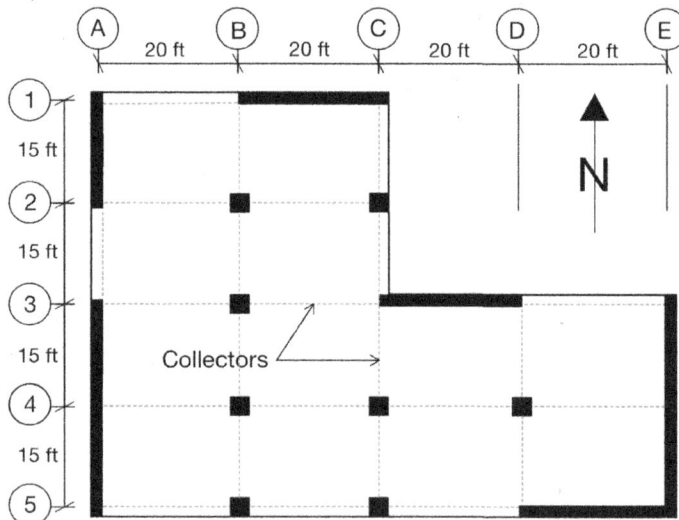

Figure 4-8 Example of Collectors Location within the Diaphragm

Consider Figure 4-8, as an example to the above discussion. This figure shows a floor plan with an irregularity. The diaphragm between lines 1 and 3 does not connect to the shear wall at line 3 for a distance of 40 *ft*. The same issue applies to the diaphragm between lines 3 and 5. To address this problem, a structural member is normally added along line 3, between points A and C, to which the diaphragm can be attached. This structural element collects the diaphragm loads at those identified locations and transfer them back to the shear wall at line 3. This member is referred to as a "collector" (as identified in the previous section) or as a "drag line" or "drag strut."

4.4.6.   Ties and Continuities Around Openings

When openings in diaphragms create sub-diaphragms, the main diaphragm's chord reinforcement must extend to resist the forces acting on the sub-diaphragm chords and to collect shear forces where the diaphragm depth is reduced (see Figure 4-9).

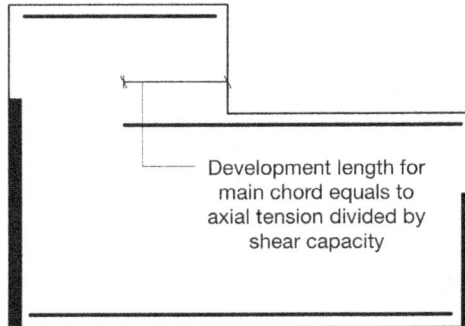

Figure 4-9 Irregular Diaphragm with Lateral Load in S-N Direction

The sub-chord and collector reinforcement should be extended into the adjacent diaphragm to effectively transfer the axial force through shear. The required development length is determined by dividing the axial force in the sub-chord by the shear capacity (force per unit length) of the main diaphragm.

In designs with wood and metal deck diaphragms, framing members often function as continuity elements, acting as sub-chords and collectors at discontinuities. These members also transfer out-of-plane wall forces to the main diaphragm when the diaphragm itself cannot directly resist the anchorage force, as shown in Figures 3-6 and 3-7 of Chapter 3.

4.5. Seismic Load Combinations and Directions

4.5.1.   Strength Design (SD)/LRFD Effects and Combinations

There are two main load combinations (6 and 7) per ASCE 7 Standard Section 2.3.6 that deal with seismic loads, under the LRFD/SD method and they are as follows:

Combination 6: $1.2D + E_v + E_h \ (or \ E_{mh}) + L + 0.2S$

Combination 7: $0.9D - E_v + E_h \ (or \ E_{mh})$

Where:

D is the effect of dead load.

L is the effect of live load.

S is the effect of snow load.

E_h & E_v represent the effects of seismic loads from the horizonal and vertical load applications respectively. These are not the loads themselves. For instance, the effect of (E_h) can result in moment or shear forces, while the effect of (E_v) can result in axial forces. In this context, the vertical effect of an earthquake can generate horizontal forces when applied to a system of braces.

E_{mh} is the effect of horizontal load with overstrength (Ω_o) added to it.

E_v is mainly taken as $(0.2S_{DS}D)$ – see comment at the end of this chapter (**).

The horizonal load effect of earthquakes is normally taken as $(E_h = \rho Q_E)$ unless the effects of overstrength shall be included in which case we use $(E_{mh} = \Omega_o Q_E)$.

The redundancy factor (ρ) and overstrength factor (Ω_o) were defined in Section 2.3.5 of Chapter 2 of this book. Their use and respective values, where applicable, are mostly provided in Tables 2-1 and 2.-2 of this book for both horizontal & vertical irregularities – **and below** ⇓. The overstrength factor (Ω_o) values are given for each Structural Force Resisting System (SFRS) in Table 12.2-1 or any other similar table in the ASCE 7 Standard.

Q_E on the other hand, is the effect of horizontal seismic force from base shear (V) or from the diaphragm design force (F_{px}) or any other force on any element (F_p).

⇒ **The overstrength factor (Ω_o):** as identified above and included in the horizontal load effects (E_{mh}) is only used with few elements or few scenarios and those are:

1. Foundation and other elements providing overturning resistance at the base of cantilever column SFRS elements, as per ASCE 7 Standard, Section 12.2.5.2.

2. Structural elements supporting discontinuous walls or frames of structures with horizontal irregularity Type 4 or vertical irregularity Type 4, as outlined in ASCE 7 Standard, Section 12.3.3.3.

3. Collector elements for Seismic Design Categories C, D, E, and F, in accordance with ASCE 7 Standard, Section 12.10.2.1.

4. Batter piles and their connections per ASCE 7 Standard, Section 12.13.8.4.

5. Pile anchorage design for uplift and rotational resistance, as per ASCE 7 Standard, Section 12.13.8.5.

6. Egress stairs and ramps when part of the building with no sliding or ductile connections provided between the stairs and the building in accordance with ASCE 7 Section 13.5.10. Overstrength in this case should be used and taken as $\Omega_o \geq 2.5$.

⇒ **The redundancy factor (ρ):** its value is permitted to be taken as 1.0 per ASCE 7 Standard Section 12.3.4.1 under specific conditions, including structures assigned to Seismic Design Category B or C, design of non-structural components, and design of nonbuilding structures that are not similar to buildings, the design of collectors or other elements when overstrength (Ω_o) is already in effect, also the design for out-of-plane walls including their anchorage. Apart from this, and specifically for SDC D, E and F, along with Type 1b horizontal irregularity, the value of (ρ) should be taken as 1.3.

4.5.2. Allowable Stress Design (ASD) Effects and Combinations
There are three main load combinations (8, 9 and 10) per ASCE 7 Standard Section 2.4.5 that deal with seismic loads under the ASD method, and they are as follows:

Combination 8: $1.0D + 0.7E_v + 0.7E_h \ (or \ 0.7E_{mh})$

Combination 9: $1.0D + 0.525E_v + 0.525E_h \ (or \ E_{mh}) + 0.75L + 0.75S$

Combination 10: $0.6D - 0.7E_v + 0.7E_h \ (or \ E_{mh})$

The rest of the definitions and procedures provided in Section 4.5.1 apply to this section as well.

4.5.3. Cantilever Load Combinations

When dealing with horizontal cantilevers in Seismic Design Categories D, E or F, those should be designed for a net upward force of 0.2 times the dead load in addition to the above combinations.

4.5.4. Direction of Loading Combinations

ASCE 7 Section 12.5 provides guidelines for determining the direction of seismic loading to ensure structures are adequately designed to resist seismic forces. For most buildings, seismic forces are assumed to act horizontally in both horizontal directions. The dominant direction of seismic forces is typically determined by the building's layout and site characteristics.

- For SDC B:

  ✓ The design can consider applying loads independently in each direction, without considering orthogonal interaction effects.

- For SDC C, D, E, and F: the design must consider the maximum of:

  ✓ The method used in SDC B.

  ✓ Forces in both directions, typically applying 100% of the load in one direction and 30% in the other direction. Accidental torsion is included for the 30% direction.

  This method can result in up to 16 different load combinations, including load application in both directions, reversals of the applied directions, and accidental torsion in the 100% direction. If accidental torsion is also considered for the 30% direction, the total number of combinations increases to 32.

If the structure has type 5 horizontal irregularity with a non-parallel SFRS, the above method may still be used. Alternatively, pairs of orthogonal directions can be applied to the structure, and the linear response history procedure to be used to design the structure.

4.5.5. The Simplified Method

The load combinations for the simplified method are explained in Section 12.14.3.1 of the ASCE 7 Standard. These are largely similar to the combinations outlined above, with a few notable adjustments. We will highlight the key changes here as follows:

➢ The horizonal load effect is taken as ($E_h = Q_E$), with overstrength ($E_{mh} = 2.5Q_E$).
➢ Vertical load effect is ($E_v = 0.2S_{DS}D$) – see comment at the end of this chapter (**).
➢ The vertical load effect (E_v) is taken as zero when $S_{DS} < 0.125$, or when determining demand on the soil structure interface with an uplift condition for a foundation.

The above is applicable to load combinations 6 and 7 for the SD method, as described in Section 4.5.1 here, or combinations 8, 9 and 10 for the ASD method per Section 4.5.2 here.

However, when overstrength is to be included, some modifications shall apply to the simplified method. For instance, the Strength Design (SD) method (i.e., LRFD), load combinations 5 (*) and 7 (*) are affected as follows:

Combination 5: $(1.2 + 0.2S_{DS})D + E_{mh} + L + 0.2S$ ....where $E_{mh} = 2.5Q_E$

Combination 7: $(0.9 - 0.2S_{DS})D + E_{mh}$ ....where $E_{mh} = 2.5Q_E$

Seismic Analysis Procedures

As for the ASD method, the above applies to the following load combinations (*):

Combination 5: $(1.0 + 0.14S_{DS})D + 0.7E_{mh}$ ....where $E_{mh} = 2.5Q_E$

Combination 6b: $(1.0 + 0.105S_{DS})D + 0.525E_{mh} + 0.75L + 0.75S$ ......where $E_{mh} = 2.5Q_E$

Combination 8: $(0.6 - 0.14S_{DS})D + 0.7E_{mh}$ ....where $E_{mh} = 2.5Q_E$

For the combinations mentioned above, it is important to refer to key notes on page 117 of the ASCE 7 Standard, which primarily address live loads (L), fluid loads (F) and lateral earth pressure loads (H). For instance, the standard allows live loads to be reduced by half if the occupancy load is less than $100\ psf$, except for garages and public assembly areas. Additionally, the standard provides guidance on how fluid loads (F) or lateral earth loads (H) should be applied in combination with the live loads, as necessary.

As for the direction of loading for the simplified method, it is similar to the approach used for SDC B in the conventional method. In both cases, the loading effects are applied separately in each orthogonal direction, and the combination of effects from the two directions is not considered.

(*) For clarity, you may have noticed that the numbering system in this method appears to be inconsistent compared to the one presented in the previous section. This discrepancy arises because the simplified method **with overstrength** references ASCE 7 Sections 2.3.2 and 2.4.1, not 2.3.6 and 2.4.5.

(**) S_{DS} for the calculation of E_v can be taken as $1.0 \geq 0.7 \times S_{DS,actual}$ provided: no irregularities, stories ≤ 5 with mezzanine considered as a story, $T \leq 0.5\ sec$, $\rho = 1$, not Site Class E or F, or Risk Category III or IV. For more details refer to ASCE 12.8.1.3.

4.6. Chapter References
The following references were used in this chapter. The reference(s) in italics is/are required for the exam, as noted in the preface of this book.

[1] ACI 318. Building Code Requirements for Structural Concrete and Commentary, 2019.

[2] *American Society of Civil Engineers, ASCE/SEI 7-16 Minimum Design Loads for Buildings and Other Structures.*

[3] *AWC SDPWS-2021 Special Design Provisions for Wind and Seismic.*

[4] CRSI 2019. Design Guide for Reinforced Concrete Diaphragms. First Edition. Concrete Reinforcing Steel Institute.

[5] Design of Wood Structures ASD/LRFD 7th Edition, 2014 by Donald E. Breyer, Kenneth J. Fridley and Kelly E. Cobeen. McGraw-Hill, Inc. New.

[6] Moehle, J., Hooper, J. & Meyer, T. R. (2010). Seismic Design of Cast-in-Place Concrete Diaphragms, Chords, and Collectors: A Guide for Practicing Engineers. NEHRP Seismic Design Technical Brief No.3. National Institute of Standards and Technology. U.S. Department of Commerce.

[7] PCI Design Handbook: Precast and Prestressed Concrete 7th edition.

Problems & Solutions

The following questions have been carefully designed to provide comprehensive coverage of the material presented in this book, while also being educational in nature, as you will notice. Additionally, you may come across a few questions not directly covered in this chapter; these have been intentionally included to ensure broader coverage of the content.

➢ Problem 4.1: Center of Rigidity for a Rigid Diaphragm
➢ Problem 4.2: Load Distribution (1)
➢ Problem 4.3: Load Distribution (2)
➢ Problem 4.4: Flexible Diaphragm (1)
➢ Problem 4.5: Flexible Diaphragm (2)
➢ Problem 4.6: Rigid Diaphragm (1)
➢ Problem 4.7: Rigid Diaphragm (2)
➢ Problem 4.8: Rigid Diaphragm (3)
➢ Problem 4.9: Chord Design (1)
➢ Problem 4.10: Chord Design (2)
➢ Problem 4.11: Chord Design (3)
➢ Problem 4.12: Chord Design (4)
➢ Problem 4.13: Chord Design Issues
➢ Problem 4.14: Shear Demand
➢ Problem 4.15: Collectors Design
➢ Problem 4.16: Load Distribution with Moment Resisting Frames
➢ Problem 4.17: Tie Down Force
➢ Problem 4.18: Collector Load
➢ Problem 4.19: Drag Force
➢ Problem 4.20: Loading Direction
➢ Problem 4.21: Load Combination
➢ Problem 4.22: Allowable Design
➢ Problem 4.23: Cantilever Load
➢ Problem 4.24: Redundancy Factor
➢ Problem 4.25: Load Combination using the Simplified Method

PROBLEM 4.1 *Center of Rigidity for a Rigid Diaphragm*

On the plan below, shear wall at line A is 12 *in* thick and D is 8 *in* thick. All walls are made of reinforced concrete for a single-story building which is 8 *ft* tall.

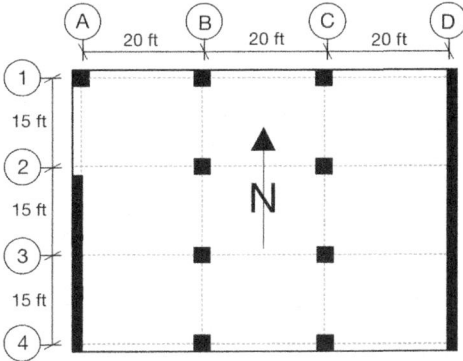

Based on the above information, calculate the location of Center of Rigidity for the walls in the South-North direction using axis A as datum.

(A) 45 *ft*

(B) 42 *ft*

(C) 36 *ft*

(D) 30 *ft*

PROBLEM 4.2 *Load Distribution (1)*

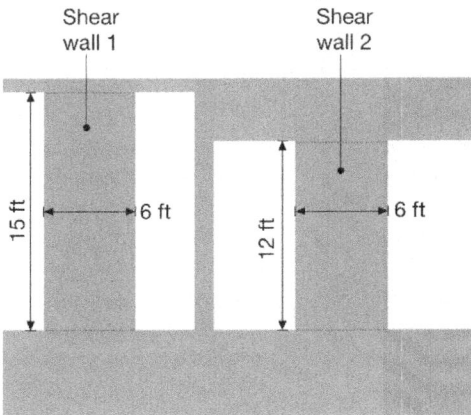

The above is an elevation view for two concrete shear walls located in the same single-story building and they both support a flexible diaphragm, where shear wall No.

1 and 2 are cantilevers (fixed from bottom only).

Assuming the thickness of the two shear walls is 10 *in*, ignoring the stiffness of the column shown, the lateral load portion on those two walls is 75 *kip*, what is the load portion transferred to wall 1?

(A) 12.4 *kip*

(B) 26.3 *kip*

(C) 50.2 *kip*

(D) Insufficient information

PROBLEM 4.3 *Load Distribution (2)*

This elevation view is for two concrete shear walls located in the same single-story building and they both support a rigid diaphragm, where shear wall 1 is cantilever, and shear wall 2 is fixed from both ends.

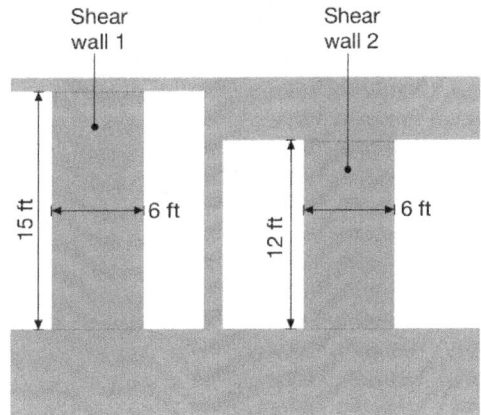

Assuming the thickness of the two shear walls is 10 *in*, ignoring the stiffness of the column shown, the lateral load portion on those two walls is 75 *kip*, what is the load portion transferred to wall 1?

(A) 12.4 *kip*

(B) 24.8 *kip*

(C) 50.2 *kip*

(D) Insufficient information

PROBLEM 4.4 *Flexible Diaphragm (1)*

In this plan, the floor diaphragm is flexible, and the lateral load in the East-West direction is 420 *kip*. All shear walls are made of masonry, have the same thickness, and are 14 *ft* tall cantilever walls in a single-story structure.

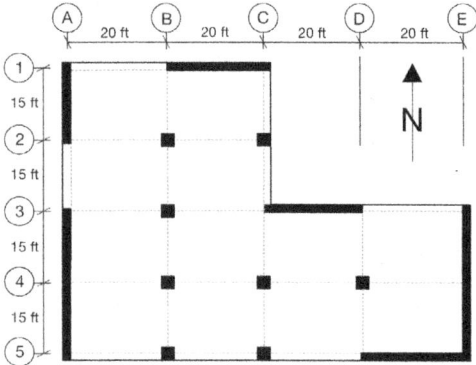

Based on this, ignoring columns' stiffness, what is the lateral force transferred to wall at axis 3?

    (A) 140 *kip*

    (B) 210 *kip*

    (C) 280 *kip*

    (D) 320 *kip*

PROBLEM 4.5 *Flexible Diaphragm (2)*

In this plan, the floor diaphragm is flexible, and the lateral load in the South-North direction is 260 *kip*. All shear walls are made of masonry, have the same thickness, and are 14 *ft* tall cantilever walls in a single-story structure.

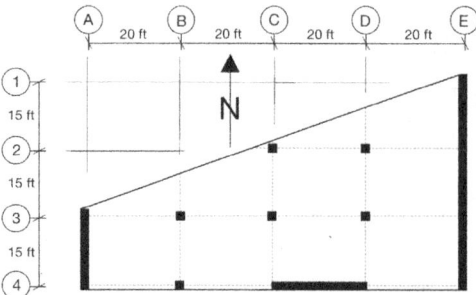

Based on this, ignoring columns' stiffness, what is the lateral force transferred to wall E?

    (A) 195.5 *kip*

    (B) 65.5 *kip*

    (C) 151.5 *kip*

    (D) 108.5 *kip*

PROBLEM 4.6 *Rigid Diaphragm (1)*

A three-story building assigned to Seismic Design Category B consists of five shear walls resisting in the short direction and four in the long direction spaced as shown.

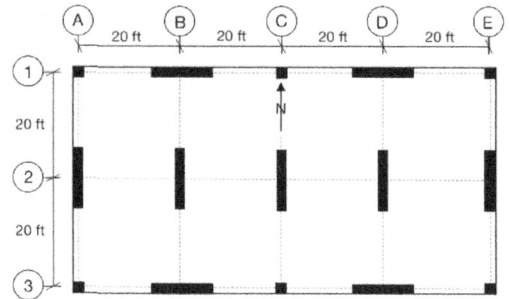

All shear walls' have the same material, length, thickness and height (hence rigidity). The Center of Mass coincides with center of rigidity, and both are located at the geometric center of all those reinforced concrete floors.

The lateral seismic force experienced at the second floor is 200 *kip* in the North direction.

Ignoring columns' rigidities, what is the seismic force transferred to the wall at axis D of the second floor due to accidental torsion alone?

    (A) 1.2 *kip*

    (B) 2.9 *kip*

    (C) 6.7 *kip*

    (D) 7.5 *kip*

PROBLEM 4.7 *Rigid Diaphragm (2)*

The below building has a lateral force applied to it as shown. The building has four shear walls with relative rigidities as depicted in the figure below.

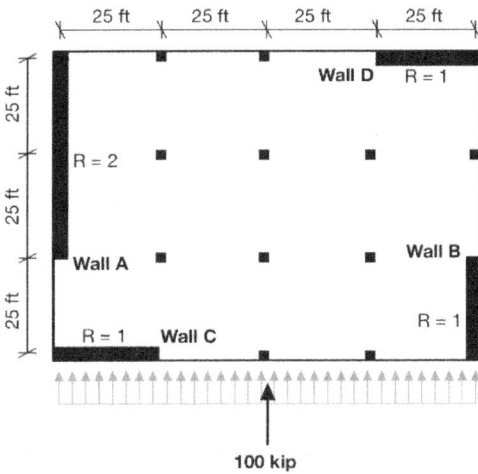

100 kip

Ignoring the stiffness of columns or any other framing system, what is the maximum potion of shear transferred to shear wall D (*)?

(A) 8.6 kip ←

(B) 6.6 kip →

(C) 8.6 kip →

(D) 6.6 kip ←

(*) Check the extras after the solution where we solve for the rest of the walls.

PROBLEM 4.8 *Rigid Diaphragm (3)*

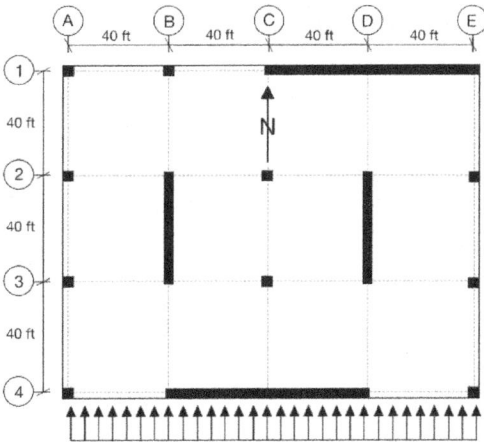

Given a load applied in the South-North direction, determine the governing shear force transferred to wall B. Assume all walls are made of reinforced concrete, have the same thickness, and that the floor is rigid

and walls are 8 ft tall. The lateral load applied is 5 kip/ft.

(A) 400.0 kip

(B) 469.5 kip

(C) 430.5 kip

(D) 414.5 kip

PROBLEM 4.9 *Chord Design (1)*

The below is a flexible diaphragm and is supported laterally with the use of three shear walls. The lateral load has been calculated as shown on the provided sketch.

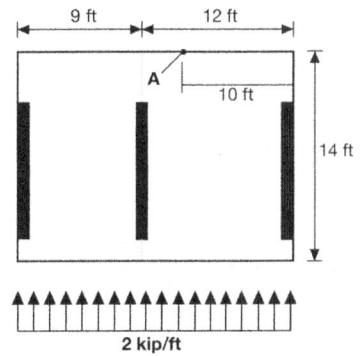

2 kip/ft

What is the tension force on the flexible diaphragm's chord at point A?

(A) 1.5 kip

(B) 3.3 kip

(C) 7.5 kip

(D) 20.0 kip

PROBLEM 4.10 *Chord Design (2)*

The below is a rigid diaphragm and is supported laterally with the use of four shear walls as shown.

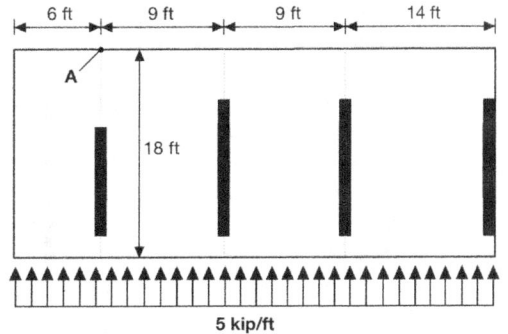

5 kip/ft

What is the compression force on the rigid diaphragm's chord at point A?

(A) 1.5 *kip*

(B) 5.0 *kip*

(C) 7.5 *kip*

(D) More information is required.

PROBLEM 4.11 *Chord Design (3)*

The below is a flexible diaphragm and is supported laterally with the use of four shear walls as shown.

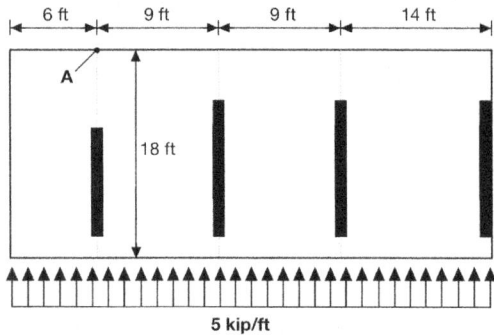

What is the compression force on the flexible diaphragm's chord at point A?

(A) 1.5 *kip*

(B) 5.0 *kip*

(C) 7.5 *kip*

(D) More information is required.

PROBLEM 4.12 *Chord Design (4)*

The below is a flexible diaphragm and is supported laterally with the use of shear walls. The lateral load in the South-North direction is a total of 200 *kip*.

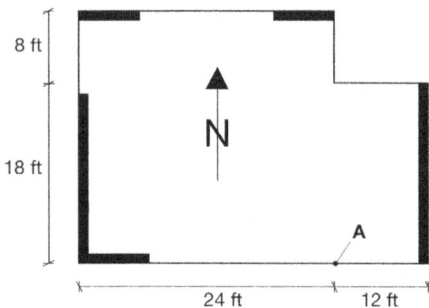

What is the maximum compression force on the flexible diaphragm's chord that could be experienced at point A?

(A) 45 *kip*

(B) 30 *kip*

(C) 25 *kip*

(D) 15 *kip*

PROBLEM 4.13 *Chord Design Issues*

In seismic design categories D, E, and F, diaphragms with large slab openings can create thinner sub-diaphragms, which are subject to potential issues. Which of the following is most critical when designing those sub-diaphragms?

(A) The maximum aspect ratio for sub-diaphragms should not exceed 2.5 : 1 (length to width) to prevent instability and excessive flexibility.

(B) Extra tension may be present at chords, this may require additional reinforcement.

(C) Sub-diaphragms formed due to large openings can be prone to buckling, which should be carefully considered in the design process.

(D) All the above.

PROBLEM 4.14 *Shear Demand*

The below is a flexible diaphragm and is supported laterally with the use of three shear walls.

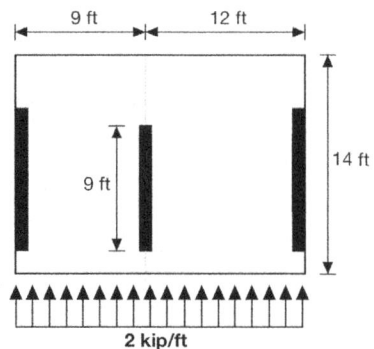

In order to attach the diaphragm to the left of the middle wall, what should be the capacity of the shear dowels provided to the left of this wall in pound per linear foot (plf)?

(A) 500 plf

(B) 1,000 plf

(C) 1,500 plf

(D) 2,000 plf

PROBLEM 4.15 *Collectors Design*

Which of the following statements is true regarding the drag forces in collectors?

(A) Collectors may be in compression with the magnitude of compression varying depending on location.

(B) Collectors may be in tension with the magnitude of tension varying depending on location.

(C) Collectors can experience both compression and tension in the same collector, depending on the geometry of the system.

(D) All the above.

PROBLEM 4.16 *Load Distribution with Moment Resisting Frames*

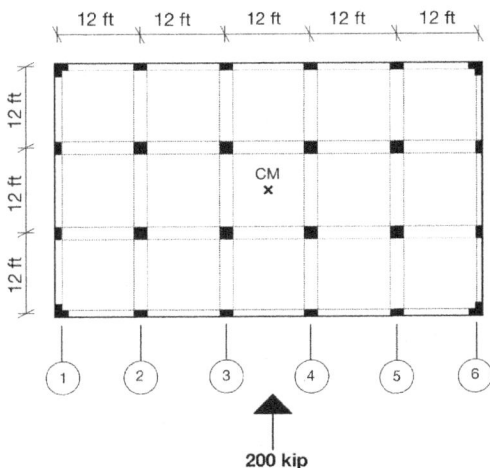

A four-story building consists of six reinforced concrete moment frames in the short direction and four in the long direction spaced as shown.

The interior frames have double the stiffness of the exterior frames, and the Center of Mass (CM) coincides with the Center of Rigidity (CR). The lateral seismic force experienced at the third floor is 200 kip.

What is the seismic force applied on frame 4 of the third floor generated from accidental torsion, ignoring the frames' stiffness orthogonal to loading (*)?

(A) 5.6 kip

(B) 2.2 kip

(C) 6.7 kip

(D) 3.3 kip

(*) Check extras in the solution section.

PROBLEM 4.17 *Tie Down Force*

The below are two wooden shear walls placed at the same line of action with a lateral force transferred to this line of 10 kip.

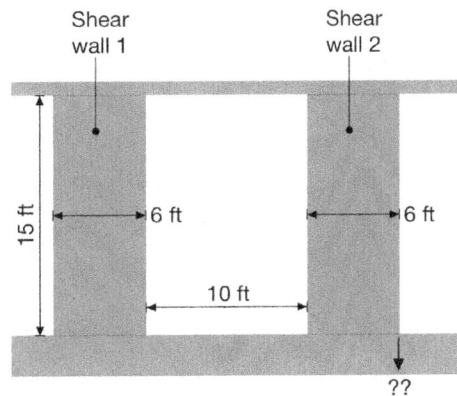

What is the tie down force required at shear wall 2 as shown in this sketch?

(A) 6.25kip

(B) 12.5 kip

(C) 18.75 kip

(D) 25.0 kip

PROBLEM 4.18 *Collector Load*

On the flexible diaphragm shown below, the lateral force in the South-North direction is a total of 180 *kip*.

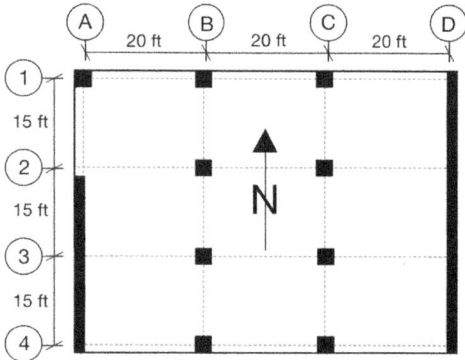

What is the drag force on a collector placed at axis A between axis 1 and 2?

(A) 15 *kip*

(B) 30 *kip*

(C) 45 *kip*

(D) 90 *kip*

PROBLEM 4.19 *Drag Force*

On the flexible diaphragm shown below, the lateral force in the West-East direction is a total of 120 *kip*.

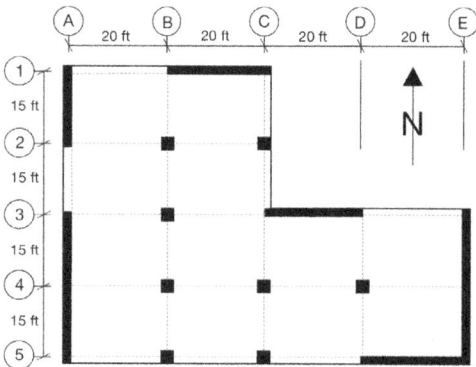

What is the drag force on a collector placed at axis 3 between axis A and C?

(A) 10 *kip*

(B) 20 *kip*

(C) 40 *kip*

(D) 80 *kip*

(*) Check extras in the solution section.

PROBLEM 4.20 *Loading Direction*

According to ASCE 7, for Seismic Design Category C how are seismic forces typically applied to a building in two horizontal directions?

(A) Seismic forces are applied equally in both horizontal directions (100% in both directions).

(B) Seismic forces are applied at 50% in one direction and 50% in the perpendicular direction.

(C) Seismic forces are applied at 100% in one direction and 30% in the perpendicular direction.

(D) Seismic forces are applied at 100% in both directions, with no reduction in one direction.

PROBLEM 4.21 *Load Combination*

A concrete footing, which is part of a braced system, to be designed using the Strength Design (SD) method. The footing is assigned to Seismic Design Category E with redundancy factor $\rho = 1.3$ and $S_{DS} = 0.75$ along with the following loads:

Load Type	kip
Dead Load	5.0
Live Load	6.0
Roof Load	1.5
Effect of horizontal seismic force (Q_E)	3.0
Wind Load	2.0

The combined load, that include earthquake loading, that shall be used for this purpose is:

(A) 16.7 *kip*

(B) 15.0 *kip*

(C) 16.0 *kip*

(D) 17.7 *kip*

PROBLEM 4.22 *Allowable Design*

how should the shear capacities for wooden walls and diaphragms be adjusted for seismic design in the case of the Allowable Stress Design (ASD) method?

(A) The allowable shear capacity must be multiplied by a reduction factor of 2.8.

(B) The allowable shear capacity must be divided by a reduction factor of 2.8.

(C) The nominal shear capacity must be multiplied by a reduction factor of 0.5.

(D) The nominal shear capacity must be divided by a reduction factor of 2.8.

PROBLEM 4.23 *Cantilever Load*

What upward force should horizontal cantilevers be designed to withstand in the relevant design combinations for Seismic Design Category D in addition to other load combinations?

(A) 0.2 times the dead load.

(B) 0.8 times the dead load.

(C) 1.1 times the dead load.

(D) 1.2 times the dead load.

PROBLEM 4.24 *Redundancy Factor*

For which of the following conditions is the redundancy factor permitted to be taken as 1.0?

(A) The design of nonstructural components.

(B) Structures assigned to Seismic Design Categories B, C & D.

(C) The design of nonbuilding structures similar to buildings.

(D) The design of collector elements without considering overstrength.

PROBLEM 4.25 *Load Combination using the Simplified Method*

A concrete footing, designed to resist overturning at the base of a cantilever column, is designed using the Strength Design method. The building's SFRS is a cantilever column system with some bracing elements that converts horizontal loads to vertical forces.

Based on the building's characteristics, it qualifies for the use of the simplified method.

The building is assigned to Seismic Design Category B with redundancy factor $\rho = 1.0$ and $S_{DS} = 0.1$. The building is a small office, with a live load of $50\,psf$.

The following loads are considered:

Load Type	kip
Dead Load	5.0
Live Load	6.0
Fluid Load	1.0
Effect of horizontal seismic force (Q_E)	2.0

The combined load, that includes earthquake loading, that shall be used for this purpose is:

(A) 7.3 kip

(B) 10.3 kip

(C) 12.3 kip

(D) 15.3 kip

SOLUTION 4.1

Aspect ratio for wall A is $\frac{h}{l} = \frac{8}{30} = 0.27 < 0.3$. Per Table 4-1 of this chapter, use area as indication for relative rigidity with $A = 30 \times \frac{12}{12} = 30\,ft^2$ (*).

Aspect ratio for wall D is $\frac{h}{l} = \frac{8}{45} = 0.18 < 0.3$. Per Table 4-1 of this chapter, use area as indication for relative rigidity with $A = 45 \times \frac{8}{12} = 30\,ft^2$ (*).

Assuming the distance from datum (axis A) to the Center of Rigidity is (x):

$$x = \frac{0 \times 30 + 60 \times 30}{30 + 30}$$

$$\rightarrow x = 30\,ft$$

Correct Answer is (D)

(*) Observe that we consistently used the area as an indicator to the relative rigidities for the two walls because the aspect ratios for both the walls $h/l < 0.3$.

SOLUTION 4.2

For flexile diaphragm load distribution, check Figure 4-1 in this chapter, and provided those two concrete walls fall on the same line of action, you need to distribute loads based on their relative rigidities as follows:

$$0.3 \leq (h_1/l_1 = 2.5) \leq 3.0$$

$$0.3 \leq (h_2/l_2 = 2.0) \leq 3.0$$

This means that the stiffness of the two walls should include both flexural and shear stiffness per Table 3-5.

For simplicity, Tables 3-6 (*) for cantilever walls can be used as follows:

$$\rightarrow k_1 = 0.014 \times \left(\frac{10}{12}\right)$$

$$\rightarrow k_2 = 0.026 \times \left(\frac{10}{12}\right)$$

$$Wall\ 1 = \frac{0.014}{0.014 + 0.026} \times 75 = 26.3\,kip$$

Correct Answer is (B)

(*) In some resources, you may find that the rigidity in this table multiplied by 10. If consistency is maintained throughout, both approaches are acceptable.

SOLUTION 4.3

In reference to Figure 4-1 of this chapter, and for rigid diaphragms, it is necessary to determine both the Center of Mass (CM) and the Center of Rigidity (CR) to complete the calculation. As a result, this question lacks the required information to arrive at a solution.

Correct Answer is (D)

SOLUTION 4.4

In reference to Figure 4-1 of this chapter, these are flexible two uniformly shaped diaphragms and the walls in question are parallel. Based on this, tributary area in the direction of loading from each diaphragm can be used.

Initially we need to determine the force distributed to the diaphragms at portions between axes 1 to 3, and 3 to 5.

Area 1 to 3 $= 40 \times 30 = 1,200\,ft^2$
33.33% of total diaphragm area
Force taken by this portion $= 140\,kip$

Area 3 to 5 $= 80 \times 30 = 2,400\,ft^2$
66.67% of total diaphragm area
Force taken by this portion $= 280\,kip$

$$Wall\ 3 = \frac{140}{2} + \frac{280}{2} = 210\,kip$$

Correct Answer is (B)

SOLUTION 4.5

In reference to Figure 4-1 of this chapter, this is a flexible non-uniformly shaped diaphragm and the walls in question are parallel. Based on this, the simply supported beam method shall be used.

Initially, we need to determine where the Center of Mass (CM) is located as it determines the location of the lateral force resultant – there are other method that can be used, this one is the shortest method.

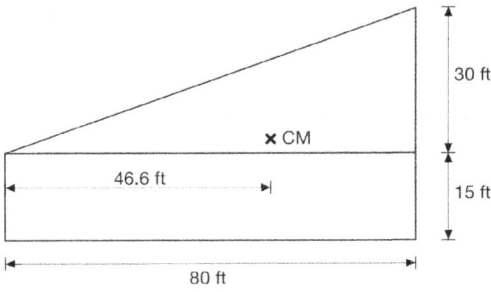

By dividing the total area into a rectangle and a triangle, the Center of Mass (CM) is determined by taking datum at axis A as follows:

$$CM = \frac{80 \times 15 \times \left(\frac{80}{2}\right) + 0.5 \times 80 \times 30 \times \left(\frac{2}{3} \times 80\right)}{80 \times 15 + 0.5 \times 80 \times 30} = 46.6 \; ft$$

Based on this information the following simple beam can be constructed:

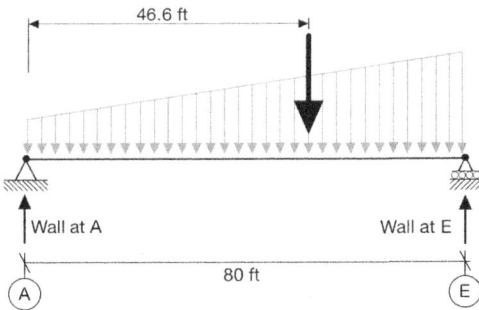

Forces transferred to wall at axis E can be calculated by taking moment around A:

$$Wall \; E = \frac{260 \times 46.6}{80} = 151.5 \; kip$$

Correct Answer is (C)

SOLUTION 4.6
ASCE 7-16 Section 12.8.4.2 sets accidental torsion at 5% of the perpendicular dimension that the seismic force is applied to.

Accidental torsion is calculated as follows:

$$M_{ta} = 0.05 \times (80 \; ft) \times 200 \; kip$$

$$= 800 \; kip.ft$$

Given that the question only asks for the torsional moment contribution, the second half of Equation 1a of Section 4.3.3 is used here where all rigidities, since equal, are denoted as R, and M_{ta} replaces M_T:

$$F_{y,D} = \frac{M_{ta}(x)R_{y,D}}{\sum R_y(x)^2 + \sum R_x(y)^2}$$

$$= \frac{800 \; kip.ft \times 20 \; ft \times R}{\left(\begin{array}{c} 40^2 + 20^2 + 0^2 + 20^2 + 40^2 \\ + \\ 20^2 + 20^2 + 20^2 + 20^2 \end{array}\right) \times R}$$

$$= \frac{800 \; kip.ft \times 20 \; ft \times R}{5,600 \; ft^2}$$

$$= 2.9 \; kip$$

Correct Answer is (B)

SOLUTION 4.7
This is a rigid diaphragm, the requested wall is orthogonal to the direction of loading, and since accidental torsional moment (M_{ta}) shall be considered per ASCE 7 Standard, we can use either Equation 2 or 2a of Section 4.3.3 of this chapter.

First, start by finding the location of Center of Rigidity (CR) knowing that the Center of Mass (CM) will be located in the geometric centroid of this floor.

Taking the center of wall A as datum:

$$X_{CR} = \frac{0 \times 2 + 1 \times 100}{1 + 2} = 33.33 \; ft$$

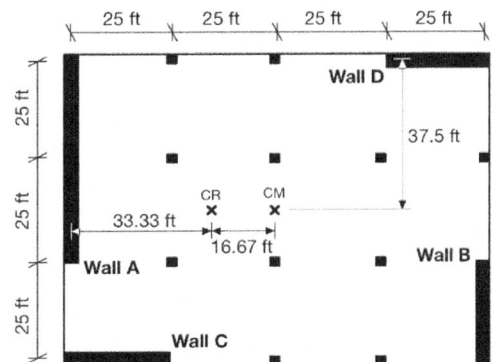

This makes eccentricity:

$$e_x = 50 - 33.33 = 16.67\,ft$$

The inherent torsional moment (M_T):

$$M_T = 16.67\,ft \times 100\,kip = 1{,}667\,kip.ft$$

The accidental torsional moment (M_{ta}):

$$M_{ta} = 0.05 \times 100\,ft \times 100\,kip = 500\,kip.ft$$

Assume $x-axis$ for the long direction and $y-axis$ for the short, also, using Equation 2a, and adding the accidental moment to the inherent torsional moment as it generates the maximum resultant:

$$F_{x,D} = \frac{(M_T+M_{ta})(y)R_{x,D}}{\sum R_y(x)^2 + \sum R_x(y)^2}$$

$$= \frac{(1{,}667+500)(37.5)\times 1}{\left(\begin{array}{c}2\times 33.33^2 + 1\times(100-33.33)^2 \\ + \\ 1\times 37.5^2 + 1\times 37.5^2\end{array}\right)}$$

$$= \frac{(1{,}667+500)(37.5)\times 1}{9{,}479.2}$$

$$= 8.6\,kip \leftarrow$$

Correct Answer is (A)

Question extras:

Let's solve for the rest of the walls, mainly walls A and B taking into account accidental torsional moment using the same method presented in Example 4.7 of this chapter which enables us to find the force that governs.

Using Equation 1a for this matter – also observe the following sketch to understand whether you need to add the accidental torsional moment or deduct it to generate maximum results:

$$F_{y,A} = \frac{R_{y,A}}{\sum R_y} \times F_y \mp \frac{(M_T \mp M_{ta})(x)R_{y,A}}{\sum R_y(x)^2 + \sum R_x(y)^2}$$

$$= \frac{2}{2+1} \times 100 - \frac{(1{,}667 - 500)\times 33.33 \times 2}{9{,}479.2}$$

$$= 58.5\,kip.ft$$

For simplicity both signs should be either negative or positive

For wall A, the 100 kip force pushes the wall forward, while the torsional moment causes the wall to rotate in such a way that it pushes the wall backward. This results in the negative sign in the equation. Therefore,

subtracting the accidental torsional moment will yield the maximum result.

As for wall B in the following calculation, the torsional moment force contribution acts in the same direction of the 100 kip force, and hence the positive sign and the addition of the accidental moment to the torsional moment.

$$F_{y,B} = \frac{R_{y,B}}{\sum R_y} \times F_y \mp \frac{(M_T \mp M_{ta})(x)R_{y,B}}{\sum R_y(x)^2 + \sum R_x(y)^2}$$

$$= \frac{1}{2+1} \times 100 + \frac{(1{,}667+500)\times(100-33.33)\times 1}{9{,}479.2}$$

$$= 48.6\,kip.ft$$

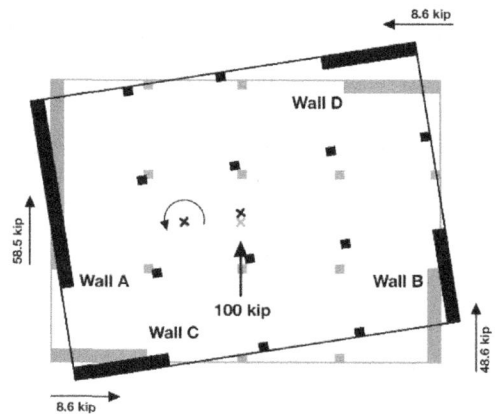

The below table provides a summary for all transferred forces broken down where the sign in this case represents direction:

Wall	Shear force	Due to torsional moment	Due to accidental torsion	Total
	(kip)	(kip)	(kip)	(kip)
A	+ 66.67	− 11.72	+ 3.5	+ 58.5
B	+ 33.33	+ 11.72	+ 3.5	+ 48.6
C	−	− 6.6	− 2.0	− 8.6
D	−	+ 6.6	+ 2.0	+ 8.6

SOLUTION 4.8

This is a rigid diaphragm, and the requested wall is aligned with the applied load direction. Equation 1 or 1a of Section 4.3.3 can be used in this case.

Due to the even distribution of walls around the floor, with all having the same thickness and length, the Center of Rigidity (CR) is located at the geometric center of the floor. Since the floor is a uniform rectangle with no irregularities, the Center of Mass also coincides with the Center of Rigidity, with both positioned at the diaphragm's centroid. Hence no inherent torsional moment is considered, only the accidental per ASCE 7 Section 12.8.4.2 calculated as follows:

$$M_{ta} = 0.05 \times 160 \times (160 \times 5) = 6,400 \; kip.ft$$

Where $F_y = 5.0 \; ft \times 160 \; kip/ft = 800 \; kip$

Use Equation 1a of this chapter as follows:

$$F_{y,B} = \frac{R_{y,B}}{\Sigma R_y} \times F_y + \frac{M_T(x)R_{y,B}}{\Sigma R_y(x)^2 + \Sigma R_x(y)^2}$$

In order to solve this equation, we need to determine wall rigidities. Using the aspect ratio of the shortest wall $\frac{h}{l} = \frac{8}{40} = 0.2 < 0.3$, in which case, and in reference to Table 4-1 of this chapter, wall length can be used for relative rigidity for all walls.

$$F_{y,B} = \frac{40}{40+40} \times 800 + \frac{6,400 \times 40 \times 40}{2 \times (40 \times 40^2) + 2 \times (80 \times 60^2)}$$

$$= 400 + 14.5$$

$$= 414.5 \; kip$$

Correct Answer is (D)

SOLUTION 4.9

We can refer to the shear and moment diagrams for simple beams and flexible diaphragms provided in Figure 4-2 of this chapter.

$$M_x = 0.5wx(L - x)$$

$$= 0.5 \times 2 \times 10 \times (12 - 10)$$

$$= 20 \; kip.ft$$

$$T_A = M/D$$

$$= 20 \; kip.ft/14 \; ft$$

$$= 1.43 \; kip$$

Most standards use 0.95 of the diaphragm's depth as lever arm, this makes $T = 1.5 \; kip$.

Correct Answer is (A)

SOLUTION 4.10

Although solving for a continuous beam requires longer steps, the location of point A falls at the cantilever part of this continuous beam. The relevant moment can be calculated using the lever arm to the load resultant – i.e., 3 ft.

With a resultant on the cantilever part of $6 \; ft \times 5 \; kip/ft = 30 \; kip$.

$$M_A = 30 \; kip \times 3 \; ft$$

$$= 90 \; kip.ft$$

$$C_A = M_A/D$$

$$= 90 \; kip.ft/18 \; ft$$

$$= 5 \; kip$$

Correct Answer is (B)

SOLUTION 4.11

This question can be approached in a similar fashion to question 4.10, there are some requirements that must be met for open and cantilever wood flexible diaphragms.

While these requirements are not covered in this chapter, Chapter 6, second part of Section 6.3.3, points to them, and they are also outlined in SDPWS 2021 Section 4.2.6.

Correct Answer is (B)

SOLUTION 4.12

In solution 4.4 and 4.5 we created a distributed load based on the flexible diaphragm's area contribution. Taking a similar approach here, see below:

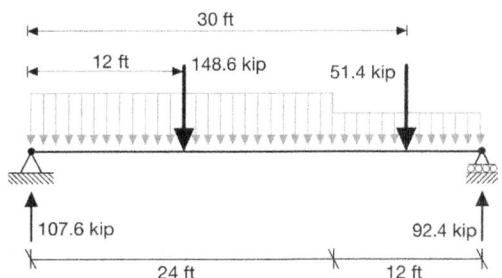

Below is the process used to generate this loading diagram:

Total area of diaphragm: $A_T = 840\ ft^2$
Area left $A_{left} = 26 \times 24 = 624\ ft^2$ (74.3%)
Area right $A_{right} = 216\ ft^2$ (25.7%)

Based on the above area contribution, the lateral load 200 kip can be divided as follows (*):

$$P_{left} = 74.3\% \times 200 = 148.6\ kip$$

$$P_{right} = 25.7\% \times 200 = 51.4\ kip$$

Find the reaction to the right, which is the lateral force transferred to the wall in that location:

$$R_{right} = \frac{148.6 \times 12 + 51.4 \times 30}{36} = 92.4\ kip$$

Cut the diaphragm at point A to find moment at this location as follows:

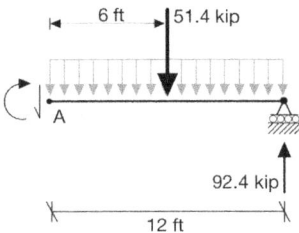

$$M_A = 92.4 \times 12 - 51.4 \times 6 = 800.4\ kip.ft$$

The maximum compression force can be calculated using the shortest lever arm at location A:

$$C_A = \frac{M_A}{D} = \frac{800.4\ kip.ft}{18\ ft} \approx 45\ kip$$

Correct Answer is (A)

(*) If you don't have time to solve a similar problem during the exam with the provided level of detail, approximate the solution by distributing the load uniformly across the beam, then use $\left(M = \frac{wl^2}{8}\right)$. This should yield a result close to the correct value.

SOLUTION 4.13

All the options highlight key design considerations for diaphragms with large openings in seismic design categories D, E, and F. The ASCE 7 Standard specifies a maximum aspect ratio of 2.5:1 for sub-diaphragms to prevent instability and other issues (See ASCE 7 Section 12.11.2.2.1). Additionally, extra tension reinforcement may be necessary due to the reduced lever arm in thinner sub-diaphragms, and the risk of buckling should be carefully addressed to avoid structural failure.

Correct Answer is (D)

SOLUTION 4.14

This is a flexible diaphragm. Figure 4-2 of this chapter can be used to determine the maximum shear force to the left of this wall as follows:

$$V = 0.5wL = 0.5 \times 2 \times 9 = 9\ kip$$

$$v = \frac{9,000\ lb}{9\ ft} = 1,000\ plf$$

Correct Answer is (B)

SOLUTION 4.15

In reference to Sections 4.4.4 and 4.4.5 of this book, drag forces in collectors are not fixed to a single state. Depending on the geometry and design of the system, collectors may experience compression (Option A), tension (Option B), or both simultaneously (Option C). The distribution of forces is influenced by factors such as the relative location and arrangement of the collectors in relation to the shear walls and diaphragm. Therefore, all the statements are correct.

Correct Answer is (D)

SOLUTION 4.16

This question was included in our publication *The Essential Guide to Passing the*

Structural Civil PE Exam and is repeated here because of the important concepts it covers.

ASCE 7-16 Section 12.8.4.2 sets accidental torsion at 5% from the direction of the perpendicular dimension that the seismic force is applied to multiplied by the seismic force.

$$M_{ta} = 0.05 \times (5 \times 12 \, ft) \times 200 \, kip$$
$$= 600 \, kip.ft$$

The forces distribution due to this moment is proportional to the rotational stiffness of each frame.

Given that the Center of Mass (CM) coincides with the Center of Rigidity (CR), forces in frames 1, 2 and 3 will be opposite in direction and equal in magnitude to forces in frames 4, 5 and 6.

Since the question is asking for the effect of accidental torsion alone, the second part of Equation 1a as presented in Section 4.3.3 of this chapter will be used as follows:

$$F_{y,4} = \frac{M_{ta}(x) R_{y,4}}{\sum R_y(x)^2 + \sum R_x(y)^2}$$

Ignoring the horizontal direction (i.e., the $x-axis$) as requested in the question provides slightly conservative results, and reduces the equation to the following:

$$F_{y,4} = \frac{M_{ta}(x) R_{y,4}}{\sum R_y(x)^2}$$

Where:

$$R_1 = R_6 = k$$
$$R_2 = R_3 = R_4 = R_5 = 2k$$

x is the distance from the center of each frame to CR:

$$x_1 = x_6 = 30 \, ft$$
$$x_2 = x_5 = 18 \, ft$$
$$x_3 = x_4 = 6 \, ft$$

$$F_{y,i} = \frac{600}{\left(\begin{array}{c} R_1 x_1^2 + R_2 x_2^2 + R_3 x_3^2 \\ + \\ R_4 x_4^2 + R_5 x_5^2 + R_6 x_6^2 \end{array} \right)} \times R_{y,i} x_i$$

$$= \frac{600}{\left(\begin{array}{c} 900(k) + 324(2k) + 36(2k) \\ + \\ 36(2k) + 324(2k) + 900(k) \end{array} \right)} \times R_{y,i} x_i$$

$$= \frac{R_{y,i} x_i}{5.4k}$$

$$F_{y,4} = \frac{R_{y,4} x_4}{5.4 \times k} = \frac{(2k) \times 6}{5.4 \times k} = 2.22 \, kip \; (= F_{y,3})$$

Correct Answer is (B)

Question extras:

The rest of the lateral forces' distribution are calculated using the same method and are shown in the first figure below:

$$F_{y,1} = \frac{k_1 x_1}{5.4 \times k} = \frac{k \times 30}{5.4 \times k} = 5.6 \, kip \, (= F_{y,6})$$

$$F_{y,2} = \frac{k_2 x_2}{5.4 \times k} = \frac{(2k) \times 18}{5.4 \times k} = 6.7 \, kip \, (= F_{y,5})$$

5.6 kip   6.7 kip   2.2 kip   2.2 kip   6.7 kip   5.6 kip

Frames resistance due to the horizontal force alone and without the torsional effect is proportional to each frame's stiffness. Using the first part of Equation 1a, See below:

$$F_{y,i} = \frac{R_{y,i}}{\sum R_y} \times F_y$$

$$\sum R_y = k + 2k + 2k + 2k + 2k + k = 10 \times k$$

$$F_{y,1} = F_{y,6} = \frac{k}{10 \times k} \times 200 = 20 \ kip$$

$$F_{y,2,3,4,5} = \frac{2k}{10 \times k} \times 200 = 40 \ kip$$

The following figure represents the lateral forces distribution without the torsional effect:

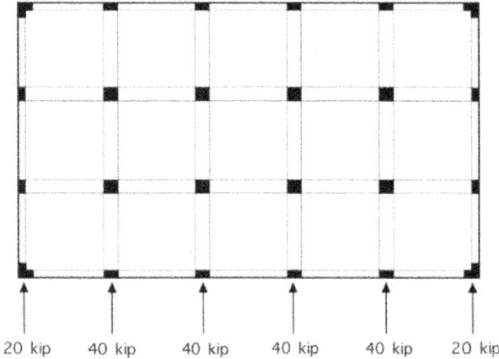

20 kip   40 kip   40 kip   40 kip   40 kip   20 kip

The following figure represents all forces' distribution combined:

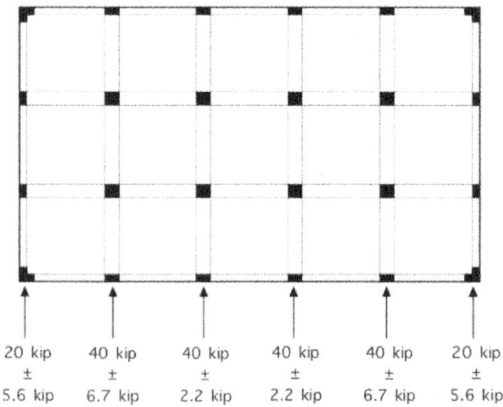

20 kip	40 kip	40 kip	40 kip	40 kip	20 kip
±	±	±	±	±	±
5.6 kip	6.7 kip	2.2 kip	2.2 kip	6.7 kip	5.6 kip

SOLUTION 4.17

In reference to Figure 4-1 of this chapter, the two walls have the same length, and because of this, each wall will take half the load which is 5 kip.

Shear wall 2 is illustrated below with the tie down requirement, which can be calculated using the summation of moment at the bottom left edge of this wall.

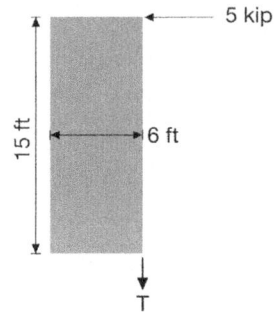

$$T = \frac{5 \times 15}{6} = 12.5 \ kip$$

Correct Answer is (B)

SOLUTION 4.18

In reference to Section 4.4.4 of this chapter, the total lateral force – given this is a flexible and uniformly shaped diaphragm – is divided to 90 kip on axis A and 90 kip on axis D.

For axis A, the length of this side is 45 ft with a unit shear $= \frac{90 \ kip}{45 \ ft} = 2 \ kip/ft$

Collector total force $= 2\frac{kip}{ft} \times 15 \ ft = 30 \ kip$

Correct Answer is (B)

SOLUTION 4.19

In reference to Section 4.4.4 of this chapter, the total lateral force – given this is a flexible and uniformly shaped diaphragm – is divided based on the respective areas as follows:

$$Area \ 1 \ to \ 3 = 1,200 \ ft^2 \ (33.33\%)$$

$$Area \ 3 \ to \ 5 = 2,400 \ ft^2 \ (66.67\%)$$

The unit shear on axis 3 at the southern side of the diaphragm

$$= \frac{\left(\frac{120 \times 66.67\%}{80}\right)}{2} = 0.5 \ kip/ft$$

The unit shear on axis 3 at the northern side of the diaphragm

$$= \frac{\left(\frac{120 \times 33.33\%}{40}\right)}{2} = 0.5 \ kip/ft$$

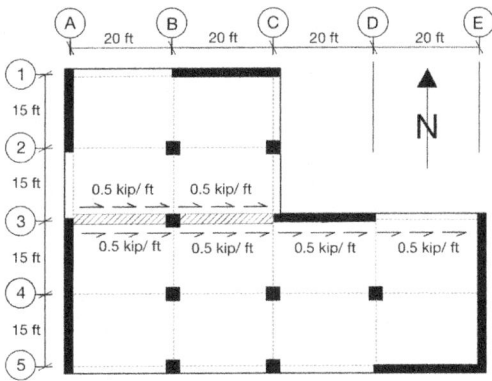

The collector drag force

$$= (0.5 + 0.5)\frac{kip}{ft} \times 40\,ft = 40\,kip \;(*)$$

Correct Answer is (C)

Question extras:
The following is the method of constructing the shear profile.

You need to find the shear on wall C-D using methods presented in Figure 4-1 of this chapter. The shear on this wall is half the lateral load from diaphragm 1 to 3 (i.e., 20 kip), and the other half from 3 to 5 (i.e., 40 kip). A total of 60 kip distributed over the length of the wall of 20 kip, which gives a unit shear on that wall of 3 kip/ft.

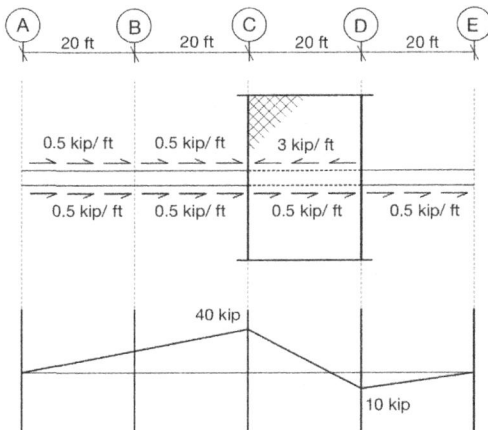

From this profile, you can observe that the drag force on collector D-E is in tension as it pulls wall C-D in the West to East direction, while collector A-C is in compression as it pushes the wall.

SOLUTION 4.20
In reference to Section 4.5.4 of this chapter, Seismic forces should be applied at 100% in one direction and 30% in the perpendicular direction for SDC C.

Correct Answer is (C)

SOLUTION 4.21
ASCE 7-16 code Section 2.3.6, or Section 4.5.1 in this chapter, is referred to in this solution.

Horizontal and vertical seismic effects are calculated as follows:

$$E_h = \rho\,Q_E = 1.3 \times 3 = 3.9$$

$$E_v = 0.2\,S_{DS}D = 0.2 \times 0.75 \times 5 = 0.75$$

The below is the full load combinations, load combination 1 to 5 are provided as extras as they have not been requested in this question.

	Combination based on allowable stress	Result (kip)
1	1.4 D 1.4×5	7.0
2	1.2 D + 1.6 L + 0.5 (L_r or S or R) 1.2×5 + 1.6×6 + 0.5 (1.5 or 0 or 0)	16.4
3	1.2 D + 1.6 (L_r or S or R) + (L or 0.5W) 1.2×5+ 1.6(1.5 or 0 or 0) + (6 or 0.5×2)	14.4
4	1.2 D + 1.0 W + L + 0.5 (L_r or S or R) 1.2×5 + 1 × 2 + 6 + 0.5 (1.5 or 0 or 0)	14.8
5	0.9 D + 1.0 W 0.9×5 + 1.0×2	6.5
6	1.2 D + E_v + E_h + L (*) + 0.2 S 1.2×5 + 0.75 + 3.9 + 6.0 + 0	16.7
7	0.9 D − E_v + E_h 0.9×5 − 0.75 + 3.9	7.7

The design shall proceed using the maximum load generated which belongs to load combination 6 with 16.7 kip.

Correct Answer is (A)
(*) You are permitted to multiply Live Load by (1/2) provided it is less than 100 psf except for garages and public assemblies.

SOLUTION 4.22

As presented in Section 4.4.3 of this chapter and also outlined in Table 4.1.4 of the SDPWS 2021. The allowable shear capacities for wooden walls and diaphragms for seismic design must be divided by a reduction factor of 2.8 for the Allowable Stress Design (ASD) method.

Correct Answer is (B)

SOLUTION 4.23

As presented in Section 4.5.3 of this chapter, and also outlined in Section 12.4.4 of the ASCE 7 Standard. When dealing with horizontal cantilevers in Seismic Design Categories D, E or F, those should be designed for an upward force of 0.2 times the dead load in addition to the load combinations.

Correct Answer is (A)

SOLUTION 4.24

In reference to Section 4.5.1 page 148 in this chapter, or Section 12.3.4.1 of the ASCE 7 Standard, the only applicable scenario here is option (A), which is the design of non-structural components.

Correct Answer is (A)

SOLUTION 4.25

ASCE 7-16 code Section 12.14.3.1, or Section 4.5.5 in this chapter, is referred to in this solution.

It is important to observe that overstrength should be used instead of the redundancy factor, as this is a foundation element supporting a cantilever column against overturning (see page 148 of this book).

In accordance with Section 12.2.5.2 of the ASCE 7 Standard (as mentioned in Section 4.5.1 of this chapter).

$$E_{mh} = 2.5Q_E = 2 \times 2.5 = 5.0 \ kip$$

Moreover, since fluid loads are included, and per the notes provided in page 117 of the ASCE 7 Standard, those should be added to the dead load. Also, given that the live load is not for a garage or public use, and is less than $100 \ psf$, a factor of 0.5 can be applied to it.

Combination 5:

$$= (1.2 + 0.2S_{DS})D + E_{mh} + 0.5L + 0.2S$$
$$= (1.2 + 0.2 \times 0.1) \times (5 + 1) + 5 + 0.5 \times 6 + 0$$
$$= 15.3 \ kip$$

Combination 7:

$$= (0.9 - 0.2S_{DS})D + E_{mh}$$
$$= (0.9 - 0.2 \times 0.1) \times (5 + 1) + 5$$
$$= 10.3 \ kip$$

The controlling load in this case is $15.3 \ kip$.

Correct Answer is (D)

CHAPTER

5

Seismic Forces: Nonbuilding Structures

5.1. Purpose of this Chapter

Seismic forces applied to non-building structures are just as important as those applied to typical buildings, and often even more critical due to the essential operational functions they serve. For example, systems like sprinklers and generators in hospitals may be crucial for post-earthquake operations.

This chapter discusses methods for calculating seismic forces on nonstructural components and nonbuilding structures that are either similar to buildings or not similar to buildings. It also provides guidance on how to distinguish between these categories and outlines the different methods used for each, highlighting the variations in parameters and requirements. Also, for a chapter overview diagram, refer to Figure 5-6, which provides a general-purpose process diagram.

5.2. Nonbuilding Structures and Nonstructural Components

5.2.1. Overview

Seismic structural design generally falls into three categories. Those are the design of (1) normal building structures, (2) nonstructural components, and (3) nonbuilding structures. While previous chapters of this book focused solely on the first category, the other categories are equally, if not more, important depending on their location and use before or after an earthquake.

Nonstructural components can be defined as architectural, mechanical, plumbing, or electrical elements that are permanently attached to structures. These components can be classified as either nonbuilding structures or nonstructural components. Table 5-1 differentiates between the two, providing examples as addressed in the ASCE 7 Standard, and indicating which chapter of the standard deals with each component.

5.2.2. Nonstructural Components

The nonstructural components primarily include architectural, mechanical, plumbing, and electrical elements, some of which are illustrated in Figure 5-1 here. Detailed guidelines for calculating the seismic design force for each of these components can be found in Chapter 13 of the ASCE 7 Standard. This chapter provides clear instructions on how to determine the seismic forces for various nonstructural components. Specific response modification factors and component amplification factors are provided in Table 13.5-1 for architectural components, and Table 13.6-1 for mechanical, plumbing, and electrical components. The following sections of this chapter will outline the details of these calculations. Also remember that the Seismic Design Category of the component is similar to that of the structure it is attached to.

Table 5-1 Nonstructural Components and Nonbuilding Structures in ASCE 7 Standard

Nonstructural Components ASCE 7 Chapter 13		Both Chapter 13 and 15 of ASCE 7	Nonbuilding Structures Not Similar to Buildings ASCE 7 Chapter 15
Access floors	Escalator components	Billboards	Amusement structures
Air conditioning units	Evaporators	Bins	Elevated hoppers
Air distribution boxes	Furnaces	Chimneys	Monuments
Air handlers	Generators	Conveyors	Cast-in-place Silos
Air separators	Heat exchangers	Cooling towers	
Battery racks	HVAC	Signs	
Boilers	Inverters	Stacks	
Cabinet heaters	Lighting fixtures	Tanks	
Cabinets	Manufacturing	Towers	
Cable trays	equipment	Vessels	
Ceiling	Motor control centers		
Chillers	Motors		
Communications	Panel boards		
equipment	Parapets		
Compressors	Penthouse (when not		
Computers	part of the building		
Ductwork	frame)		
Electrical conduit	Piping		
Elevators	Plumbing Process		
Engines	equipment		
Switch gear	Pumps		
Transformers	Walls and wall panels		
Tubing	Water heaters		
Turbines	Vibration isolated		
Veneers	systems		

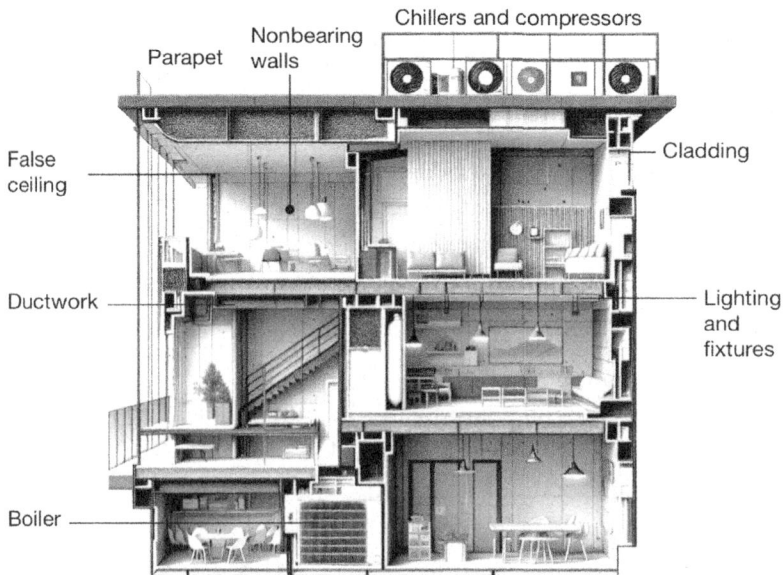

Figure 5-1 Example for Nonstructural Components

5.2.3.   Nonbuilding Structures

Nonbuilding structures generally share similarities with buildings in terms of structural dynamics and response, but they differ in their specific structural characteristics. These structures are classified into two categories: (1) similar to buildings, and (2) not similar to buildings.

Examples of nonbuilding structures similar to buildings include steel storage racks, pipe racks, and towers for tanks or vessels. The design of these structures follows the requirements outlined in Chapter 15 of the ASCE 7 Standard, along with relevant sections from the previous chapters discussed earlier in this book. In contrast, nonbuilding structures that are not similar to buildings include earth-retaining structures, chimneys, and telecommunication towers. These structures have distinct design requirements, also covered in Chapter 15 of the ASCE 7 Standard.

Figure 5-2 Example for a Nonbuilding Structure (Oil Refinery) with Tanks and Pipe Racks

5.3. Design for Nonstructural Components

A nonstructural component, as defined in Chapter 13 of the ASCE 7 Standard, is a component that is attached to the main structure and weighs less than 25% of the effective weight of the structure. For nonbuilding structures which weigh > 25% reference should be made to Chapter 15 of the ASCE 7 Standard (or Section 5.4 here).

5.3.1.   The Importance Factor

The importance factor (I_p) is a key coefficient in this chapter and is typically taken as 1.5 for certain conditions, particularly when the component is critical for life safety and must remain operational after an earthquake. Examples of such components include egress stairways and sprinkler systems for fire protection. The importance factor for those should always be taken as $I_p = 1.5$ regardless of the risk category they belong to.

Additional components that require the factor of $(I_p = 1.5)$ are listed in Section 13.1.3 of the ASCE 7 Standard, which includes components containing toxic or highly toxic materials, those attached to a Risk Category IV structure, or components that contain hazardous substances.

Also, you need to observe the list of exemptions in the same section, interesting to know for example that mechanical equipment in SDC B are fully exempted, while they are also exempted for SDC C when $(I_p = 1.0)$, or when they weigh less than (20 *lb*).

5.3.2.   Reference Documents and Testing Requirements

It is important to understand that many manufacturers of equipment, cabinets, chillers, and other components are required to provide reference documents for their products. These documents often include proof of seismic testing, such as shake table tests, and sometimes certifications, before installation.

For specific equipment, the manufacturer can submit their documentation to the relevant authority having jurisdiction over the structure. It is important to note that the seismic forces used to design the equipment should not be less than those specified in the ASCE 7 Standard Section 13.3.1 (and Section 5.3.3 of this chapter). Additionally, seismic interactions with any connected elements must be considered, including potential drifts, deflections, or displacements. The anchorage requirements must also comply with the standard, which, as you will see in the next section, involve multiplying the design force by the relevant overstrength factor (Ω_o).

If the reference documents use the Allowable Stress Design (ASD) method, the calculated forces from Section 13.3.1 of the ASCE 7 Standard shall be multiplied by (0.7). The dead load, earthquake loads, and operating loads should be added to the values provided in the reference, with relative displacement taken into account. The standard allows in this case NOT to use the load combinations covered in its Chapter 2.

Regarding certifications and special certifications, manufacturers must submit a special certification for designated seismic systems assigned to Seismic Design Categories C through F. This certification should confirm that the equipment, which must remain operable after an earthquake, can do so based on an approved shake table test. Alternatively, experience data can be accepted. Additionally, if the equipment contains hazardous materials with an importance factor of $(I_p = 1.5)$, the manufacturer must provide both an analysis and a shake table test certification. Experience data may also be accepted in such cases.

Lastly, it is essential to consider the interrelationship between components to prevent consequential damage. The failure of one component, whether essential or non-essential, should not cause the failure of essential components. This ensures that the failure of one part of the system does not compromise the overall functionality of the equipment or the safety of the structure.

5.3.3.   Seismic Design Force

This section primarily covers three types of nonstructural components: architectural, mechanical (such as plumbing), and electrical. With this in mind, you will see that the focus is on calculating the relevant seismic forces (F_p). These forces can be determined using the equation presented in Section 13.3.1 of the ASCE 7 Standard.

$$F_p = \frac{0.4a_p S_{DS} W_p}{\left(\frac{R_p}{I_p}\right)}\left(1 + 2\frac{z}{h}\right)$$

$$0.3 S_{DS} I_p W_p \le F_p \le 1.6 S_{DS} I_p W_p$$

If the manufacturer elected to use an alternative testing method apart from the acceptable ICC-ES AC 156, then the maximum limit of (F_p) increases to $(3.2 I_p W_p)$.

Where:

ICC-ES AC 156 is a guideline issued by the International Code Council Evaluation Service (ICC-ES) for the seismic qualification of nonstructural components. The full title is ICC-ES AC 156: Acceptance Criteria for Seismic Qualification by Shake-Table Testing of Nonstructural Components.

a_p is the component's amplification factor which varies from 1.0 to 2.5 and can be collected from Table 13.5-1 and 13.6-1 of the ASCE 7 Standard – included in this chapter's appendix for ease of reference. It is also important to observe how this amplification factor increases for some cantilever component when braced above the component's Center of Mass compared to below it – e.g., check chimneys in Table 13.5-1 in this chapter's appendix.

R_p is the response modification factor which varies from 1.0 to 12.

W_p is the operating weight, which is the weight of the component during operation, for instance with fluids or other material necessary for its operation.

For suspended ceilings, (W_p) includes the ceiling grid, tiles or panel, light fixtures and shall not be taken less than $4\ psf$. For access floors, it shall include the weight of all equipment fastened to it and 25% of equipment not fastened to it.

z/h is the ratio of the height of the component to the average height of the structure. This indicates that if the component was at base, then:

$$F_p = \frac{0.4a_p S_{DS} W_p}{(R_p/I_p)}$$

And if the component was at the roof, then:

$$F_p = 3 \times \frac{0.4a_p S_{DS} W_p}{(R_p/I_p)}$$

Most importantly, remember that when anchorage force to masonry or concrete is to be calculated, this force is to be multiplied by the overstrength factor (Ω_o). While in attachments, response modification factor is taken as a maximum of $R_p = 6$.

Additionally, the resultant force shall be applied orthogonally and independently in the two directions along with the service or operating load.

Finally, and do not overlook that the vertical seismic load for the component is calculated as $(\mp 0.2 S_{DS} W_p)$.

Seismic Forces: Nonbuilding Structures

Example 5.1 Roof Air Conditioning Unit

The Carrier 30XV is a large air conditioning unit with dimensions of $7.25\ ft$ (width) by $7.25\ ft$ (depth) and a height of $8.42\ ft$, while it weighs $5\ kip$. Assume its center of gravity is at the geometric centroid of the unit. This unit is installed on the roof of a Category IV hospital. Given a short acceleration of $S_{DS} = 0.75$ and knowing that the unit is supported by four supports located at each of the four bottom edges, what is the seismic force contribution to each support in the E – W direction?

Based on the description of the building that this air conditioning unit is placed on top, the importance factor $I_p = 1.5$, and based on Table 3.6-1 for an air conditioning unit, $a_p = 2.5$ and $R_p = 6$. With the unit mounted on the roof:

$$F_p = 3 \times \frac{0.4 a_p S_{DS} W_p}{(R_p / I_p)} = 3 \times \frac{0.4 \times 2.5 \times 0.75 \times 5}{(6/1.5)} = 2.81\ kip$$

$$\left(0.3 S_{DS} I_p W_p = 1.7\ kip\right) \leq F_p \leq \left(1.6 S_{DS} I_p W_p = 9.0\ kip\right) \rightarrow OK$$

Based on the below diagram, the lateral force (F_p) that is creating moment at base shall be stabilized by a vertical couple created through supports. Due to symmetry, all reactions are equal in magnitude and opposite in direction as shown below (*). Use R_v for vertical and R_h for horizontal reactions per support:

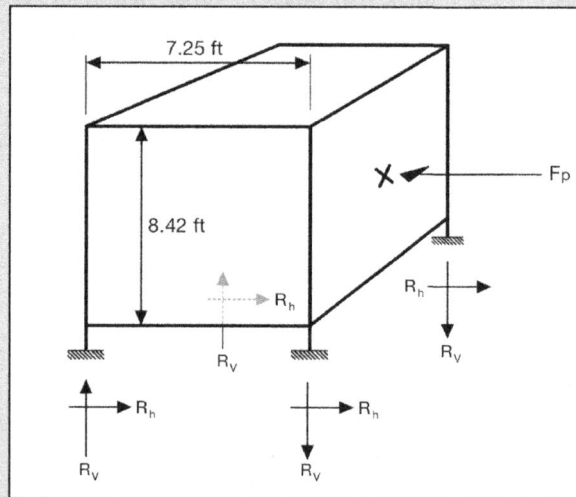

$$2.81\ kip \times \frac{8.42\ ft}{2} = (2R_v) \times 7.25\ ft$$

$$\rightarrow R_v = 0.8\ kip$$

$$R_h = \frac{2.81}{4} = 0.7\ kip$$

(*) In structural engineering, it is essential to consider additional factors, such as the flexibility of supports, to make a final determination of their values since this is an indeterminate structure. However, this is beyond the scope of this book and the exam.

5.3.4.   Displacement and Relative Displacement

With the design of certain components – such as piping, windows, stairs, cladding or sprinkler systems – flexibility is more important that strength, and displacements between structures, or the relative displacement (D_p) within the same structure, both play an important role in the design process. See Figure 5-2 and Figure 5-3 for illustration of those displacements.

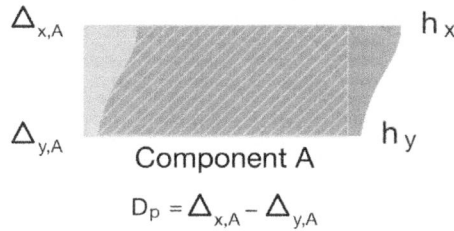

$$\Delta_{x,A} \qquad\qquad\qquad\qquad h_x$$

$$\Delta_{y,A} \qquad\qquad\qquad h_y$$

Component A

$$D_p = \Delta_{x,A} - \Delta_{y,A}$$

Figure 5-3 Relative Displacement for a Cladding Component

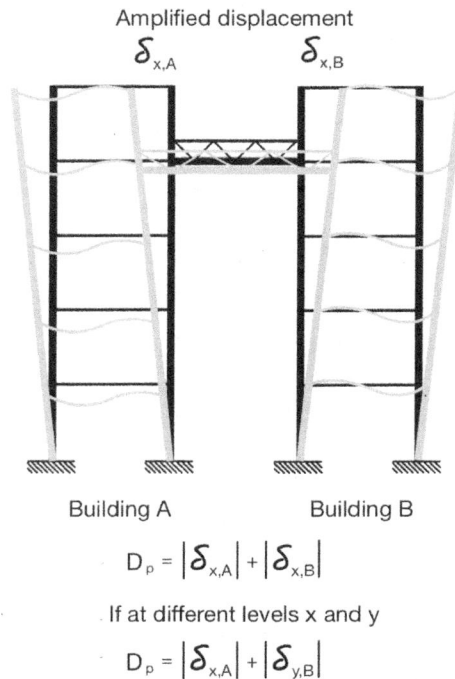

Amplified displacement

$$\delta_{x,A} \qquad\qquad \delta_{x,B}$$

Building A   Building B

$$D_p = \left|\delta_{x,A}\right| + \left|\delta_{x,B}\right|$$

If at different levels x and y

$$D_p = \left|\delta_{x,A}\right| + \left|\delta_{y,B}\right|$$

Figure 5-4 Displacement Between Two Buildings

The seismic displacement (D_{pI}) in this case relies on the importance factor (I_e) taken from Section 11.5.1 of the ASCE 7 (<u>not the component importance factor I_p</u>) and is calculated as follows:

$$D_{pI} = I_e D_p$$

5.3.5.   Component Period

The period for a nonstructural component is calculated in a similar fashion to the method explained earlier in Section 3.2.4 here, provided that the stiffness of the component can be determined, and that the component can be represented with a mass and spring model similar to Figure 3-1 of Chapter 3 of this book. The component period (T_p) can be calculated using the following equation:

$$T_p = 2\pi \sqrt{\frac{W_p}{K_p g}}$$

Where (W_p) is the operating weight of the component in lb or kip, (K_p) in this case is the total stiffness of contributing elements in the desired direction, and $g = 32.174 \, ft/sec^2$.

Component period is used to classify components as either rigid when $T_p < 0.06 \, sec$, or flexible when $T_p \geq 0.06 \, sec$.

5.3.6.   Nonstructural Components Requirements

This section is not intended to provide a detailed summary for the ASCE 7 Standard Chapter 13; rather, the following tables provides high-level requirements for some of the key components along with some of the foundational concepts. You are advised to refer to relevant sections in the standard for more detailed requirements.

Architectural Components, Supports and Attachments

Table 5-2 below provides some high-level requirements on a selection of architectural components, supports and attachments as detailed in ASCE 7 Standard Section 13.5.

Table 5-2 Architectural Components, Supports and Attachments' Requirements

Nonstructural Component	Requirement(s)	Exception(s)
Exterior wall elements and connections (such as cladding)	Connection design should be based on relative seismic displacement (D_{pl}) or $(0.5 \, in)$, whichever is greater.	-
Glass *Such as glass in glazed walls, glazed storefronts or glazed partitions.*	The relative seismic displacement at which glass fallout occurs $$\Delta_{fallout} \geq 1.25 D_{pl}$$ Or 0.5 in Whichever is greater	When clearance between glass and frame $\geq 1.25 D_{pl}$ Fully tempered monolithic glass in risk category I, II and III located $< 10 \, ft$ above walking surface. Annealed or heat tempered.
Suspended ceilings	Operating weight (W_p) shall not be taken less than $4 \, psf$. For areas $> 2,500 \, ft^2$ in SDC D to F, seismic separation joint (or full height partition that reaches the ceiling) to be provided with ratio of long to short dimension < 4.	Area $<144 \, ft^2$ surrounded by walls and soffits. Screw or nail attached gypsum board on one level that are connected by laterally braced walls.
Access floors	Operating weight (W_p) includes the weight of all equipment fastened to it and 25% of equipment not fastened to it.	
Partitions	Partitions $> 6 \, ft$ high and tied to the ceiling shall be laterally braced to the building structure.	Partitions $< 9 \, ft$ AND Linear wt. $< 10 \, lb \times height$ AND Seismic force $< 5 \, psf$

Egress stairs and ramps	Design is based on relative seismic displacement (D_{pI}) along with diaphragm deformation.	If no sliding or ductile connections are provided, the stiffness and strength of those stairs should be included in the overall building structural model. The stairs in this case should include the relevant overstrength of the building. (See Section 4.5.1 page 148 of Chapter 4 of this book). Overstrength (Ω_o) in this case should be $\geq$ 2.5.
	Sliding connections with slotted holes should permit movement of (D_{pI}) or $(0.5\ in)$, whichever is greater.	
	Sliding components without end stops should be designed to accommodate $(1.5D_{pI})$ or $(1.0\ in)$, whichever is greater.	

Mechanical and Electrical Components, Supports and Attachments

Section 13.6 and the related commentary sections of the ASCE 7 Standard provide comprehensive requirements for the seismic design of mechanical and electrical nonstructural components, as well as their supports and attachments. These components include HVACR equipment, electrical components (e.g., batteries and their distribution systems such as conduits, cable trays, and raceways), ducting systems, tubing, fire protection sprinkler systems and sprigs, boilers and pressure vessels, elevators, and roof panels.

A few key concepts are common across these components. First, it is important to understand that seismic safety for these components is addressed at two levels. The **usual safety level** applies to components that are not required to operate after an earthquake but must not pose a life safety hazard if detached. The focus here is on ensuring proper support and attachment. The **higher level** applies to components that must remain operational after an earthquake. These are classified as **designated seismic systems** and must be designed to withstand seismic forces, in addition to requiring proper support and attachment.

Mechanical and electrical components can be grouped into three categories:

1. Airside components (e.g., fans, air handlers, and air distribution boxes) are relatively light and flexible, with an amplification factor of $a_p = 2.5$.
2. Wet-side components (e.g., boilers, chillers, and water heaters) are more rigid, with a lower amplification factor of $a_p = 1.0$.
3. Rugged components (e.g., engines, turbines, pumps, and compressors) are rigid but somewhat ductile.

Vibration is an important aspect to mechanical equipment subjected to seismic loading as vibration can either damage the equipment itself, or the wall or slab that it is attached to. For that, there are distinctions on how vibrating components should be supported. **Neoprene pads** for instance are more effective at damping vibrations compared to **springs**, as the ductility factor is typically higher when using Neoprene, and hence they have higher response modification factor (R_p). Moreover, for vibrating components with a nominal clearance (or air gap) of less than $0.25\ in$ between the supporting frame and the restraint, the seismic design force should be taken as $(2F_p)$.

When fastening sheet metal plates for those vibrating components, **Belleville** washers are preferred over standard washers because their larger surface area helps distribute seismic and vibrating stresses more evenly across the metal sheet, reducing the risk of tearing especially for sheets as thin as 0.18 *inches*, or 7 gauge. See Figure 5-5 for a typical washer connection.

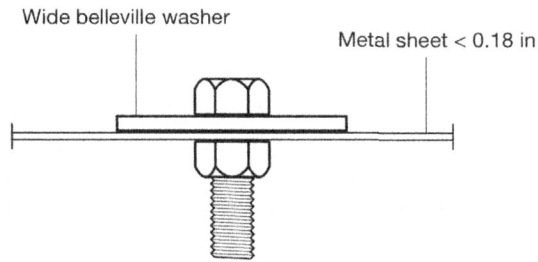

Figure 5-5 Belleville Washer used with Thin Metal Sheet

The commentary sections of the ASCE 7 Standard also highlight the importance of accommodating seismic relative displacements for components that could be damaged or that could cause damage to other components. This is especially relevant for systems fastened at multiple locations within the structure, such as ducts, cable trays, elevator guide rails, and piping systems.

5.4. Design for Nonbuilding Structures

The methods outlined in Chapter 3 and 4 in this book (or Section 12 of the ASCE 7 Standard), which discussed normal building structures, can be applied here as well to evaluate the design forces for nonbuilding structures, whether or not they are similar to buildings, with some key differences that will be addressed here. Additionally, it is important to highlight the excluded structures specified in the ASCE 7 Standard, including vehicular bridges, electrical transmission towers, hydraulic structures, buried utility lines, and nuclear reactors.

5.4.1.   The Seismic Force Resisting Systems (SFRS) for Nonbuilding Structures

Nonbuilding structures require a lateral Seismic Force Resisting System (SFRS) similar to that of typical buildings. These systems are outlined in Tables 15.4-1 and 15.4-2 of the ASCE 7 Standard, which address nonbuilding structures similar to buildings and those that are not, respectively. For convenience, these tables are also provided in the appendix of this chapter.

For systems applicable to nonbuilding structures similar to buildings, you will notice that Table 15.4-1 of the ASCE 7 Standard presents systems that closely resemble those listed for typical buildings in Table 12.2-1 of the ASCE Standard. For example, ordinary, intermediate, and special moment frames are listed in a manner similar to Table 12.2-1, with one key difference: height.

Height is directly tied to the Seismic Design Category (SDC) of the structure. For example, in the case of ordinary moment frames for typical buildings, height limitations exist, and certain SDCs do not permit these systems. This limitation also applies to nonbuilding structures similar to buildings, though there is some flexibility allowed. This flexibility is likely due to the unique nature of nonbuilding structures, where taller elements like pipe racks may be necessary. This leeway is reflected in the lateral force requirements, which can be adjusted by reducing the response modification factor (R). Table 5-3, below, provides an example taken from Table 15.4-1, illustrating this adjustment through a comparison of a steel ordinary moment frame system with an unlimited height requirement.

Table 5-3 Example of SFRS for Ordinary Moment Frame with Height Permissions

| Nonbuilding Structure Type | R | Ω_o | C_d | Structural Height Seismic Design Category | | | | |
				B	C	D	E	F
Steel ordinary moment frames	3.5	3	3	NL	NL	NP[b,i]	NP[b,i]	NP[b,i]
With permitted height increase	2.5	2	2.5	NL	NL	100	100	NP[b,i]
With unlimited height	1	1	1	NL	NL	NL	NL	NL

NP: Not Permitted, NL: Not Limited

[b] Steel ordinary and intermediate moment frames are permitted in pipe racks up to 65 ft where moment joints of field connections are constructed of bolted end plates.

[i] Steel ordinary and intermediate moment frames are permitted in pipe racks up to 35 ft.

5.4.2.   Base Shear

Base shear for nonbuilding structures is calculated in a similar fashion to the base shear for typical buildings per Section 3.3.2 of this book with few variations to the limiting coefficients as summarized in Table 5-5 & 5-6 for specific AWWA (American Water Works Association), API (American Petroleum Institute) and ACI (American Concrete Institute) designs. Rigidity of the structure plays an important role in determining these forces as described in Table 5-4 below.

Table 5-4 Base Shear Equations for Nonbuilding Structures

Rigid Nonbuilding Structures $T < 0.06$	Not Rigid Nonbuilding Structures $T \geq 0.06$
$V = 0.3 S_{DS} W I_e$	$V = C_s W \quad \rightarrow \quad C_s$ from Table 5-5

Where (R) is the response modification factor provided by Table 15.4-1 or 15.4-2 of ASCE 7, (I_e) is the importance factor and (W) is the operating weight.

Table 5-5 Seismic Response Coefficient (C_s) Limits for Nonbuilding Structures

Equation	Condition for Use
$C_s = \dfrac{S_{DS}}{\left(\frac{R}{I_e}\right)}$	Used when $T \leq T_s$ C_s calculated here cannot exceed C_s calculated below when $T_s < T \leq T_L$
$C_s = \dfrac{S_{D1}}{T\left(\frac{R}{I_e}\right)}$	Used when $T_s < T \leq T_L$ .... Where $T_s = S_{D1}/S_{DS}$
$C_s = \dfrac{S_{D1}T_L}{T^2\left(\frac{R}{I_e}\right)}$	Used when $T > T_L$
$C_{s,min} = 0.044 S_{DS} I_e \geq 0.03$	Used when R is taken from Table 15.4-2 (not similar to buildings)
$C_{s,min} = \dfrac{0.8 S_1}{\left(\frac{R}{I_e}\right)}$	Used when $S_1 \geq 0.6g$

Table 5-6 Seismic Response Coefficient (C_s) for Tanks, Vessels, Stacks and Chimneys

Equation	Condition for Use
$C_{s,min} = 0.044 S_{DS} I_e \geq 0.01$	Use when R is taken from Table 15.4-2
$C_{s,min} = \dfrac{0.5 S_1}{\left(\frac{R}{I_e}\right)}$	Used when $S_1 \geq 0.6g$

When nonbuilding structures that are not similar to buildings are supported by other structures which they are not part of their SFRS, the following shall apply – also refer to Figure 5-6 for a graphical summary, or Section 15.3 of the ASCE 7 for more details:

1. If the nonbuilding structure weighs less than 25% of the seismic weight of the supporting structure, **it can be designated as a nonstructural component**, and the provisions of Chapter 13 of the ASCE 7 Standard shall apply (Section 5.3 here).

2. If the nonbuilding structure weight $\geq$ 25% of the seismic weight and has a period of less than 0.06 seconds (i.e., rigid):

    → It is considered as a rigid element.

    → Use the R factor of the support structure for the support structure.

    → Use requirements of ASCE 7 Chapter 13 (nonstructural components), use R_p from Table 15.4-2 (not similar to buildings) and $a_p = 1$ for the rigid element.

3. If the nonbuilding structure weighs more than 25% of the seismic weight and has a period greater than 0.06 seconds, both the nonbuilding structure and the supporting structure **must be modeled together** in a combined model, **the R factor used is the lesser of the values for the nonbuilding structure and the supporting structure**.

Sometimes there are components to nonbuilding structures that their response modification factor is not found in Chapter 13 of the ASCE 7. In this case, their response modification factor is taken from Chapter 15.

5.4.3.   Vertical Seismic Force Distribution

Vertical seismic forces distribution for nonbuilding structures is determined in a similar fashion to the methods used for the normal building structures, and those are explained in Section 3.3.3 of this book.

5.4.4.   Accidental Torsional Moment for Nonbuilding Structures

Accidental torsional moment, unlike typical buildings, are not always included in the calculations for the nonbuilding structures, especially for rigid nonbuilding structures, when proper calculation and modeling have been conducted, and all masses of elements are accounted for in the analysis. For more details on this, check the ASCE Standard item 5 page 149.

5.4.5.   Consideration for Vertical Ground Acceleration

Nonbuilding structures typically have different shapes compared to regular buildings, with elements such as racks, pipes, or other components extending horizontally much more than in typical buildings. With this in mind, these horizontal projections require special consideration when applying load directions to determine the most critical design case. Vertical ground acceleration was discussed in Chapter 1 of this book and in Section 11.9 of the ASCE 7 Standard. Based on this, the following load combinations should be used, and worst-case effects are used:

1- Use 100% of loads in on direction, 30% in the orthogonal direction, and 30% in the vertical direction. (This combination is also used to evaluated overturning and stability as well)

2- Use 30% of loads in on direction, 30% in the orthogonal direction, and 100% in the vertical direction.

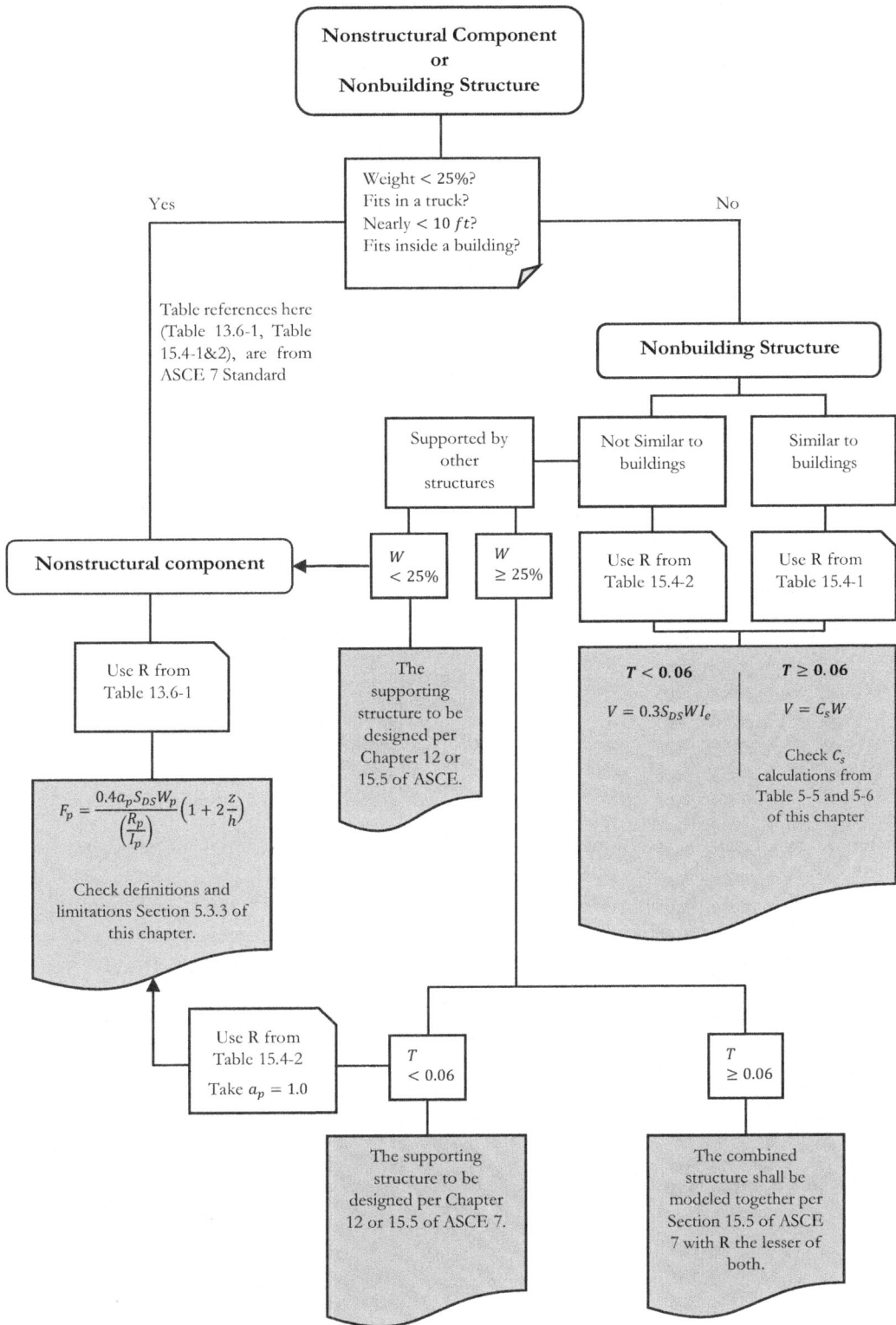

Figure 5-6 Design Process for Nonstructural Components and Nonbuilding Structures

5.4.6.   Fundamental Period for Nonbuilding Structures

The fundamental period or the approximate period outlined in Section 3.2.4 of this book cannot be applied to nonbuilding structures. A more accurate method involves calculating the lateral load elastic deflections (δ_i) generated by the lateral forces (f_i) at each level of the nonstructural building. These deflections are summed across all levels, as detailed in Section 15.4.4 of the ASCE 7 Standard.

$$T = 2\pi \sqrt{\frac{\sum w_i \delta_i^2}{g \sum f_i \, \delta_i}}$$

A quicker and more conservative method for calculating (T) is explained in Section ASCE 7 C15.4.4 as follows:

$$T = T_s = \frac{S_{D1}}{S_{DS}}$$

Example 5.2 Fundamental Period Calculation for a Storage Rack

A 17 ft tall steel storage rack with three levels that is assigned a Seismic Design Category C has been modelled using the forces presented in the following table, and generated displacements as shown.

Level	Force (kip)	Level Displacement (in)	Effective Seismic Weight (kip)
1	22	0.7	35
2	27	1.2	35
3	17	2.5	28

While S_{DS} = 1.1 and S_{D1} = 0.55, what is the fundamental period for this nonbuilding structure?

In reference to Section 5.4.6 of this book, or Section 15.4.4 of the ASCE 7 Standard, the fundamental period method, or the approximate fundamental period method used in typical buildings cannot be used for nonbuilding structures. Instead, for nonbuilding structures, the following equation shall be used:

$$T = 2\pi \sqrt{\frac{\sum w_i \delta_i^2}{g \sum f_i \delta_i}}$$

$$= 2\pi \sqrt{\frac{35\times0.7^2+35\times1.2^2+28\times2.5^2}{32.17\frac{ft}{sec^2}\left(\times\frac{12\ in}{1\ ft}\right)\times(22\times0.7+27\times1.2+17\times2.5)}}$$

$$= 0.52 \ sec$$

Alternatively, the following conservative method can be used:

$$T = T_s = \frac{S_{D1}}{S_{DS}} = \frac{0.55}{1.1} = 0.5 \ sec$$

5.4.7.   Nonbuilding Structures Requirements

This section is not intended to provide a detailed summary for the ASCE 7 Standard Chapter 15; rather, the following sections and tables provide high-level requirements for some of the key components along with some of the foundational concepts. You are advised to refer to relevant sections in the standard for more detailed requirements.

Nonbuilding Structures Similar to Buildings

Some specific requirements for nonbuilding structures that are similar to buildings are provided in sufficient details in Section 15.5 of the ASCE 7 Standard. The referenced section from ASCE covers several structures: (1) Pipe Racks, (2) Storage Racks, (3) Electrical Power-Generating Facilities, (4) Structural Towers for Tanks and Vessels, and (5) Piers and Wharfs that are accessible to the public. See Table 5-7 for some detail on this, for more detail refer to ASCE 7 Standard.

Table 5-7 Nonbuilding Structures Similar to Buildings Requirements

Nonbuilding Structure Not Similar to Buildings	Requirement(s)	Exception(s) /Comment(s)
Pipe racks	Use amplified deflection to assess pounding of piping system similar to section 2.4.1 of this book: $$\delta_x = \frac{C_d \delta_{xe}}{I_e}$$ Friction resulting from gravity loads not to be considered to resist seismic forces.	
Storage racks	Steel storage racks when anchored to concrete or masonry, overstrength of $\Omega_o = 2$ is used. Cantilever steel storage racks when anchored to concrete or masonry, overstrength is used per relevant tables, and redundancy factor in load combinations should be $\rho = 1$. Load combination (racks should be designed for each of those combinations): (1) weight of rack with all levels loaded at 67% of rated capacity, (2) weight of racks along with the highest level loaded at 100% of rated capacity.	Steel storage racks supported at base can use $R = 4$. Importance factor when open to public such as warehouse retails $I_e = 1.5$.
Structural towers for tanks and vessels	When supported on a grillage beam[12], vertical reaction from weight should be increased by 20% to account for nonuniform support.	
Pier and wharfs	The design shall account for the effects of liquefaction and shall consider all applicable marine loading combinations such as mooring, berthing, wave and current on piers and wharfs.	Piers and wharfs that are not open to pubic are excluded from ASCE 7 Standard.

---

[12] A grillage beam is a system of intersecting beams arranged in a grid pattern to distribute loads evenly across a foundation or structure typically used for heavy-load foundations.

Nonbuilding Structures Not Similar to Buildings

Some specific requirements for nonbuilding structures that are not similar to buildings are provided in sufficient details in Section 15.6 & 15.7 of the ASCE 7 Standard. The referenced sections from ASCE cover several structures: (1) Earth Retaining Structures, (2) Chimneys and Stacks, (3) Amusement Structures, (4) Special Hydraulic Structures, (5) Secondary Containment Systems, (6) Telecommunication Towers, (7) Steel Tubular Support Structures for Onshore Wind Turbine Generator Systems, (8) Ground-Supported Cantilever Walls and Fences, and (9) Tanks and Vessels.

Table 5-8 below outlines few specific requirements for some of the above-mentioned nonbuilding structures that are not similar to building, while Table 5-9 and Table 5-10 provide some key requirements for Tanks and Vessels.

Table 5-8 Nonbuilding Structures Not Similar to Buildings Requirements

Nonbuilding Structure Similar to Buildings	Requirement(s)[13]	Exception(s)
Earth retaining structures	Requirements applies to SDC D, E & F. Designed using the (MCE_G) presented in Chapter 1 here, or ASCE 7 Section 11.8.3. Risk Category should be similar to adjacent structures.	Excludes SDC A, B & C
Chimneys and stacks	Separation between liner and chimney equals to (C_d) multiplied by the differential lateral drift. When made of concrete for SDC D, E & F: - Splices for vertical bars shall be staggered with no more than 50% of the splices in one section. - where openings > 10%, $\Omega_o = 1.5$ is used for design purposes.	
Special hydraulic structures	Designed for out-of-phase movement for the hydrodynamic fluid forces.	
Secondary containments systems	Examples: impoundment dikes and walls. Load combination: (1) 100% of MCE when empty, (2) two-thirds of MCE when full, (3) in case of a similar magnitude aftershock, and the containment is highly toxic or explosive, 100% of MCE shall be considered when full as well. *(MCE: Maximum Considered Earthquake)* Freeboard: used to determine height of impoundment. Check Problem 5.11 here for a freeboard example.	
Ground-supported cantilever walls and fences	Applicable to heights > 6 ft. Plain concrete or masonry or ordinary plain Autoclave Aerated Concrete (AAC)[14] not permitted in SDC C, D, E & F.	

---

[13] Redundancy factor in load combinations should be $\rho = 1$.

[14] AAC concrete is lightweight, porous, precast material made from cement, lime, water, and an aerating agent. Cured under high pressure and temperature in an autoclave. It is strong, insulated, and fire-resistant, commonly used in walls, floors, and roofs.

Tanks and Vessels

Tanks and vessels are used to store liquids, gases or granular material. They are given a special attention in ASCE 7 due to their critical functions before and after an earthquake, the type of material they store, and their importance to life safety or toxicity (if these contain toxic material), all of which require specific design considerations to maintain their functionality.

During an earthquake, two primary forces act on these structures, as shown in Figure 5-7. Impulsive loads result from the rigid-body motion of the liquid near the tank's base, with forces concentrated at the bottom which occur at higher frequencies. Convective loads are caused by the sloshing of the liquid, with forces concentrated at the top and occurring at lower frequencies. Impulsive loads are generally more significant in smaller tanks, while convective loads dominate in larger tanks. These forces are usually combined using different methods, depending on the relationship between the periods of the impulsive and convective modes.

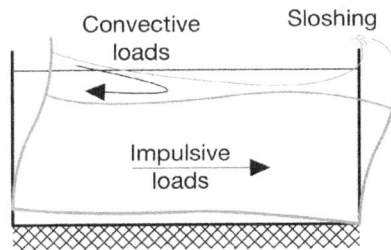

Figure 5-7 Convective and Impulsive Loads in Tanks and Vessels

Table 5-9 Requirements for Tanks and Vessels Storing Liquids

Tanks and Vessels Component	Requirement(s)	Comments/ Equation
Vertical acceleration (S_{av}) and content density	Noncylindrical: increase liquid unit load (γ_L) by:	$0.4S_{av}\gamma_L$
	Cylindrical tanks with diameter (D_i), hoop stresses per unit height (N_h) at height (y) in a liquid height of (H_L):	$N_h = \frac{S_{av}}{R}\gamma_L (H_L - y)\left(\frac{D_i}{2}\right)$
	Vertical inertia:	$0.4S_{av}W$
Pipe attachments	When attached above base, multiply minimum displacement of ASCE 7 Table 15.7-1 by (C_d) to avoid connection rupture. ASCE 7 Table 15.7-1 does not account for foundation displacement.	Applies to tanks storing granular material as well.
Ground supported storage tanks for liquids	Base shear is a combination of impulsive (V_i) and convective (V_c) components: $V = V_i + V_c$ Problem 5.24 provides a calculated example. See ASCE 7 page 156 for more details.	$V_i = \frac{S_{ai}}{R/I_e}W_i$ $V_c = \frac{S_{ac}I_e}{1.5}W_c$
Sliding resistance for flat bottom steel tanks	Maximum base shear (V): W is the effective seismic weight of tank, roof and content after reduction for coincident vertical earthquake.	$V < W\tan 30°$

Table 5-10 Requirements for Tanks and Vessels Storing Granular Material

Tanks and Vessels Component	Requirements & Key Considerations
Effective mass and load path	Loss of intergranular friction during shaking increases lateral pressure and hoop stresses.
	All lateral loads are of impulsive type.
	Once compacted, temperature changes increase hoop stresses.
	Intergranular friction transfers seismic shear directly to foundation.
	S_{DS} is used to determine base shear: $$V = \frac{S_{DS} W_{effective}}{R/I_e}$$
Effective density reduction	When height to diameter ratio $h/D < 2$ energy is lost through intergranular movement, which reduces the mass subject to horizontal acceleration.
	Based on this, use a reduction factor of: 0.8 for most granular material. 0.9 for high moduli of elasticity such as aggregate and metal ore.
Lateral sliding	Tanks with steel bottom that can resist lateral load with friction, no anchorage is required.
	Bottomless tanks require anchorage to be provided. Granular material stored on ground in bottomless tanks.

When tanks are elevated on a tower, whether storing liquid, gas, or granular materials, the stored material is treated as a rigid mass acting at the volumetric center of gravity of the tank. In certain instances, the effects of fluid-structure forces are permitted to be considered, particularly in the case of high sloshing fundamental period ($T_c > 3T$), which is beyond the scope of this book.

5.5. Chapter References

The following references were used in this chapter. The reference(s) in italics is/are required for the exam, as noted in the preface of this book.

[1] *American Society of Civil Engineers, ASCE/SEI 7-16 Minimum Design Loads for Buildings and Other Structures.*

[2] Bachman, R. E., and Dowty, S. M. (2008). Nonstructural component or nonbuilding structure? Building Safety J., April-May.

[3] Malhotra, P. K. (2005). Sloshing loads in liquid-storage tanks with insufficient freeboard." Earthq. Spectra, 21(4), 1185–1192.

Chapter Appendix

This appendix contains the following information:

➢ Table 13.5-1 of ASCE 7 Standard
   Seismic Coefficients for Architectural Components

➢ Table 13.6-1 of ASCE 7 Standard
   Seismic Coefficients for Mechanical and Electrical Components

➢ Table 15.4-1 of ASCE 7 Standard
   Seismic Coefficients for Nonbuilding Structures Similar to Buildings

➢ Table 15.4-2 of ASCE 7 Standard
   Seismic Coefficients for Nonbuilding Structures Not Similar to Buildings

All tables provided in this appendix are copied here with permission from ASCE.

ASCE Table 13.5-1 Seismic Coefficients for Architectural Components

Architectural Component	a_p^a	R_p	Ω_o^b
Interior nonstructural walls and partitions[c]			
Plain (unreinforced) masonry walls	1	1½	1½
All other walls and partitions	1	2½	2
Cantilever elements (unbraced or braced to structural frame below its center of mass)			
Parapets and cantilever interior nonstructural walls	2½	2½	2
Chimneys where laterally braced or supported by the structural frame	2½	2½	2
Cantilever elements (braced to structural frame above its center of mass)			
Parapets	1	2½	2
Chimneys	1	2½	2
Exterior nonstructural walls[c]	1[b]	2½	2
Exterior nonstructural wall elements and connections[b]			
Wall element	1	2½	NA
Body of wall panel connections	1	2½	NA
Fasteners of the connecting system	1¼	1	1
Veneer			
Limited deformability elements and attachments	1	2½	2
Low-deformability elements and attachments	1	1½	2
Penthouses (except where framed by an extension of the building frame)	2½	3½	2
Ceilings			
All	1	2½	2
Cabinets			
Permanent floor-supported storage cabinets more than 6 ft (1,829 mm) tall, including contents	1	2½	2
Permanent floor-supported library shelving, book stacks, and bookshelves more than 6 ft (1,829 mm) tall, including contents	1	2½	2
Laboratory equipment	1	2½	2
Access floors			
Special access floors (designed in accordance with Section 13.5.7.2)	1	2½	2
All other	1	1½	1½
Appendages and ornamentations	2½	2½	2
Signs and Billboards	2½	3	2
Other rigid components			
High-deformability elements and attachments	1	3½	2
Limited-deformability elements and attachments	1	2½	2
Low-deformability materials and attachments	1	1½	1½
Other flexible components			
High-deformability elements and attachments	2½	3½	2½
Limited-deformability elements and attachments	2½	2½	2½
Low-deformability materials and attachments	2½	1½	1½
Egress stairways not part of the building seismic force-resisting system	1	2½	2
Egress stairs and ramp fasteners and attachments	2½	2½	2½

[a] A lower value for a_p shall not be used unless justified by detailed dynamic analysis. The value for a_p shall not be less than 1. The value of $a_p = 1$ is for rigid components and rigidly attached components. The value of $a_p = 2½$ is for flexible components and flexibly attached components.
[b] Overstrength where required for nonductile anchorage to concrete and masonry. See Section 12.4.3 for seismic load effects including overstrength.
[c] Where flexible diaphragms provide lateral support for concrete or masonry walls and partitions, the design forces for anchorage to the diaphragm shall be as specified in Section 12.11.2.

Seismic Forces: Nonbuilding Structures

ASCE Table 13.6-1 Seismic Coefficients for Mechanical and Electrical Components

Components	a_p^a	R_p^b	Ω_o^c
MECHANICAL AND ELECTRICAL COMPONENTS			
Air-side HVACR, fans, air handlers, air conditioning units, cabinet heaters, air distribution boxes, and other mechanical components constructed of sheet metal framing	2½	6	2
Wet-side HVACR, boilers, furnaces, atmospheric tanks and bins, chillers, water heaters, heat exchangers, evaporators, air separators, manufacturing or process equipment, and other mechanical components constructed of high-deformability materials	1	2½	2
Air coolers (fin fans), air-cooled heat exchangers, condensing units, dry coolers, remote radiators and other mechanical components elevated on integral structural steel or sheet metal supports	2½	3	1½
Engines, turbines, pumps, compressors, and pressure vessels not supported on skirts and not within the scope of Chapter 15	1	2½	2
Skirt-supported pressure vessels not within the scope of Chapter 15	2½	2½	2
Elevator and escalator components	1	2½	2
Generators, batteries, inverters, motors, transformers, and other electrical components constructed of high-deformability materials	1	2½	2
Motor control centers, panel boards, switch gear, instrumentation cabinets, and other components constructed of sheet metal framing	2½	6	2
Communication equipment, computers, instrumentation, and controls	1	2½	2
Roof-mounted stacks, cooling and electrical towers laterally braced below their center of mass	2½	3	2
Roof-mounted stacks, cooling and electrical towers laterally braced above their center of mass	1	2½	2
Lighting fixtures	1	1½	2
Other mechanical or electrical components	1	1½	2
VIBRATION-ISOLATED COMPONENTS AND SYSTEMS^b			
Components and systems isolated using neoprene elements and neoprene isolated floors with built-in or separate elastomeric snubbing devices or resilient perimeter stops	2½	2½	2
Spring-isolated components and systems and vibration-isolated floors closely restrained using built-in or separate elastomeric snubbing devices or resilient perimeter stops	2½	2	2
Internally isolated components and systems	2½	2	2
Suspended vibration-isolated equipment including in-line duct devices and suspended internally isolated components	2½	2½	2
DISTRIBUTION SYSTEMS			
Piping in accordance with ASME B31 (2001, 2002, 2008, and 2010), including in-line components with joints made by welding or brazing	2½	12	2
Piping in accordance with ASME B31, including in-line components, constructed of high- or limited-deformability materials, with joints made by threading, bonding, compression couplings, or grooved couplings	2½	6	2
Piping and tubing not in accordance with ASME B31, including in-line components, constructed of high-deformability materials, with joints made by welding or brazing	2½	9	2
Piping and tubing not in accordance with ASME B31, including in-line components, constructed of high- or limited-deformability materials, with joints made by threading, bonding, compression couplings, or grooved couplings	2½	4½	2
Piping and tubing constructed of low-deformability materials, such as cast iron, glass, and nonductile plastics	2½	3	2
Ductwork, including in-line components, constructed of high-deformability materials, with joints made by welding or brazing	2½	9	2
Ductwork, including in-line components, constructed of high- or limited-deformability materials with joints made by means other than welding or brazing	2½	6	2
Ductwork, including in-line components, constructed of low-deformability materials, such as cast iron, glass, and nonductile plastics	2½	3	2
Electrical conduit and cable trays	2½	6	2
Bus ducts	1	2½	2
Plumbing	1	2½	2
Pneumatic tube transport systems	2½	6	2

^{a} A lower value for a_p is permitted where justified by detailed dynamic analyses. The value for a_p shall not be less than 1. The value of a_p equal to 1 is for rigid components and rigidly attached components. The value of a_p equal 2 ½ for flexible components and flexibly attached components.

^{b} Components mounted on vibration isolators shall have a bumper restraint or snubber in each horizontal direction. The design force shall be taken as $2F_p$ if the nominal clearance (air gap) between the equipment support frame and restraint is greater than 0.25 in. (6 mm). If the nominal clearance specified on the construction documents is not greater than 0.25 in. (6 mm), the design force is permitted to be taken as F_p.

^{c} Overstrength as required for anchorage to concrete and masonry. See Section 12.4.3 for seismic load effects including overstrength.

ASCE Table 15.4-1 Seismic Coefficients for Nonbuilding Structures Similar to Buildings

Nonbuilding Structure Type	Detailing Requirements in ASCE 7	R	Ω_o	C_d	Structural System and Structural Height h_n limits (ft)[a]				
					Seismic Design Category				
					B	C	D[b]	E[b]	F[c]
Steel storage racks	15.5.3.1	4	2	3.5	NL	NL	NL	NL	NL
Steel cantilever storage racks hot-rolled steel									
Ordinary moment frame (cross-aisle)	15.5.3.2	3	3	3	NL	NL	NP	NP	NP
Ordinary moment frame (cross-aisle)[d]	15.5.3.2	2.5	2	2.5	NL	NL	NL	NL	NL
Ordinary braced frame (cross-aisle)	15.5.3.2	3	3	3	NL	NL	NP	NP	NP
Ordinary braced frame (cross-aisle)[d]	15.5.3.2	3.25	2	3.25	NL	NL	NL	NL	NL
Steel cantilever storage racks cold-formed steel[e]									
Ordinary moment frame (cross-aisle)	15.5.3.2	3	3	3	NL	NL	NP	NP	NP
Ordinary moment frame (cross-aisle)	15.5.3.2	1	1	1	NL	NL	NL	NL	NL
Ordinary braced frame (cross-aisle)	15.5.3.2	3	3	3	NL	NL	NP	NP	NP
Building frame systems:									
Steel special concentrically braced frames		6	2	5	NL	NL	160	160	100
Steel ordinary concentrically braced frame		$3\frac{1}{4}$	2	$3\frac{1}{4}$	NL	NL	35[f]	35[f]	NP[f]
With permitted height increase		$2\frac{1}{2}$	2	$2\frac{1}{2}$	NL	NL	160	160	100
With unlimited height		1.5	1	1.5	NL	NL	NL	NL	NL
Moment resisting frame systems:									
Steel special moment frames		8	3	5.5	NL	NL	NL	NL	NL
Special reinforced concrete moment frames[g]		8	3	5.5	NL	NL	NL	NL	NL
Steel intermediate moment frames:		4.5	3	4	NL	NL	35[h,i]	NP[h,i]	NP[h,i]
With permitted height increase		2.5	2	2.5	NL	NL	160	160	100
With unlimited height		1.5	1	1.5	NL	NL	NL	NL	NL
Intermediate reinforced concrete moment frames:		5	3	4.5	NL	NL	NP	NP	NP
With permitted height increase		3	2	2.5	NL	NL	50	50	50
With unlimited height		0.8	1	1	NL	NL	NL	NL	NL
Steel ordinary moment frames:		3.5	3	3	NL	NL	NP[h,i]	NP[h,i]	NP[h,i]
With permitted height increase		2.5	2	2.5	NL	NL	100	100	NP[h,i]
With unlimited height		1	1	1	NL	NL	NL	NL	NL
Ordinary reinforced concrete moment frames:		3	3	2.5	NL	NP	NP	NP	NP
With permitted height increase		0.8	1	1	NL	NL	50	50	50

[a] NL = no limit and NP = not permitted.

[b] See Section 12.2.5.4 for a description of seismic force-resisting systems limited to structures with a structural height, h_n, of 240 ft (73.2 m) or less.

[c] See Section 12.2.5.4 for seismic force-resisting systems limited to structures with a structural height, h_n, of 160 ft (48.8 m) or less.

[d] The column-to-base connection shall be designed to the lesser of M_n of the column or the factored moment at the base of the column for the seismic load case using the overstrength factor.

[e] Cold-formed sections that meet the requirements of AISC 341, Table D1.1, are permitted to be designed in accordance with AISC 341.

[f] Steel ordinary braced frames are permitted in pipe racks up to 65 ft (20 m).

[g] In Section 2.3 of ACI 318, the definition of "special moment frame" includes precast and cast-in-place construction.

[h] Steel ordinary moment frames and intermediate moment frames are permitted in pipe racks up to 65 ft (20 m) where the moment joints of field connections are constructed of bolted end plates.

[i] Steel ordinary moment frames and intermediate moment frames are permitted in pipe racks up to 35 ft (11 m).

ASCE Table 15.4-2 Seismic Coefficients for Nonbuilding Structures Not Similar to buildings

Nonbuilding Structure Type	Detailing Requirements in ASCE 7[c]	R	Ω_o	C_d	Structural System and Structural Height h_n limits (ft)[a,b] Seismic Design Category				
					B	C	D	E	F
Elevated tanks, vessels, bins, or hoppers:									
On symmetrically braced legs (not similar to buildings)	15.7.10	3	2[d]	2.5	NL	NL	160	100	100
On unbraced legs or asymmetrically braced legs (not similar to buildings)	15.7.10	2	2[d]	2.5	NL	NL	100	60	60
Horizontal, saddle-supported welded steel vessels	15.7.14	3	2[d]	2.5	NL	NL	NL	NL	NL
Flat-bottom ground-supported tanks:	15.7								
Steel or fiber-reinforced plastic:									
Mechanically anchored		3	2[d]	2.5	NL	NL	NL	NL	NL
Self-anchored		2.5	2[d]	2	NL	NL	NL	NL	NL
Reinforced or prestressed concrete:									
Reinforced nonsliding base		2	2[d]	2	NL	NL	NL	NL	NL
Anchored flexible base		3.25	2[d]	2	NL	NL	NL	NL	NL
Unanchored and unconstrained flexible base		1.5	1.5[d]	1.5	NL	NL	NL	NL	NL
All other		1.5	1.5[d]	1.5	NL	NL	NL	NL	NL
Cast-in-place concrete silos that have walls continuous to the foundation	15.6.2	3	1.75	3	NL	NL	NL	NL	NL
All other reinforced masonry structures not similar to buildings detailed as intermediate reinforced masonry shear walls	14.4.1[e]	3	2	2.5	NL	NL	50	50	50
All other reinforced masonry structures not similar to buildings detailed as ordinary reinforced masonry shear walls	14.4.1	2	2.5	1.75	NL	160	NP	NP	NP
All other nonreinforced masonry structures not similar to buildings	14.4.1	1.25	2	1.5	NL	NP	NP	NP	NP
Concrete chimneys and stacks	15.6.2	2	1.5	2.0	NL	NL	NL	NL	NL
Steel chimneys and stacks	15.6.2	2	2	2	NL	NL	NL	NL	NL
All steel and reinforced concrete distributed mass cantilever structures not otherwise covered herein, including stacks, chimneys, silos, skirt-supported vertical vessels; single-pedestal or skirt-supported.	15.6.2								
Welded steel	15.7.10	2	2[d]	2	NL	NL	NL	NL	NL
Welded steel with special detailing[f]	15.7.10, 5 a&b	3	2[d]	2	NL	NL	NL	NL	NL
Prestressed or reinforced concrete	15.7.10	2	2[d]	2	NL	NL	NL	NL	NL
Prestressed or reinforced concrete with special detailing	15.7.10	3	2[d]	2	NL	NL	NL	NL	NL
Trussed towers (freestanding or guyed), guyed stacks, and chimneys	15.6.2	3	2	2.5	NL	NL	NL	NL	NL
Steel tubular support structures for onshore wind turbine generator systems	15.6.7	1.5	1.5	1.5	NL	NL	NL	NL	NL
Cooling towers:									
Concrete or steel		3.5	1.75	3	NL	NL	NL	NL	NL
Wood frames		3.5	3	3	NL	NL	NL	50	50
Telecommunication towers:	15.6.6								
Truss: Steel		3	1.5	3	NL	NL	NL	NL	NL
Pole: Steel		1.5	1.5	1.5	NL	NL	NL	NL	NL
Wood		1.5	1.5	1.5	NL	NL	NL	NL	NL
Concrete		1.5	1.5	1.5	NL	NL	NL	NL	NL
Frame: Steel		3	1.5	1.5	NL	NL	NL	NL	NL
Wood		1.5	1.5	1.5	NL	NL	NL	NL	NL
Concrete		2	1.5	1.5	NL	NL	NL	NL	NL
Amusement structures and monuments	15.6.3	2	2	2	NL	NL	NL	NL	NL
Inverted pendulum type structures (except elevated tanks, vessels, bins, and hoppers)	12.2.5.3	2	2	2	NL	NL	NL	NL	NL
Ground-supported cantilever walls or fences	15.6.8	1.25	2	2.5	NL	NL	NL	NL	NL
Signs and billboards		3.0	1.75	3	NL	NL	NL	NL	NL
All other self-supporting structures, tanks, or vessels not covered above or by reference standards that are not similar to buildings		1.25	2	2.5	NL	NL	50	50	50

[a] NL = no limit and NP = not permitted.

[b] For the purpose of height limit determination, the height of the structure shall be taken as the height to the top of the structural frame making up the primary seismic force-resisting system.

[c] If a section is not indicated in the detailing requirements column, no specific detailing requirements apply.

[d] See Section 15.7.3.a for the application of the overstrength factors, Ω_o, for tanks and vessels.

[e] Detailed with an essentially complete vertical load-carrying frame.

[f] Sections 15.7.10.5.a and 15.7.10.5.b shall be applied for any risk category.

Chapter 5 Appendix

Problems & Solutions

The following questions have been carefully designed to provide comprehensive coverage of the material presented in this book, while also being educational in nature, as you will notice. Additionally, you may come across a few questions not directly covered in this chapter; these have been intentionally included to ensure broader coverage of the content.

- ➤ Problem 5.1: Loading Direction (1)
- ➤ Problem 5.2: Component Special Certification
- ➤ Problem 5.3: The Allowable Stress Design Method for Components
- ➤ Problem 5.4: Seismic Force for an Equipment when Relocated
- ➤ Problem 5.5: Parapet Out-of-Plane Seismic Force
- ➤ Problem 5.6: Seismic Relative Displacement Between Two Buildings
- ➤ Problem 5.7: Component Anchor
- ➤ Problem 5.8: Sprinkler Clearance
- ➤ Problem 5.9: Suspended Ceilings
- ➤ Problem 5.10: Mechanical and Electrical Components
- ➤ Problem 5.11: Tank Freeboard
- ➤ Problem 5.12: Loading Direction (2)
- ➤ Problem 5.13: Seismic Parameters
- ➤ Problem 5.14: Height Limitation
- ➤ Problem 5.15: Buildings Not Covered in the ASCE 7 Standard
- ➤ Problem 5.16: Accidental Torsional Moment for a Nonbuilding Structure
- ➤ Problem 5.17: Nonbuilding Structure Supported by a Another Structure
- ➤ Problem 5.18: Ground Supported Storage Tanks for Granular Material
- ➤ Problem 5.19: Nonbuilding Structure Design Force Calculation (1)
- ➤ Problem 5.20: Nonbuilding Structure Design Force Calculation (2)
- ➤ Problem 5.21: Tanks and Vessels Support
- ➤ Problem 5.22: Load Combinations for Storage Racks
- ➤ Problem 5.23: Load Combinations for Secondary Containment Systems
- ➤ Problem 5.24: Tank Impulsive Load
- ➤ Problem 5.25: Ground-Supported Storage Tanks for Granular Material

PROBLEM 5.1 *Loading Direction (1)*

For Seismic Design Category C how are seismic forces typically applied to a nonstructural component in two horizontal directions?

(A) Seismic forces are applied independently in both horizontal directions along with the service and operating loads.

(B) Seismic forces are applied at 50% in one direction and 50% in the perpendicular direction along with the service and operating loads.

(C) Seismic forces are applied at 100% in one direction and 30% in the perpendicular direction along with the service and operating loads.

(D) Seismic forces are applied at 100% in both directions simultaneously, with no reduction in one direction, along with the service and operating loads.

PROBLEM 5.2 *Component Special Certification*

Which of the following statements about seismic special certification for designated seismic systems equipment is true?

(A) Manufacturers are required to submit special certifications for all components regardless of SDC.

(B) Certification components in SDC C through F must confirm operability after an earthquake through a shake table test or experience.

(C) Only shake table test is accepted for certification for SDC C through F.

(D) If equipment contains hazardous materials no additional certification is required.

PROBLEM 5.3 *The Allowable Stress Design Method for Components*

Which of the following is true regarding the reference documents with allowable stress design for nonstructural components?

(A) The calculated forces must be multiplied by 0.7, and the load combinations from ASCE 7 must always be applied.

(B) The calculated forces must be multiplied by 0.7, and the load combinations from ASCE 7 need not to be used.

(C) The calculated forces must be multiplied by 1.0, and the load combinations from ASCE 7 must always be applied.

(D) ASD requires that only dead load and earthquake loads be added to the calculated forces, with no consideration of relative displacement.

PROBLEM 5.4 *Seismic Force for an Equipment when Relocated*

A generator that is currently located $15\,ft$ above the base level of a Category IV hospital, with an importance factor of 1.5, is to be relocated to the roof of the $90\,ft$ building. The generator was originally designed to withstand a horizontal seismic force of $15\,kip$ at its current location. What is the new horizontal seismic force the generator will be subjected to once relocated to the roof?

(A) 15.0 kip

(B) 22.5 kip

(C) 27.75 kip

(D) 33.75 kip

PROBLEM 5.5 *Parapet Out-of-Plane Seismic Force*

A 6 ft tall 12 in thick parapet made of 150 pcf reinforced concrete is located at a tall building's roof with an importance factor of $I_p = 1.0$.

With short spectral acceleration of $S_{DS} = 0.65$, what is the moment for a 1 ft wide strip of this parapet at its base generated from horizontal, out-of-plane, seismic forces?

    (A) 350 lb. ft per ft

    (B) 700 lb. ft per ft

    (C) 2,100 lb. ft per ft

    (D) 2,800 lb. ft per ft

PROBLEM 5.6 *Seismic Relative Displacement Between Two Buildings*

What is the seismic relative displacement for Category III buildings shown below with architectural features installed between the roofs of two adjacent buildings?

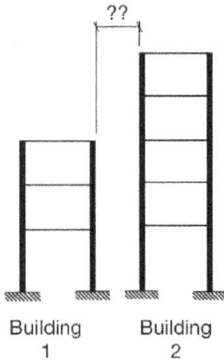

	Amplified displacement	
Floor No	**Building 1**	**Building 2**
1	1.3 in	1.5 in
2	2.3 in	2.5 in
3 (or roof of bldg. 1)	3.1 in	3.4 in
4	–	4.2 in
Roof of bldg. 2	–	5.1 in

    (A) 18.7 in

    (B) 12.3 in

    (C) 10.3 in

    (D) 8.1 in

PROBLEM 5.7 *Component Anchor*

For the below plumbing works, the linear operating weight of the pipe that is hung from the basement concrete ceiling is 20 lb/ft and it is anchored at 3 ft intervals. The building is assigned to Seismic Design Category B with $I_p = 1.0$ and $S_{DS} = 0.8$.

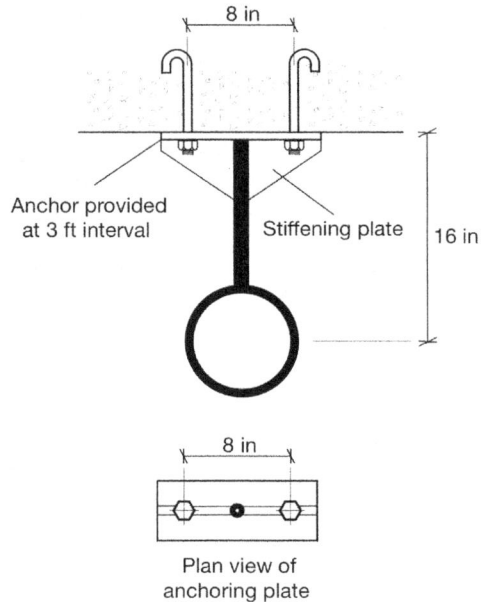

Using the ASD load combination No. 8 from the ASCE 7 Standard, what is the maximum tension force that any of the two bolts of the above shown anchor should be designed for?

    (A) 40 lb

    (B) 55 lb

    (C) 70 lb

    (D) 85 lb

PROBLEM 5.8 *Sprinkler Clearance*

The clearance between a flexible hose sprinkler system and any other distribution system in the same vicinity is:

    (A) 3 in

    (B) 5 in

    (C) 9 in

    (D) *Not Applicable*

PROBLEM 5.9 *Suspended Ceilings*

What is the area for a suspended ceiling's acoustic tiles that can go without providing a seismic separation joint to comply with the requirements of the ASCE 7 Standard for a building assigned to SDC D?

(A) $1,250\ ft^2$

(B) $1,750\ ft^2$

(C) $2,500\ ft^2$

(D) $3,000\ ft^2$

PROBLEM 5.10 *Mechanical and Electrical Components*

Which statement is correct regarding the seismic design of mechanical and electrical nonstructural components?

(A) Wet-side components, such as boilers and chillers, are more flexible than airside components.

(B) Components that must remain operational after an earthquake are classified as designated seismic systems and must be designed to withstand seismic forces.

(C) Neoprene pads have a lower response modification factor due to their higher ductility compared to spring base.

(D) All of the above.

PROBLEM 5.11 *Tank Freeboard*

A $25\ ft$ deep, $80\ ft$ diameter, circular tank is placed above ground with soil class B has been assigned a risk category IV with the following spectral parameters $S_S = 0.25$, $S_1 = 0.1$, $T_L = 6\ sec$ and a sloshing period of $T_c = 2.5\ sec$. The freeboard requirement for this tank is:

(A) $1.1\ ft$

(B) $2.7\ ft$

(C) $3.3\ ft$

(D) $4.4\ ft$

PROBLEM 5.12 *Loading Direction (2)*

When determining the critical load cases for nonbuilding structures with horizontal projections, what would be one possible loading direction to be considered?

(A) Seismic forces are applied at 100% in one direction, 30% in the orthogonal direction, and 30% in the vertical direction.

(B) Seismic forces are applied at 50% in one direction, 50% in the orthogonal direction, and 50% in the vertical direction.

(C) Seismic forces are applied at 30% in one direction, 100% in the orthogonal direction, and 100% in the vertical direction.

(D) Seismic forces are applied at 100% in one direction, 100% in the orthogonal direction, and 100% in the vertical direction.

PROBLEM 5.13 *Seismic Parameters*

Which of the following Seismic Design Categories is/are exempted from seismic design requirements for nonstructural components?

(A) SDC A

(B) SDC B

(C) SDC C

(D) SDC A + SDC B

PROBLEM 5.14 *Height Limitation*

The height limit (with no permitted height increase) for a steel intermediate moment frame for a pipe rack in SDC D, where field moment connections are made of bolted end plates is:

(A) $35\ ft$

(B) $65\ ft$

(C) $100\ ft$

(D) $160\ ft$

PROBLEM 5.15 *Buildings Not Covered in the ASCE 7 Standard*

The following buildings' seismic design is not covered in the ASCE 7 Standard?

(A) Electrical transmission towers

(B) Cast in place silos

(C) Hydraulic structures

(D) A + C

PROBLEM 5.16 *Accidental Torsional Moment for a Nonbuilding Structure*

Given that a proper analysis is performed, and all masses of both structural and nonstructural elements are accounted for in the modeling and analysis, which of the following conditions applies to accidental torsion for nonbuilding structures that are similar to buildings?

(A) Accidental torsion requirements need to be considered when $R \leq 3.5$, and eccentricity between CM and CR is larger than 5%.

(B) Accidental torsion requirements need to be considered when $R \leq 3.5$ eccentricity between CM and CR is lesser than 5%.

(C) Accidental torsion requirements need to be considered when $R > 3.5$, and eccentricity between CM and CR is larger than 5%.

(D) Accidental torsion requirements need to be considered when $R > 3.5$ eccentricity between CM and CR is lesser than 5%.

PROBLEM 5.17 *Nonbuilding Structure Supported by Another Structure*

A nonbuilding structure that is not similar to buildings is supported by another structure. If the weight of the nonbuilding structure is 30% of the total seismic weight of both the supporting structure and the nonbuilding structure, and its fundamental period is $T = 0.08 \, sec$, which of the following methods apply to this case?

(A) The supporting structure to be designed per Chapter 12 or Section 15.5 of ASCE 7.

(B) The supporting structure to be designed per Chapter 12 or Section 15.5 of ASCE 7 with R of the nonbuilding structure.

(C) The combined structure shall be modeled together per Section 15.5 of ASCE 7 with R the lesser of both.

(D) The combined structure shall be modeled together per Chapter 15.5 of ASCE 7 with R the larger of both.

PROBLEM 5.18 *Ground Supported Storage Tanks for Granular Material*

For ground supported storage tanks that holds granular material, it is allowed to reduce the effective density of the material that it holds by how much?

(A) 10%

(B) 15%

(C) 20%

(D) 25%

PROBLEM 5.19 *Nonbuilding Structure Design Force Calculation (1)*

A mechanically anchored, prestressed concrete, with reinforced nonsliding base, ground supported tank is lifted and placed on a 35 ft tall steel concentrically ordinary frame in an SDC C. The fundamental period of the tank is 0.05 sec and the supporting frame is 0.75 sec and $T_L = 2 \, sec$.

The operating weight for this tank is 800 kip, while the weight of the supporting structure is 100 kip.

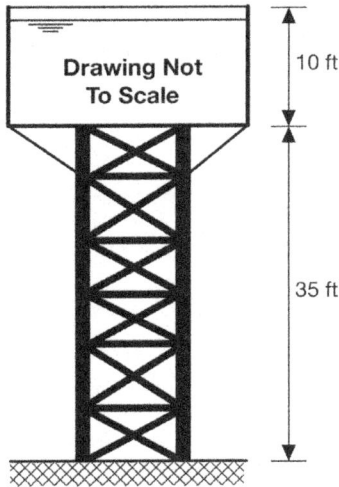

With an importance factor of 1.5 along with $S_{DS} = 0.55$ & $S_{D1} = 0.35$, What is the seismic horizontal force the attachment of the tank to the supporting structure should be designed for (*)?

(A) 100 *kip*

(B) 200 *kip*

(C) 300 *kip*

(D) 400 *kip*

(*) Extra solution for the design of base shear for the supporting structure is provided in the solution section.

PROBLEM 5.20 *Nonbuilding Structure Design Force Calculation (2)*

A prestressed concrete with reinforced nonsliding base and ground supported tank in an SDC C. The fundamental period of the tank is 0.05 *sec*.

The operating weight for this tank is 800 *kip*.

With an importance factor of 1.5 along with $S_{DS} = 0.55$, What is the seismic base shear for this tank?

(A) 100 *kip*

(B) 200 *kip*

(C) 300 *kip*

(D) 400 *kip*

PROBLEM 5.21 *Tanks and Vessels Support*

Under the category of tanks and vessels, and when those are supported on grillage beams, the vertical reaction caused by weight and overturning shall be adjusted. How should this rection be adjusted?

(A) Decrease by 20%

(B) Increase by 20%

(C) Decrease by 10%

(D) Increase by 10%

PROBLEM 5.22 *Load Combinations for Storage Racks*

Which of the following load combination should be considered as part of the design load combinations for storage racks?

(A) Weight of racks plus every storage level loaded to 50% of their rated load capacity.

(B) Weight of racks plus every storage level loaded to 67% of their rated load capacity.

(C) Weight of racks plus the highest rated rack loaded to 100%.

(D) Weight of racks plus the highest rated rack loaded to 67%.

PROBLEM 5.23 *Load Combinations for Secondary Containment Systems*

Which of the following load combination effects should be considered for the design of secondary containment systems?

(A) 100% of MCE (*) when empty and two thirds of MCE when full.

(B) Two thirds MCE (*) when empty and two thirds of MCE when full.

(C) 100% of MCE (*) when empty and 100% of MCE when full.

(D) Two thirds MCE (*) when empty and 100% of MCE when full.

(*)MCE: Maximum Considered Earthquake ground motion.

PROBLEM 5.24 *Tank Impulsive Load*
A reinforced concrete with reinforced nonsliding base and ground supported tank is located in an SDC B with $S_{D1} = 0.65$ and $S_{DS} = 1.0$ along with an importance factor of $I_e = 1.0$.

The fundamental period of the tank structure and the impulsive component from the liquid content has been determined via modelling as 0.09 *sec*.

The impulsive component of the contained liquid along with the tank other elements (i.e., roof, equipment, etc.) weighs 800 *kip*.

What is the base shear caused by the impulsive component from the weight of the tank and liquid content?

(A) 100 *kip*
(B) 200 *kip*
(C) 300 *kip*
(D) 400 *kip*

PROBLEM 5.25 *Ground-Supported Storage Tanks for Granular Material*
Which of the following statement is true when it comes to lateral load determination for ground-supported storage tanks for granular material?

(A) Loss of intergranular friction during shaking increases lateral pressure and hoop stress.

(B) Once compacted, temperature changes increase hoop stresses.

(C) Intergranular friction transfers seismic shear directly to foundation.

(D) All the above.

SOLUTION 5.1

According to ASCE 7 Standard Section 13.3.1.1 (Section 5.3.3 here), horizontal forces are applied independently in both horizontal directions along with the service and operating loads.

Correct Answer is (A)

SOLUTION 5.2

According to ASCE 7 Standard Section 13.2.2 (Section 5.3.2 here), manufacturers must submit certification for designated seismic systems equipment in Seismic Design Categories C through F to confirm that the equipment will remain operable after an earthquake.

Correct Answer is (B)

SOLUTION 5.3

According to ASCE 7 Standard Section 13.1.8 (Section 5.3.2 here), The calculated forces must be multiplied by 0.7. The allowable stress design combination shall consider dead, live, operation and earthquake load in addition to those in the reference documents, and the load combinations from ASCE 7 Chapter 2 need not to be used.

Correct Answer is (B)

SOLUTION 5.4

The horizontal seismic force for structural components depends on the height they are placed at in relation to the height of the building (z/h) as discussed in Section 5.3.3 of this chapter.

$$F_p = \frac{0.4 a_p S_{DS} W_p}{\left(\frac{R_p}{I_p}\right)}\left(1 + 2\frac{z}{h}\right)$$

While the first part of this equation will not change for components placed in the same building with no changes in their S_{DS}, a_p, R_p

or, W_p, the second part with $\left(1 + 2\frac{z}{h}\right)$ is the only one affected with this relocation.

Seismic force before equipment relocation:

$$15\ kip = \frac{0.4 a_p S_{DS} W_p}{\left(\frac{R_p}{I_p}\right)}\left(1 + 2 \times \frac{15\ ft}{90\ ft}\right)$$

$$\rightarrow \frac{0.4 a_p S_{DS} W_p}{\left(\frac{R_p}{I_p}\right)} = 11.25\ kip$$

Seismic force after equipment relocation:

$$F_p = \frac{0.4 a_p S_{DS} W_p}{\left(\frac{R_p}{I_p}\right)}\left(1 + 2\frac{z}{h}\right)$$

$$= 11.25 \times \left(1 + 2 \times \frac{90\ ft}{90\ ft}\right)$$

$$= 33.75\ kip$$

Correct Answer is (D)

SOLUTION 5.5

We need to determine the weight of this parapet, based upon which, we can determine its horizontal force. Its weight for a $1\ ft$ wide strip and per vertical linear ft is:

$$w = 150\frac{lb}{ft^3} \times 1\ ft \times \frac{12\ in}{12\ in/ft} = 150\ lb/ft$$

Referring to Table 13.5-1 for architectural components, "parapet" braced below its centroid is $a_p = 2.5$ and $R_p = 2.5$.

With the parapet at the roof, force F_p is calculated as follows:

$$F_p = 3 \times \frac{0.4 a_p S_{DS} W_p}{\left(\frac{R_p}{I_p}\right)}$$

$$= 3 \times \frac{0.4 \times 2.5 \times 0.65 \times 150}{\left(\frac{2.5}{1.0}\right)}$$

$$= 117\ lb/ft$$

The following upper and lower criteria should be maintained for F_p:

$$(0.3 S_{DS} I_p W_p = 29.3) \leq F_p \leq (1.6 S_{DS} I_p W_p = 156)$$

$$\rightarrow ok$$

Based on the below uniform distribution diagram, the moment generated at the base of this architectural component is:

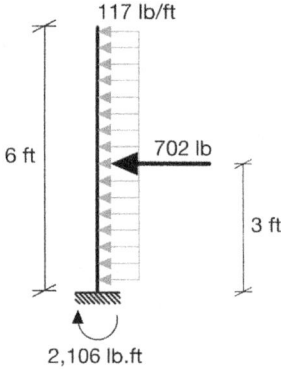

117 lb/ft

6 ft   702 lb

3 ft

2,106 lb.ft

$$M = 117\frac{lb}{ft} \times 6\,ft \times 3\,ft = 2{,}106\,lb.ft\;per\;ft$$

Correct Answer is (C)

SOLUTION 5.6

In reference to Section 13.3.2 of the ASCE 7 Standard, or Section 5.3.4 of this chapter:

$$D_p = |\delta_{x1}| + |\delta_{y2}|$$

Where level 'x' of building 1 represents the roof of that building, and level 'y' of building 2 represents the roof of building 2.

$$D_p = 3.1 + 5.1 = 8.2\,in$$

Also, using an importance factor $I_e = 1.25$ for Category III structure as noted in Table 1-5 of Chapter 1 here:

$$D_{pl} = I_e D_p = 1.25 \times 8.2 = 10.25\,in$$

Correct Answer is (C)

SOLUTION 5.7

Equations in Section 13.3.1 of ASCE 7 will be used, where the operating weight is $W_p = 3\,ft \times 20\,lb/ft = 60\,lb$. Moreover, the following is collected from Table 13.6-1 for plumbing works: $a_p = 1, R_p = 2.5\,\&\,\Omega_o = 2.0$, while $Z = 0$ given the location of this anchor is below base level – i.e., at the basement.

$$F_p = \frac{0.4a_pS_{DS}W_p}{\left(\frac{R_p}{I_p}\right)}\left(1+2\frac{z}{h}\right)$$

$$= \frac{0.4\times1\times0.8\times60}{\left(\frac{2.5}{1}\right)}\left(1+2\times\frac{0}{h}\right)$$

$$= 7.7\,lb$$

The vertical component will either be in the upward or the downward direction, the worst case is the downward direction, which shall be reflected in the load combination identified in this question.

$$= \mp 0.2S_{DS}W_p = 0.2 \times 0.8 \times 60 = 9.6\,lb$$

Load combination No. 8 shall include overstrength in it since this is anchored to concrete, where $E_{mh} = \Omega_o E_h$, as follows:

$$D + 0.7E_v + 0.7E_{mh}$$

D is the dead load $= W_p$ in this case, E_v is the effect of seismic vertical loads, which is generated from the 9.6 lb identified earlier, and E_{mh} is the effect on the bolt from the horizontal load F_p including overstrength.

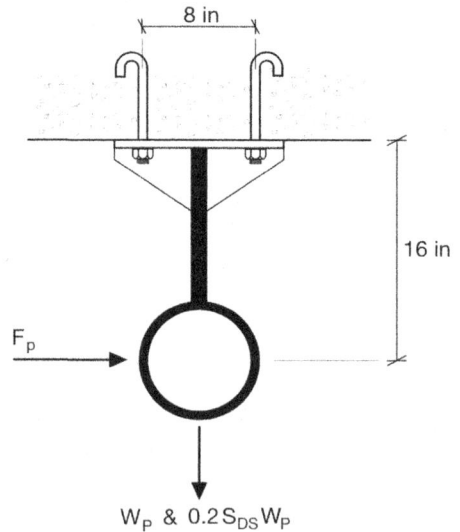

8 in

16 in

F_p

W_P & $0.2\,S_{DS}\,W_P$

From the above diagram, the bolt on the left side will be subjected to the maximum tension force. The effect of the vertical component on the bolt to the left is

$$E_v = 9.6/2 = 4.8\,lb \downarrow$$

The effect the horizontal force on the bolt to the left can be calculated by taking moment around the bolt to the right, or by using the method of couples.

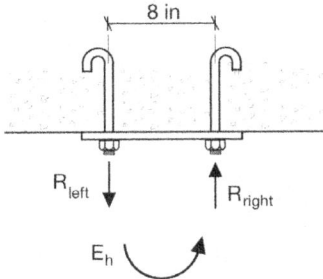

For R_{left} only:

$$8\ in \times E_h = 16\ in \times 7.7\ lb$$

$$E_h = 15.4\ kip \downarrow$$

Overstrength to be included as follows (See comment "c" under ASCE 7 Table 13.6-1):

$$E_{mh} = \Omega_o E_h = 2.0 \times 15.4 = 30.8\ lb \downarrow$$

ASD Load combination 8 is as follows (all forces are pointing downwards):

$$R_{left} = D + 0.7E_v + 0.7E_{mh}$$
$$= \frac{60}{2} + 0.7 \times 4.8 + 0.7 \times 30.8$$
$$= 54.9\ lb$$

Correct Answer is (B)

SOLUTION 5.8
According to ASCE 7 Standard Section 13.2.3.1, the exception section mentions that flexible hose sprinkler system has no requirements for clearance with any other system.

Correct Answer is (D)

SOLUTION 5.9
According to ASCE 7 Standard Section 13.5.6.2.2, for SDC D through F, a seismic separation joint shall be provided when the area exceeds $2,500\ ft^2$. This information is also found in Table 5-2 in this chapter.

Correct Answer is (C)

SOLUTION 5.10
Reference is made in this question to Section 5.3.6 of this chapter, also, this information can be found in the commentary section of ASCE 7 Section 13.6.

Option A is incorrect because wet-side components (like boilers and chillers) are more rigid than airside components. For the removal of doubt compare R_p in ASCE 7 Table 13.6-1.

Option B is correct. Components that need to remain operational after an earthquake are classified as designated seismic systems, which must be designed to withstand seismic forces, in addition to proper support and attachment.

Option C is incorrect because it falsely claims that Neoprene pads have a lower response modification factor due to their higher ductility. Neoprene pads have a higher response modification factor due to their higher ductility. You can verify this information in ASCE 7 Table 13.6-1 under Vibration-Isolated Components & Systems·

This makes B the only correct statement.

Correct Answer is (B)

SOLUTION 5.11
Reference is made to the ASCE 7-16 Standard Section 11.7, Section 15.6.5.1 and Section 15.7.6.1 to determine the freeboard requirement for this tank.

Prior to that, we will determine the spectral parameters related to this tank using Sections 11.4.4 and Section 11.4.5 of the ASCE 7-16 Standard (see Table2 1-2 & 1-3 of this book) where $F_a = 0.9$ and $F_v = 0.8$ as collected from Table 11.4-1 and Table 11.4-2 of the ASCE Standard.

$$S_{DS} = \frac{2}{3} S_{MS}$$

$$= \frac{2}{3} F_a S_S$$

$$= \frac{2}{3} \times 0.9 \times 0.25$$

$$= 0.15$$

$$S_{D1} = \frac{2}{3} S_{M1}$$

$$= \frac{2}{3} F_v S_1$$

$$= \frac{2}{3} \times 0.8 \times 0.1$$

$$= 0.053$$

ASCE 7 Tables 11.6-1 & 11.6-2 (see Table 1-7 & 1-8 of this book) with the above information provides a Seismic Design Category SDC A.

Section 11.7 of ASCE 7 Standard with risk category IV and SDC A points you to Section 15.6.5.1 to determine freeboard.

Section 15.6.5.1 of the ASCE 7 Standard provides the following equation for calculating the freeboard δ_s as follows:

$$\delta_s = 0.42 D S_{ac}$$

Where D is diameter and S_{ac} is calculated using Equation 15.7-10 from ASCE 7 Standard with sloshing period $T_c < T_L$:

$$S_{ac} = \frac{1.5 S_{D1}}{T_c} \le S_{DS}$$

$$= \frac{1.5 \times 0.053}{2.5} \le S_{DS}$$

$$= 0.032 \le 0.15$$

$$\delta_s = 0.42 D S_{ac}$$

$$= 0.42 \times 80\,ft \times 0.032$$

$$= 1.1\,ft$$

You can also confirm the minimum requirement in ASCE 7 Table 15.7-3.

Correct Answer is (A)

SOLUTION 5.12
Referring to Section 5.4.5 in this book, also Section 15.1.4.1.1 of the ASCE 7 Standard, one of the loading directions that should be considered is "100% in one direction, 30% in the orthogonal direction, and 30% in the vertical direction".

Correct Answer is (A)

SOLUTION 5.13
Reference is made to Section 11.7 of the ASCE 7 Standard. This section specifies that nonstructural components in SDC A are exempt from Seismic design requirements.

Correct Answer is (A)

SOLUTION 5.14
Referring to Table 15.4-1 of the ASCE 7 Standard, while observing the table's commentary section at footnote "h", it states that steel ordinary and intermediate moment frames are "permitted in pipe racks up to $65\,ft$ where moment joints of field connections are constructed of bolted end plates"

Correct Answer is (B)

SOLUTION 5.15
Reference is made to the commentary section of the ASCE 7 Standard section C15.1.1. This information is also mentioned in this book introduction to Section 5.4.

Based on the above, the following nonbuilding structures are not covered in the ASCE 7 Standard:

(A) Electrical transmission towers

(C) Hydraulic structures

Correct Answer is (D)

SOLUTION 5.16

Reference is made to Section 15.4.1, point No. 5, of the ASCE 7 Standard. This point states that when mass locations for the structure and all elements are accounted for in the analysis, the accidental torsion requirements need not be considered for rigid nonbuilding structures, for the following:

- Not similar to buildings when $R \leq 3.5$.

- Similar to buildings when $R \leq 3.5$, also, when any of the following conditions are met:

  i. The eccentricity between CM and CR is larger than 5%.

  ii. No horizontal torsional irregularity with at least two lines of lateral resistance. One of those two lines shall be 20% away from CM both sides.

Based on the above, option A provides the most accurate representation for the above description.

Correct Answer is (A)

SOLUTION 5.17

Reference to this question can be made to Figure 5-6 of this chapter, or to Section 15.3.2.2 of ASCE 7 Standard. Based on this, Option C: "The combined structure shall be modeled together per Section 15.5 of ASCE 7 with (R) the lesser of both"

Correct Answer is (C)

SOLUTION 5.18

As explained in ASCE 7 Section C15.7.9.3.3 (also see Table 5-10), energy loss through intergranular movement and grain to grain friction reduces the mass subjected to horizontal acceleration.

A factor of 0.8 is applied to most of the granular materials' densities, and 0.9 is applied for metal ores and aggregate.

Based on this, the reduction factor in this question is 0.8.

Correct Answer is (C)

SOLUTION 5.19

Refer to the design diagram provided in Figure 5-6 of this chapter. This diagram, for a nonbuilding structure not similar to buildings, supported by other structures, and with $T < 0.06$ and its seismic weight > 25% of the overall structure and supporting structure, requires the design of its attachments to use R from Table 15.4-2 $R = 2$ and $a_p = 1.0$ per this requirement (also stated in ASCE 7 Standard Section 15.3.2.1) (*).

Per the abovementioned diagram (or the referenced section of the ASCE 7 Standard), the force equation F_p from Chapter 13, with the new location of this tank at the roof, shall be used:

$$F_p = 3 \times \frac{0.4 a_p S_{DS} W_p}{\left(\frac{R_p}{I_p}\right)}$$

$$= 3 \times \frac{0.4 \times 1.0 \times 0.55 \times 800}{\left(\frac{2.0}{1.5}\right)}$$

$$= 396 \ kip$$

The following upper and lower criteria should be maintained for F_p:

$$\left(0.3 S_{DS} I_p W_p = 198 \ kip\right) \leq F_p$$

$$\leq \left(1.6 S_{DS} I_p W_p = 1{,}056 \ kip\right)$$

$$\rightarrow ok$$

Correct Answer is (D)

(*) The supporting structure for this tank should be designed per Chapter 12 or 15.5 of ASCE 7 with R used should be for the (nonbuilding) supporting structure.

In which case, R is taken from Table 15.4-1 of the ASCE 7 Standard as $R = 3.25$ for steel ordinary concentrically braced frames with no permitted height increase.

With $T > 0.06$, the base shear for this structure V is calculated as follows:

$$V = C_s W$$

Assume $T < T_s$ just for this part:

$$C_s = S_{DS} \times \frac{I_e}{R} = 0.55 \times \frac{1.5}{3.25} = 0.25$$

$$0.044 S_{DS} I_e = 0.036 \leq C_s \rightarrow ok$$

$$V = C_s W = 0.25 \times (100 + 800) = 225 \ kip$$

SOLUTION 5.20
Reference here is made to the design diagram provided in Figure 5-6 of this chapter. This diagram, for a nonbuilding structure not similar to buildings, with $T < 0.06$ requires the base shear to be calculated as follows:

$$V = 0.3 S_{DS} W I_e$$
$$= 0.3 \times 0.55 \times 800 \times 1.5$$
$$= 198 \ kip$$

Correct Answer is (B)

SOLUTION 5.21
In reference to Table 5-7 of this chapter, or to Section 15.5.5.1 (b) of ASCE 7 Standard, and in order to account for the nonuniform support caused by the grillage beams' support, this reaction should be increased by 20%.

Correct Answer is (B)

SOLUTION 5.22
In reference to Table 5-7 of this chapter, or to Section 15.5.3.3.2 of ASCE 7 Standard, storage racks shall be designed for each of the following conditions:

a. Weight of racks plus every storge level loaded to 67% of rated capacity.

b. Weight of racks plus the highest storage level (*not the highest rated*) loaded to 100%.

Correct Answer is (B)

SOLUTION 5.23
In reference to Table 5-8 of this chapter, or to Section 15.6.5 of ASCE 7 Standard, it states that "secondary containment systems shall be designed to withstand the effects of the maximum considered earthquake ground motion where empty and two-thirds … where full".

Correct Answer is (A)

SOLUTION 5.24
Reference is made to Table 5-9 of this chapter, and to Section 15.7.6.1 of ASCE 7 Standard.

Impulsive base shear (V_i) is calculated as follows:

$$V_i = \frac{S_{ai} W_i}{R/I_e}$$

The response modification factor R for concrete vessel or tanks with nonsliding base is $R = 2$ per Table 15.4-2 of ASCE 7 Standard.

The spectral acceleration multiplier for the impulsive component $S_{ai} = S_{DS} = 1.0$ when $T_i \leq T_s$. See below:

$$T_s = \frac{S_{D1}}{S_{DS}} = \frac{0.65}{1.0} = 0.65 \ sec > T_i \ (= 0.09 \ sec)$$

$$V_i = \frac{S_{ai} W_i}{R/I_e} = \frac{1.0 \times 800}{2/1} = 400 \ kip$$

Correct Answer is (D)

SOLUTION 5.25
Reference is made to Table 5-10 of this chapter, first item, or to Section 15.7.9.1 of ASCE 7 Standard, all mentioned statements are correct.

Correct Answer is (D)

CHAPTER

6

Design Codes, Detailing and Requirements

6.1. Purpose of this Chapter

This chapter summarizes key components of several important building codes and standards, including the 2022 California Building Code (CBC), Title 24 Part 2, the 2016 TMS 402/602, and the 2021 AWC SDPWS (Special Design Provisions for Wind and Seismic), all of which are required references for the California seismic practice exam.

While this book offers an overview of the critical concepts and requirements of these codes, it also directs you to the specific chapters within these documents, encouraging you to explore them in more detail. For the best preparation, it is recommended that you review these codes at your own pace to fully grasp the necessary knowledge for the exam.

Additionally, this book includes some relevant requirements from other important codes and manuals, such as ACI 318 for concrete design and AISC 341 for steel design. Although these are not explicitly listed as exam references, they are cited in the primary codes, and familiarity with them will provide a more comprehensive understanding of the seismic detailing needed to design buildings capable of withstanding seismic forces.

6.2. California Building Code
6.2.1.   Overview

This chapter revisits key information presented in earlier chapter of this book to address specific provisions of the California Building Code (CBC). For instance, Section 1.6 of Chapter 1 introduced the four risk categories outlined in ASCE 7. In the subsequent sections of this chapter, we will discuss CBC requirements that build upon these categories, incorporating additional factors such as occupancy type and the number of occupants, in order to refine and finalize the risk categories. Moreover, while previous chapters highlighted building heights limits and provided other related requirements on buildings and nonbuilding structures, this chapter provides key variations specific to the state of California, as mandated by some of the CBC amended chapters – i.e., chapters marked with an "A" (this is further explained in Section 6.2.2).

Lastly, you can access the CBC code on their official website (see references section at the end of this chapter). To ensure you are using copyrighted material legitimately, you can subscribe to their service for a nominal fee which allows you to print relevant sections.

6.2.2.   Buildings Occupancy and Risk Category

Risk categories for buildings are determined to assess their potential risks based on their intended use and the type of occupancy they are designed to accommodate. Each occupancy is identified by a letter and a number, as explained below. The letter represents the broad occupancy group (the first letter of each occupancy type), while the number indicates a subclassification that further defines the specific activity or use. This classification system is detailed in Chapter 3 of the California Building Code and is summarized here as follows:

i. **Assembly (Group A):**
A-1: Theaters, concert halls, and similar places of assembly.
A-2: Banquet halls, nightclubs, restaurants, and similar places of assembly where food and drink are served.
A-3: Places of worship, theaters, auditoriums, art galleries, and other assembly uses not including food and drink service.
A-4: Arenas, stadiums, and places for indoor sporting events and other assembly uses.
A-5: Amusement parks, fairgrounds, and similar outdoor assembly spaces.

ii. **Business (Group B):**
Offices, professional services, banks, and other business-related uses not listed under other categories.

iii. **Educational (Group E):**
Schools for children up to the 12th grade, day care centers, and other educational facilities.

iv. **Factory and Industrial (Group F):**
F-1: Moderate hazard industrial uses (e.g., manufacturing or processing where the risk of fire is moderate).
F-2: Low hazard industrial uses (e.g., light manufacturing, storage, etc.).

v. **High Hazard (Group H):**
H-1 to H-5: Varying levels of high hazard occupancies depending on the materials used or stored (e.g., chemicals, explosives, or highly flammable materials).

Check Table 307.1 of the CBC code for more details on this group.

vi. **Institutional (Group I):**
I-1: Not used – check R-2.1 instead.
I-2: Hospitals, nursing homes, and other health care facilities where occupants are typically not capable of self-preservation.
I-3: Prisons, jails, detention centers, and similar institutions where occupants are under constant supervision and security.
I-4: Day care facilities for more than six people where the occupants require care by people other than parents and guardians.

vii. **Mercantile (Group M):**
Retail stores, shopping centers, and other similar spaces where goods are sold.

viii. **Residential (Group R):**

R-1: Hotels, motels, and similar transient lodging facilities.

R-2: Apartment buildings, condominiums, dormitories, and other residential buildings where the occupants are nontransient (i.e., resident for more than one month).

  R-2.1: Supervised environment that provides personal care.

  R-2.2: Provided by CDCR (California Department of Corrections and Rehab).

R-3: Single-family homes, duplexes, and similar small residential buildings.

  R-3.1: Government facilitated care for less than six residents.

R-4: Residential care/assisted living facilities (e.g., small group homes with non-ambulatory or elderly residents).

ix. **Storage (Group S):**

S-1: Moderate hazard storage (e.g., warehouses storing items with moderate fire risk).

S-2: Low hazard storage (e.g., storage of items with low fire risk).

x. **Utility and Miscellaneous (Group U):**

Buildings or structures that do not fall under other classifications, such as agricultural buildings, carports, greenhouses, or other accessory structures.

Based on the above description of buildings occupancy and the various use of structures, the risk categories presented in the ASCE 7 Standard – discussed here in Section 1.6 of this book – are amended in the CBC Chapter 16 and 16A. We are only presenting risk category III and IV of Chapter 16 (See Section 6.2.2 here to understand the difference between Chapter 16 and 16A of CBC 2022):

Risk Category III: Buildings and structures that represent a substantial hazard to human life in the event of failure. Examples include:

  o   Buildings with public assembly spaces with an occupant load greater than 300.
  o   Buildings containing one or more public assembly spaces each with occupant load greater than 300 and cumulative occupant load greater than 2,500.
  o   Buildings containing Group E (educational) or Group I-4 (daycare) occupancies with an occupant load greater than 250.
  o   Buildings with students above grade 12 with occupant load greater than 500.
  o   Buildings containing Group I-3 (prisons, jails, detention centers, etc.)
  o   Buildings containing Group I-2 (e.g., hospitals) with greater than 50 care recipient or not having emergency surgery or treatment facilities.
  o   Buildings with an occupant load greater than 5,000.
  o   Power-generating stations, water treatment facilities, and other public utility facilities not included in Risk Category IV.
  o   Buildings and structures containing quantities of toxic, highly toxic, or explosive materials that exceed certain thresholds and pose a substantial risk to the public if released.

Risk Category IV: Buildings and structures designated as essential facilities. Examples include:

- o Buildings containing Group I-2 (e.g., hospitals) <u>with emergency surgery or emergency treatment facilities.</u>
- o Fire, rescue, ambulance, and police stations.
- o Emergency vehicle garages and shelters.
- o Emergency preparedness, communication or operations centers.
- o Power-generating stations and other public utility facilities required as emergency backup for risk category IV structures.
- o Buildings and structures containing hazardous materials in quantities sufficient to pose a threat to the public if released which exceed quantities allowed by the code.
- o Aviation control towers, air traffic control centers and emergency hangers.
- o Buildings and structures with critical national defense functions.
- o Water storage facilities with all relevant structures which belong to them.

Where <u>multiple occupancies are provided within the same building</u> or structure, the building or the structure shall be <u>assigned with the highest risk category.</u>

The Occupant Load Factor (OLF) can be found in Table 1004.5 of the CBC and should fall between 30 to 50 square feet per person. If you subscribe to the ICC website, you can access and print a copy of this table to bring with you to the exam. Additionally, the following equation can be used to determine the OLF for malls per CBC Section 402.8.2.1. The number of occupants, derived from the OLF, is essential for determining the required means of egress for the mall.

$$OLF = 0.00007GLA + 25$$

The means of egress must be provided based on the Gross Leasable Area (GLA) of the mall, excluding anchor buildings, whether the mall is enclosed or open.

6.2.3.   The Matrix Adoption Table and the Amended Chapters in CBC 2022

In the 2022 California Building Code, you will notice that some chapters are marked with an "A" (e.g., Chapter 16A, 17A, 18A, etc.). Additionally, there is a Matrix Adoption Table at the beginning of each chapter, which provides important information about the application of each chapter. This section briefly explains what these markings mean, how they differ, and why they are used.

<u>Chapters with an "A" Designation:</u>

Chapters marked with an "A" include special provisions that apply specifically to state-regulated buildings and structures. These chapters contain additional requirements or modifications to the general provisions found in the general chapters without the "A" designation. The "A" chapters are intended to address the unique needs of state-regulated projects, ensuring they meet higher standards of safety, compliance, and performance.

<u>The Matrix Adoption Table:</u>

At the beginning of each chapter, you will find a Matrix Adoption Table, which clarifies the adoption status and applicability of the code sections to state-regulated or nonregulated buildings' agencies. To illustrate this, let us take a look at Chapter 16. Chapter 16 applies to OSHPD 3 (licensed clinics other than those listed in OSHPD 1, 1R, and 2). Chapter 16A, however, applies to OSHPD 4 (correctional treatment centers).

For more detailed information on adopting agencies abbreviations, refer to the Appendix of this chapter in this book, or the preface chapters of the CBC 2022.

Color Coding and Additional Markings:

In the code, you will also notice color coding that relates to different sections of the code. This coding indicates applicability to specific entities. Additionally, you will observe more abbreviations or markings used in the code, which are also explained in the preface chapters of the CBC.

6.2.4.   Structural Observations and Inspections Requirements

The 2022 California Building Code (CBC) outlines specific requirements for special and continuous inspections in Sections 1704 and 1705. These sections ensure that critical structural elements and systems are constructed according to the approved plans and specifications.

Section 1704 and 1704A focus on the general requirements for special inspections and tests. It mandates that special inspections be conducted by qualified individuals or agencies to verify that work complies with the approved construction documents. This includes inspections for materials, installation, fabrication, erection, and placement of components requiring special expertise. The section also emphasizes the contractor's responsibility to coordinate these inspections and ensure that they are performed as required.

Also, pay a close attention to Section 1704.6.1 which specifies that observations shall be provided to risk categories III or IV, high rises, SDC E with more than two stories above grade.

Section 1705 and 1705A detail the specific types of work that require special inspections and tests. This includes inspections for soils, foundations, concrete, masonry, steel, wood, and other critical structural elements. Each type of inspection has its own set of criteria and standards that must be met. For example, concrete construction requires inspections for reinforcing steel placement, concrete mix design, and concrete placement and curing. Similarly, structural steel inspections cover welding, bolting, and material quality.

Special attention should be given to the following tables of the CBC code which outline the special and continuous inspection requirements for various structural elements:

1705A.2.1	Required Special Inspections and Tests of Steel Construction
1705.2.3 & 1705A.2.3	Required Special Inspections of Open-Web Steel Joists and Joist Girders
1705.3 & 1705A.3	Required Special Inspections and Tests of Concrete Construction
1705.5.3 & 1705A.5.3	Required Special Inspections of Mass Timber Construction
1705.6 & 1705A.6	Required Special Inspections and Tests of Soils
1705.7 & 1705A.7	Required Special Inspections and Tests of Driven Deep Foundation Elements
1705.8 & 1705A.8	Required Special Inspections and Tests of Cast-in-Place Deep Foundation Elements
1705.13.7 & 1705A.13.7	Required Inspections of Storage Rack Systems

Special inspections are required for certain critical structural elements and systems to ensure they comply with the approved construction documents. These inspections must be conducted by qualified special inspectors who have the necessary expertise in the specific area being inspected. Special inspectors are often independent of the construction team to provide an unbiased assessment. Continuous special inspections require the special inspector to be present on-site at all times during the execution of the work that requires inspection.

Based on the above, below is a list of required special inspections for seismic design as specified in **Section 1705A.13 of the 2022 California Building Code (CBC)**:

Structural Steel: Special inspection is required for steel SFRS buildings in SDC D, E & F including structs, collectors, cords and foundation elements.

Structural Wood: When SFRS is assigned to SDC D, E & F, continuous inspection is required during field gluing of SFRS elements, and periodic inspection is required for nailing, bolting, anchoring and other fastening elements of the SFRS including wood shear walls, wood diaphragms, drag struts, braces, shear panels and hold downs.

Cold-Formed Steel Light Frames: For SDC D, E & F, periodic special inspection is required for welding operations of SFRS elements, screw attachments, bolting, anchoring and other fastening elements including shear walls, diaphragms, collectors and hold downs.

Pluming, Mechanical and Electrical Components: Most of the anchorage periodic special inspection requirements for these elements are required for SDC D, E & F.

For special inspections for general buildings, per Section 1705.13: there are exceptions. These apply to light-frame construction $< 35\,ft$, or reinforced masonry/concrete frames $< 25\,ft$, and when $S_{DS} < 0.5$. Also, detached one- or two-family dwellings under two stories when they do not have certain irregularities. Apart from that, they are required for the following cases:

Structural Steel: Special inspection is required for steel SFRS buildings in SDC B, C, D, E & F using quality assurance methods of the AISC 341.

Structural Wood: When assigned to SDC C, D, E & F, continuous inspection is required during field gluing of SFRS elements, and periodic inspection is required for nailing, bolting, anchoring and other fastening elements of the SFRS including wood shear walls, wood diaphragms, drag struts, braces, shear panels and hold downs.

Cold-Formed Steel Light Frames: For SDC C, D, E & F, periodic special inspection is required for welding operations of SFRS elements, screw attachments, bolting, anchoring and other fastening elements including shear walls, diaphragms, collectors and hold downs.

Pluming, Mechanical and Electrical Components: Most of the anchorage periodic special inspection requirements for these elements are required for SDC C, D, E & F.

6.2.5.   Structural Design Seismic Requirements

The structural design requirements of the CBC code primarily reference relevant sections of the ASCE 7-16 Standard, particularly Chapter 16. However, the CBC also includes specific modifications to ASCE 7 in Chapter 16A and Section 1617 of Chapter 16 (pertaining to community colleges), due to their importance.

In this book, we have integrated some of the key requirements for both the state regulated buildings of CBC Chapter 16A, and the community colleges of CBC Chapter 16 Section 1617, into the relevant sections or tables of this book for easier reference, and they have been identified in *Italics* or shaded throughout. The following list provides a guide to the locations where CBC requirements are integrated in this book:

- ✓ Table 2-1 of this book that summarizes prohibitions for horizontal irregularities for specific SDCs includes items from the above two CBC sections in *italics*.

- ✓ Table 2-2 of this book that summarizes prohibitions for vertical irregularities for specific SDCs includes items from the above two CBC sections in *italics*.

- ✓ ASCE 7-16 Standard Table 12.2.-1 which provides design coefficients for the various SFRS systems have been shaded as noted in the Appendix of Chapter 2 of this book.

- ✓ Sections 3.4.2 and 3.4.3 of this book have been modified to include the vertical combinations requirements relevant to the above sections of the CBC in *italics*.

- ✓ Figure 3-8 in this book which depicts the requirements for two stage analysis has been modified with the addition of the CBC requirements in *italics*.

6.2.6.   Soils and Foundations Seismic Requirements

This section introduces the seismic requirements for earth structures in Chapters 18 and 18A of the California Building Code 2022, which focus on soils and foundations. Three tables summarize key requirements for both general buildings and state-regulated structures. Table 6-1 outlines the seismic requirements from Chapter 18 for general buildings, with references to shared requirements from Chapter 18A, which are also included in Table 6-3 for state-regulated buildings. Table 6-2 addresses the nonparticipating masonry wall requirements for foundation walls in CBC Chapter 18 with reference to TMS 402.

These tables provide a concise overview of the seismic provisions for both general and state-regulated structures. For further details you can refer to the relevant sections in the code.

Table 6-1 CBC Chapter 18 Specific Soils and Foundations Seismic Design Requirements

Element	SDC	Requirement	CBC Reference
Soil investigations	C to F	Must assess the potential for liquefaction, slope instability, differential settlement and surface displacement due to faulting.	1803.5.11 & 1803A.5.11
Concrete foundation walls	A & B	When designed in accordance with Table 1807.1.6.2, shall have no less than one *No.*5 bars provided around windows, doors and similar openings.	1807.1.6.2.1
	C to F	The use of Table 1807.1.6.2 to design walls not permitted.	
Masonry foundation walls	C	See requirements of Section 7.4.3 of TMS 402 (*)	1807.1.6.3.2
	D	See requirements of Section 7.4.4 of TMS 402 (*)	
	E & F	See requirements of Section 7.4.5 of TMS 402 (*)	
Retaining walls	-	Minimum safety factor of 1.1 for sliding and overturning under seismic conditions.	1807.2.3 & 1807A.2.3

Design Codes, Detailing and Requirements

Shallow footings ties	D, E & F	For Site Class E or F, footings shall be interconnected by ties to carry a force equal to the following – where (G) is the gravity load per footing: $$Min\left\{\left(\frac{G_{larger\ footing} \times S_{DS}}{10}\right) or \left(25\% \times G_{smaller\ footing}\right)\right\}$$	1809.13 & 1809A.13
Deep foundations	D, E & F	For Site Class E or F, they shall withstand maximum imposed curvature caused by seismic motion. Except for precast prestressed, or cast-in-place with long reinforcement ratio ≥ 0.005 extending full length.	1810.2.4.1 & 1810A.2.4.1
Helical piles	D, E & F	Capacity determined by at least two project specific pre-production tests for each soil profile. At least 2% of all production piles shall be proof tested to strength. All helical pile materials that are subject to corrosion shall include at least 1/16 in corrosion allowance. Shall not be considered as carrying any horizontal loads.	1810.3.1.5.1 & 1810A.3.1.5.1
Deep foundation hooks	C, D, E & F	The end of hoops, spirals and ties shall be terminated with seismic hooks (**) and shall be turned into the confined concrete core.	1810.3.2.1.1 & 1810A.3.2.1.1
Deep foundation splices	C, D, E & F	They shall develop the lesser of: ✓ The nominal strength of the deep foundation element. ✓ The axial, shear and moment forces from the seismic load effects including overstrength (Ω_o).	1810.3.6.1 & 1810A.3.6.1
Cast-in-place deep foundation elements	C	**Longitudinal reinforcement:** Minimum of four longitudinal bars with ratio ≥ 0.0025. Their length is taken as the maximum of: ✓ One-third the element length. ✓ 10 ft. ✓ Three times the least element dimension. ✓ Distance from the top of element to crack location. **Transverse reinforcement:** ✓ Closed ties or spirals with min 3/8 in dia. ✓ Spacing < min(8 $dia\ long\ bar$ or 0.5 $least\ dim.$ or 6 in) provided within a distance of 3 times the least dimension and shall not exceed 16 dia long bar throughout the remainder of the pile.	1810.3.9.4.1
	D, E & F	**Longitudinal reinforcement:** Minimum of four longitudinal bars with ratio ≥ 0.005. Their length is taken as the maximum of: ✓ One-half the element length. ✓ 10 ft. ✓ Three times the least element dimension. ✓ Distance from the top of element to crack location. **Transverse reinforcement:** ✓ Closed ties or spirals with min $No.$ 3 bars when the element least dimension is 20 in, and $No.$ 4 bars for larger elements. ✓ Spacing < min(12 $dia\ long\ bar$ or 0.5 $least\ dim.$ or 12 in).	1810.3.9.4.2

	C	Permanent steel casing to the point of zero curvature.	
Micropiles	D, E & F	The Micropile shall be considered as an alternative system as applicable, relevant design documentation to be submitted for building officials for review and approval.	1810.3.10.4
Pile caps	C to F	Concrete deep foundations shall be anchored to pile caps. For uplift forces with concrete filled steel pipes, provide reinforcements > 0.01 the cross section of concrete fill extending into the cap a length equal to two times the required cap embedment, but not less than the development length in tension.	1810.3.11.1 & 1810A.3.11.1
Pile cap anchorage	D, E & F	When resisting uplift or rotation, restraints shall be provided to cap. Anchorage shall develop > 25% of strength in tension and soil/element friction is multiplied by 1.3. H or unfilled pipe piles inserted into the pile cap connection is designed for a tensile force > 10% of pile compression capacity.	1810.3.11.2 & 1810A.3.11.2

(*) Requirements from TMS 402 Sections 7.4.3, 4 & 5 are summarized in Table 6-2 below.
(**) A seismic hook has a 135-degree bend for enhanced anchorage in seismic regions, while a normal hook has a 90-degree bend for standard reinforcement.

Table 6-2 Nonparticipating Masonry Walls Seismic Details per TMS 402 Sections 7.4.3, 7.4.4 & 7.4.5

SDC	Orientation	Required detail (*)
C	Horizontal	**Walls > 4 in** At least two longitudinal wires of $W1.7$ ($MW11$) bed joint reinforcement spaced no more than 16 in. (**) **Walls < 4 in** At least one $W1.7$ ($MW11$) wires spaced no more than 16 in. Or At least one $No.4$ ($M\#13$) bar spaced not more than 48 in. Shall be provided within 16 in of the top and the bottom of those walls.
	Vertical	At least one $No.4$ ($M\#13$) bar spaced not more than 120 in. Shall be located within 16 in of the ends of walls.
D	Horizontal	Similar to SDC C
	Vertical	At least one $No.4$ ($M\#13$) bar spaced not more than 48 in. Shall be located within 16 in of the ends of masonry walls.
E & F	Horizontal	**When not laid in running bond** At least 0.0015 multiplied by the gross cross-sectional area of masonry. The maximum spacing of horizontal reinforcement shall be 24 in. These elements shall be fully grouted and shall be constructed of hollow open-end units or two wythes of solid units.
	Vertical	Similar to SDC D

(*) For information on participating walls, refer to the same sections of TMS 402.
(**) $W1.7$ is a 1.7 mm diameter steel wire. $M\#13$ is a 12.7 mm diameter rebar.

Table 6-3 CBC Chapter 18A Specific Soils and Foundations Seismic Design Requirements

Element	SDC	Requirement	CBC Reference
Soil investigations	C to F	Must assess the potential for liquefaction, slope instability, differential settlement and surface displacement due to faulting.	1803A.5.11 & 1803.5.11
Concrete and masonry foundation walls	-	The use of perspective design tables of Chapter 18, Section 1807.1.6, is not permitted for OSHPD 1R, 2 & 5 (i.e., acute care hospitals and renovations, skilled nursing facilities and other health care facilities)	1807.1.5 & 1807A.1.5
Retaining walls	-	Minimum safety factor of 1.1 for sliding and overturning under seismic conditions.	1807A.2.3 & 1807.2.3
Shallow footings ties	D, E & F	For Site Class E or F, footings shall be interconnected by ties to carry a force equal to the following – where (G) is the gravity load per footing: $$Min\left\{\left(\frac{G_{larger\ footing} \times S_{DS}}{10}\right) \ or \ \left(25\% \times G_{smaller\ footing}\right)\right\}$$	1809A.13 & 1809.13
Deep foundations	D, E & F	For Site Class E or F, they shall withstand maximum imposed curvature caused by seismic motion. Except for precast prestressed, or cast-in-place with long. reinforcement ratio ≥ 0.005 extending full length.	1810A.2.4.1 & 1810.2.4.1
Helical piles	D, E & F	Capacity determined by at least two project specific pre-production tests for each soil profile. At least 2% of all production piles shall be proof tested to strength. All helical pile materials that are subject to corrosion shall include at least 1/16 *in* corrosion allowance. Shall not be considered as carrying any horizontal loads.	1810A.3.1.5.1 & 1810.3.1.5.1
Deep foundation hooks	C, D, E & F	The end of hoops, spirals and ties shall be terminated with seismic hooks (*) and shall be turned into the confined concrete core.	1810A.3.2.1.1 & 1810.3.2.1.1
Deep foundation splices	C, D, E & F	They shall develop the lesser of: ✓ The nominal strength of the deep foundation element. ✓ The axial, shear and moment forces from the seismic load effects including overstrength (Ω_o).	1810A.3.6.1 & 1810.3.6.1
Cast-in-place deep foundation elements	C	Not permitted by DSA (District of the State Architect) or OSHPD (Office of Statewide Health Planning and Development).	1810A.3.9.4.1
	D, E & F	**Longitudinal reinforcement:** Minimum of four longitudinal bars with ratio ≥ 0.005. Their length is taken as the maximum of: ✓ One-half the element length. ✓ 10 *ft.* ✓ Three times the least element dimension. ✓ Distance from the top of element to crack location.	1810A.3.9.4.2 cont. next pg.

		Transverse reinforcement: ✓ Closed ties or spirals with min *No. 3* bar when the element least dimension is 20 *in*, and *No. 4* bar for larger elements. ✓ Spacing < *min*(12 *dia or* 0.5 *least dimension or* 12 *in*).	
Micropiles	D, E & F	Permanent steel casing thickness > 3/8 *in* shall be provided from the top of the Micropile to a minimum of 120% of the point of zero curvature. Capacity determined similar to helical piles or 1810A.3.3. Steel casing shall not contribute to friction.	1810A.3.10.4
Pile caps	C to F	Concrete deep foundations shall be anchored to pile caps. For uplift forces with concrete filled steel pipes, provide reinforcements > 0.01 the cross section of concrete fill extending into the cap a length equal to two times the required cap embedment, but not less than the development length in tension.	1810A.3.11.1 & 1810.3.11.1
Pile cap anchorage	D, E & F	When resisting uplift or rotation, restraints shall be provided to cap. Anchorage shall develop > 25% of strength in tension and soil/element friction is multiplied by 1.3. H or unfilled pipe piles inserted into the pile cap connection is designed for a tensile force > 10% of pile compression capacity.	1810A.3.11.2 & 1810.3.11.2

6.2.7. Other Seismic Requirements

The previous sections covered key seismic requirements found across different sections of the CBC code. Some of those key requirements have been integrated into this book as you may have noticed. One of the other requirements which was not addressed in this book is the modifications by the CBC on nonstructural components. These can be found in Sections 1617 and 1617A of the CBC. You can review these sections at your convenience to become familiar with their specific details.

Additionally, there is a requirement related to providing buildings with accelerographs for earthquake monitoring. Appendix L101 of the CBC outlines these requirements for structures located in seismic zones with an $S_1 > 0.4$. Specifically, for buildings exceeding six stories with an aggregate floor area over $60,000\ ft^2$, or those exceeding ten stories regardless of floor area, the following must be installed:

➤Three approved recording accelerographs, interconnected for common start and timing.

➤A tri-axial free-field instrument is required to be included for OSHPD 1 and 4 buildings.

These instruments should be positioned at the lowest level, mid-height, and near the top of the structure. Clear access to these instruments must be maintained at all times, and a sign should be posted stating, "MAINTAIN CLEAR ACCESS TO THIS INSTRUMENT."

In the following sections, additional requirements for wood, masonry, concrete, and steel construction from the CBC or the other required exam codes will be discussed.

6.3. Seismic Design and Detailing Requirements for Wood Structures
6.3.1.   Overview

Wood design and seismic detailing can be overwhelming because the necessary information is not located in a single reference. For example, there is often no clear explanation of the load cases presented in the SDPWS or guidance on how to use the tables in the code. Additionally, it can be unclear how to apply LRFD and ASD methods to determine shear capacities for wood structures in seismic design in combination with those tables. The distinctions between blocked and unblocked diaphragms and their effects on load cases for diaphragms are also not explained in any code and those could be tricky as you will see from the following sections. This chapter takes care of this.

This section provides key requirements and definitions from the 2022 California Building Code (CBC), and the 2021 Special Design Provisions for Wind and Seismic (SDPWS). It shows you how and where to locate the relevant information, and it offers you important details for easy reference. Most importantly, it explains the reasoning behind some of the code requirements to help you understand the "how" and "why" of wood seismic design.

6.3.2.   LRFD and ASD Methods

The Load and Resistance Factor Design (LRFD) and Allowable Stress Design (ASD) methods are both used for seismic design according to the SDPWS 2021. The LRFD method addresses uncertainties by applying load factors to design loads (refer to load combinations in Section 4.5) and then applies a resistance reduction factor to nominal material capacities[15]. Section 4.1.4 of the SDPWS 2021 specifies a reduction factor of $\theta_D = 0.5$ for nominal shear (V_n) when designing for seismic forces. In contrast, the ASD method reduces the nominal shear capacity by dividing it by a factor of 2.8. For a detailed example, see Example 6.1.

Example 6.1 Nominal Shear Strength for a Diaphragm

The wooden diaphragm shown below is supported by two concrete shear walls. A horizontal seismic load effect of $E_h = 20\ kip$ is applied to the diaphragm in the indicated direction. Vertical seismic load effects (E_v) or dead loads are not considered in this direction.

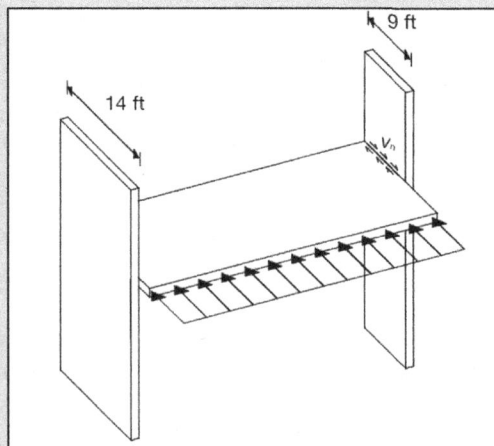

---

[15] The nominal capacity of a material in structural design is its theoretical strength calculated based on its properties and load conditions, without reduction factors.

Using Load Combination 7 for LRFD and Load Combination 10 for the ASD method, determine the maximum nominal shear force this diaphragm should resist (use the right-hand side since it is shorter hence provides maximum results).

Using the LRFD method:
Load combination 7 $(0.9D - E_v + E_h)$ is the ultimate shear (v_u), and should be less than or equal to the nominal shear strength multiplied by the seismic reduction factor (θ_D) specified in SPDWS 2021 Section 4.1.4.

$$v_u \leq \theta_D v_n$$

This is a flexible diaphragm; hence the lateral load is distributed to the two shown walls evenly – See Figure 4-2 of this book – divided by the shortest wall to generate the required (maximum) shear reaction per linear foot of the diaphragm.

$$v_u = \frac{0.5 \times (0.9D - E_v + E_h)}{9\,ft} = \frac{0.5 \times (0 - 0 + 20)}{9\,ft} = 1.11\,klf$$

$$1.11\,klf \leq 0.5 \times v_n$$

$$v_n \geq 2.22\,klf$$

Using the ASD method:
Load combination 10 $(0.6D - 0.7E_v + 0.7E_h)$ is the allowable shear (v_a), and should be less than or equal to the nominal shear strength divided by 2.8 per SPDWS 2021 Section 4.1.4.

$$v_a \leq \frac{v_n}{2.8}$$

$$V_a = \frac{0.5 \times (0.6D - 0.7E_v + 0.7E_h)}{9\,ft} = \frac{0.5 \times (0 - 0 + 0.7 \times 20)}{9\,ft} = 0.78\,klf$$

$$0.78\,klf \leq \frac{v_n}{2.8}$$

$$v_n \geq 2.20\,klf$$

Generally, ASD provides us with a more conservative results as it applies a factor of safety to the material's allowable stress, providing a larger margin of safety. In this example the two results were almost identical.

6.3.3. Diaphragms Detailing and Capacity Determination

Blocked and Unblocked Panels

The shear capacity of diaphragms is primarily governed by the capacity of the nails and their arrangement in the sheathing attached to the framing system. The sheathing material can consist of wood structural panels or lumber, and may come in various forms, such as OSB (Oriented Strand Board), plywood, or CLT (Cross-Laminated Timber). The framing system, in this case, consists of the joists. Refer to Figures 6-1 and 6-2 for further details.

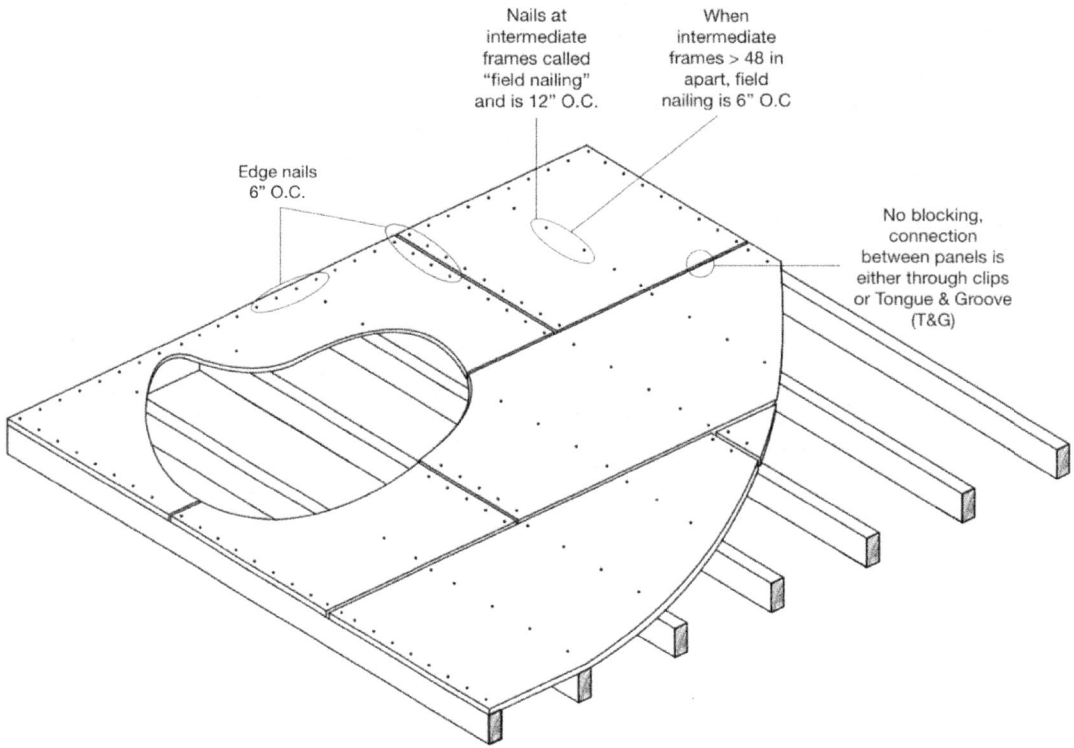

Figure 6-1 Unblocked Diaphragm Floor Nailed to Joists

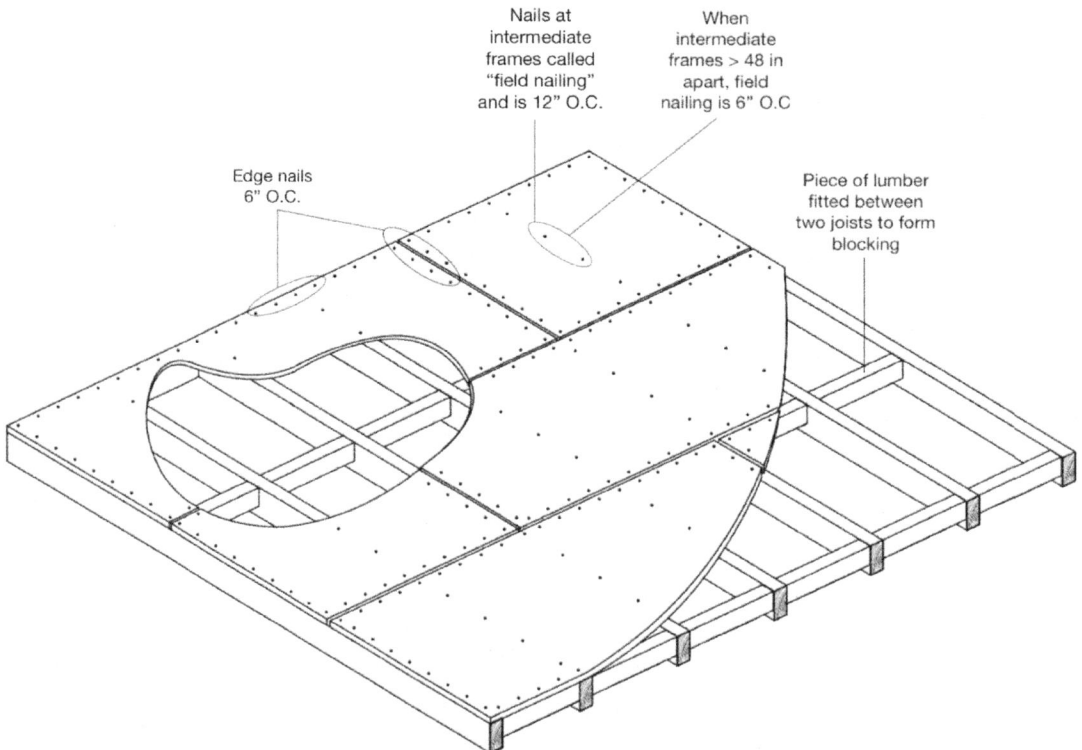

Figure 6-2 Blocked Diaphragm Floor Nailed to Joists

Wooden panels, or sheathing, can be attached to the joists using two methods: blocked and unblocked. Figure 6-1 illustrates an unblocked system, where the continuous edges of the panels are aligned against each other without any wood piece beneath them to secure the nails. Instead, the panel edges are either clipped against each other or connected using the Tongue and Groove (T&G) method. Figure 6-2 shows a blocked diaphragm, where a piece of lumber is placed beneath the panel edges between the joists and nailed every 6 *in* from each side. This method enhances shear resistance and increases the diaphragm's shear capacity.

Refer to Tables 4.2A to 4.2C in the SDPWS 2021 Standard for shear capacities based on different diaphragm blocking statuses. The load cases under each table are explained below. Note that, for similar load cases and nail capacities, the blocked diaphragm demonstrates higher shear capacity compared to the unblocked diaphragm.

SDPWS Diaphragm Load Cases

SDPWS presents six load cases (Case 1 and 3 load is presented in Figure 6-3 below). These depend on the arrangement and direction of panels in the sheathing system, and they are outlined in Tables 4.2A, 4.2B, and 4.2C of SDPWS 2021. Table 4.2A addresses blocked diaphragms, Table 4.2C covers unblocked diaphragms, and Table 4.2B accounts for multiple of rows of fasteners for blocked diaphragms (for spacing of multiple rows, refer to Figure 4C page 24 of SDPWS). These tables are designed for two sheathing grades: Structural I, which is more expensive, and "Sheathing and Single Floor" grades, which are more commonly used. They are also intended for Douglas-Fir-Larch or Southern Pine; for other species, adjust using the specific gravity (G) adjustment factor: $[1 - (0.5 - G)]$.

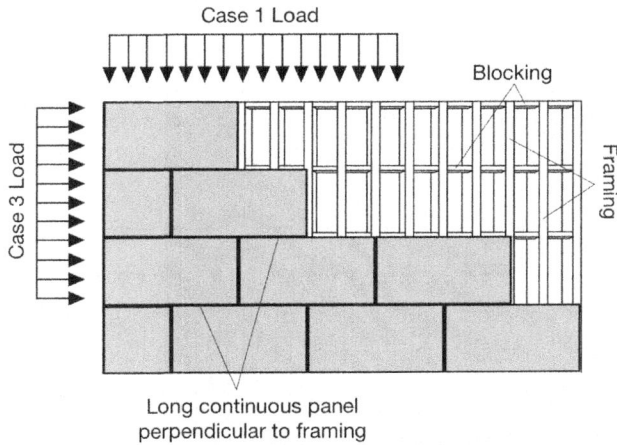

Figure 6-3 Diaphragm Detailing for Cases 1 and 3 Loads

To use the above-described load cases and tables, certain requirements must be met, as outlined in Section 4.2.8 of SDPWS. For example, the standard panel size should not be less than $4\,ft \times 8\,ft$, and the distance between nails and panel edge should be a minimum of $3/8\,in$, with nails spaced 6 *in* for blocked diaphragms or edges.

The load cases in SDPWS are based on panel orientation and the relationship between panel edges and the framing system, including whether the continuous edges are parallel or perpendicular to the framing. Loading direction is also a key factor. See below:

Design Codes, Detailing and Requirements

- **Load Case 1:** Staggered, the panels' <u>continuous edge is perpendicular</u> to the framing and the load applied in the transverse or parallel to the framing direction.

- **Load Case 3:** Staggered, the panels' <u>continuous edge is perpendicular</u> to the framing and load applied in the long direction or perpendicular to the framing direction.

- **Load Case 2:** Staggered, the panels' <u>continuous edge is parallel</u> to the framing and load applied in the transverse or perpendicular to the framing direction.

- **Load Case 4:** Staggered, the panels' <u>continuous edge is parallel</u> to the framing and load applied in the long direction or parallel to the framing direction.

- **Load Cases 5 & 6:** Non-staggered, panel edges are all continuous. The only difference is whether long panel edge is parallel or perpendicular to the framing.

Referring to the load case example presented in Figure 6-3, the continuous long edge is perpendicular to the framing and with a staggered arrangement. The panel is oriented so that the long edge is also perpendicular to the framing, which corresponds to Case 1 in the transverse direction and Case 3 in the longitudinal direction.

The SDPWS tables are designed with a cap on the unit shear capacity based on sheathing thickness and nail spacing. This means that if the sheathing thickness exceeds the ones mentioned in the table or the nail spacing falls below the minimum, the values in the table represents the maximum capacity and cannot be adjusted.

Example 6.2 Nominal Shear Strength for a Diaphragm Based on Panels' Orientation

The wooden diaphragm shown below is a blocked diaphragm with the arrangement and orientation of panels as shown. All panels are $4\,ft \times 8\,ft$ with a thickness of $\frac{1}{2}\,in$ and the $8d$ nails used are spaced $6\,in\,O.C.$ The framing system is made of Eastern White Pine with specific gravity of $G = 0.36$.

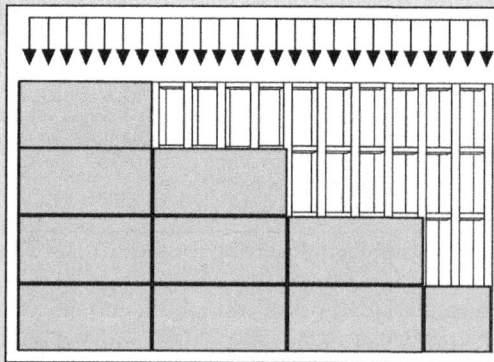

The width of the nailed face of the adjoining panels is $2\,in.$

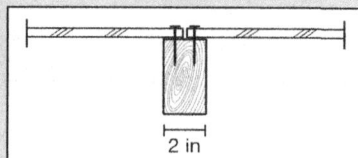

What is the nominal unit shear capacity of this sheathed diaphragm in the direction shown?

This is a non-staggered panel distribution with the long continuous edge perpendicular to the framing direction, and the load applied in the short direction, which corresponds to Case 5 in SDPWS 2021. Since this is a blocked diaphragm, Table 4.2A of the SDPWS should be used.

According to Table 4.2A, we will use the common sheathing grade other than Structural I – i.e., Sheathing and Single-Floor, as the sheathing grade was not provided, and the latter being the most used as mentioned earlier in this chapter.

With an $8d$ nail and the largest sheathing thickness $15/32\ in < 1/2\ in$, a nailed face width of $2\ in$, and $6\ in\ O.C.$ nail spacing, the unadjusted nominal unit shear capacity is:

$$v_n = 755\,plf$$

Next, apply a correction factor for the specific gravity (G) of Eastern White Pine ($G = 0.36$) per foot note 2 of the abovementioned table as specified in the question. (Also, make sure you familiarize yourself with the rest of these tables' footnotes). The adjusted value is calculated as follows:

$$v_{n,adjusted} = 755 \times (1 - (0.5 - 0.36)) = 649.3\,plf$$

Take a moment to compare how the shear capacities vary for the same loading across different diaphragm framing arrangements, whether blocked or unblocked, and with different nail capacities and spacing.

Open Front or Cantilever Diaphragms

Section 4.2.6 of the SDPWS 2021 addresses the design requirements for open structure or cantilevered diaphragms, which are diaphragms with significant openings such as windows or doors at the front. These diaphragms are subject to specific requirements to ensure they adequately transfer lateral loads to the SFRS. The design of these diaphragms must comply with the standards for blocked and unblocked diaphragms, including limitations on panel size ($4\ ft \times 8\ ft$), nail spacing of a maximum of $6\ in\ O.C.$, and edge distance $> 3/8\ in$.

Based on the above, the following aspect ratio L/W for open front cantilever diaphragm apply:

- For panel sheathed diaphragms $L/W < 1.5{:}1$.
- For diagonally sheathed with single or double layers of lumber $L/W < 1{:}1$.
- For horizontal irregularities $L/W < 0.67{:}1$, this can be relaxed to $< 1{:}1$ for single-story.

Other Requirements for Diaphragms

Take the time to familiarize yourself with the SDPWS 2021 content. For example, refer to SDPWS Table 4.2.2 for the maximum aspect ratio requirements for diaphragms (See the Notes and Common Mistakes section here), review the deflection calculations outlined in SDPWS Section 4.2.3, and check the torsional requirements specified in SDPWS Section 4.2.5. Also, familiarize yourself with Table 4.2D for lumber diaphragms.

6.3.4.   Shear Walls Detailing and Capacity Determination

The detailing and determination of unit shear capacities for wood shear walls is largely similar to that of diaphragms, with a few key differences. For example, when shear walls are constructed from wood, studs (blocked or unblocked) are used to support structural panels like plywood or Oriented Strand Board (OSB). Other materials, such as gypsum board or plaster, may also be used for paneling. Unit shear capacities for walls can be found in SDPWS Tables 4.3A to 4.3D. It is essential however, prior to finalizing those capacities, to understand some key concepts, such as the construction method and the walls' aspect ratio.

Shear Walls Construction Methods and Segmentation

There are three primary methods for erecting shear walls. Understanding these methods is crucial for determining the final shear capacity and connections to the rest of the Seismic Force Resisting System. Two of these methods are shown in Figure 6-4.

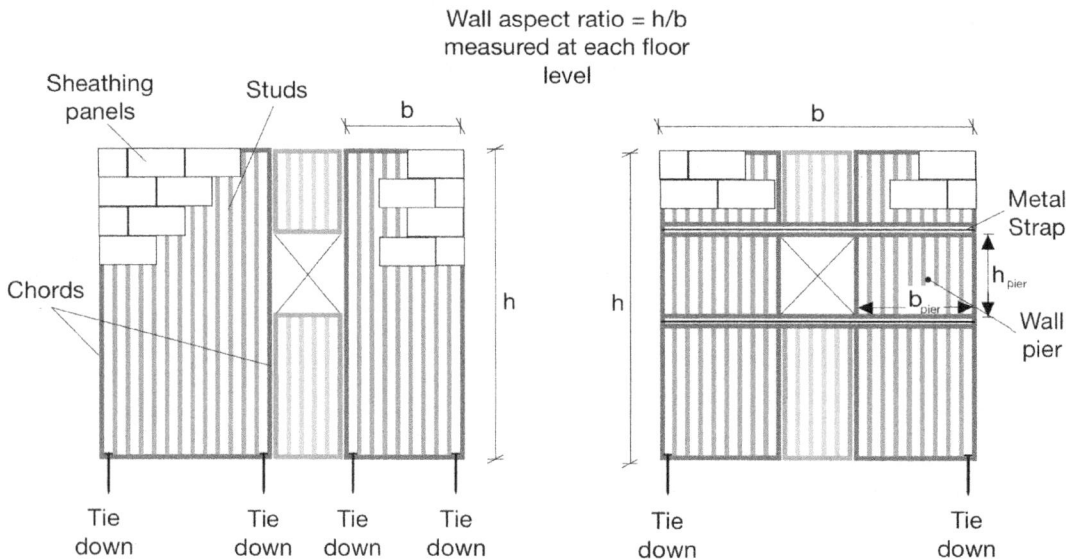

Figure 6-4 Shear Wall Construction Methods - Cantilever (left) and FTAO (right)

The first method (left in Figure 6-4) is described in SDPWS Section 4.3.2.1, this method treats each shear wall independently, excluding openings from the design process. Openings are not considered when determining the shear capacity of the wall. In this approach, each shear wall segment behaves as a cantilever diaphragm, with two chords[16], one in tension and the other in compression depending on the lateral load direction.

For this method, the aspect ratio is specified in SDPWS Table 4.3.3 and if it exceeds 2:1 for wood structural panels, the nominal shear capacity must be adjusted using the reduction factor:

$$WSP_{wood\ panel} = 1.25 - 0.125\frac{h}{b}$$

---

[16] Chord forces are computed in a similar fashion as presented in Example 2.1 of Chapter 2, with $C = T = \mp M_{base}/b$, where in this case the overturning moment is divided by the level arm (b). In case of a perforated wall (i.e., one wall piece with openings) the lever arm is reduced to account for those openings – see Equation 4.3-8 of SDPWS.

For fireboard panels (these provide fire resistance), the maximum aspect ratio is 1:1 in this case, and the reduction factor for nominal shear when this aspect ratio is exceeded is:

$$WSP_{fireboard\ panel} = 1.09 - 0.09\frac{h}{b}$$

Example 6.3 Nominal Shear Capacity for a Cantilever Structural Fireboard Wall

The figure below represents a structural fireboard sheathing blocked shear wall. The framing is a Douglas-Fir-Larch with a moisture content less than 19% during fabrication, and the sheathing is 1/2 in thick, nailed to the blocked framing system using 11 ga. galv. roofing nails spaced 4 in O.C.

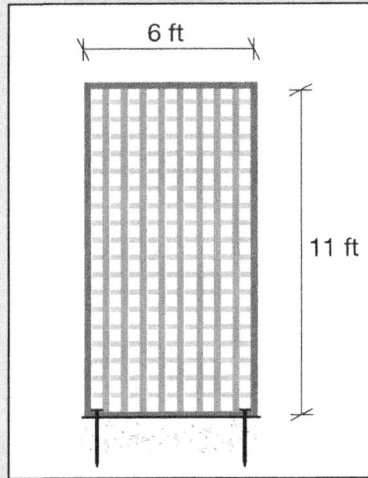

Based on this information, what is the nominal unit shear capacity for this wall?

To determine the nominal unit shear capacity for this wall, refer to SDPWS Table 4.3A, which specifies a nominal shear capacity of $v_n = 475\ plf$ for this wall configuration sheathed using structural fireboard panels.

However, due to the wall's aspect ratio of $11/6 = 1.83 > 1:1$, this value should be adjusted per SDPWS Section 4.3.3 by the relevant WSP factor.

$$WSP_{fireboard\ panel} = 1.09 - 0.09 \times \frac{11}{6} \cong 0.93$$

$$v_{n(adjusted)} = 475 \times 0.93 \cong 442\ plf$$

The second method (right in Figure 6-4) integrates openings with the shear wall and adds a coupling beam, or a metal strap, above and below the opening. This approach is known as the Force-Transfer Around Opening (FTAO) method. The aspect ratio should not exceed the limits outlined in SDPWS Table 4.3.3. The aspect ratio is measured for the entire wall as well as for each wall pier adjacent to openings.

The third method (not shown in Figure 6-4) is the perforated method. As the name suggests, it involves shear walls with perforations, and the analysis of these walls is mostly empirical.

Adjusted Nominal Shear Capacities

SDPWS 2021 Tables 4.3A to 4.3D are used to determine nominal unit shear capacities for various wall types. <u>The shear capacities provided in these tables are based on sheathing applied to one side of the wall studs.</u> For walls with sheathing on both sides, refer to the relevant adjustment coefficients in Table 6-4.

Similar to diaphragms, Tables 4.3A to 4.3D have specific requirements that must be met to be applicable. These requirements are outlined in SDPWS Section 4.3.7. While these conditions generally apply to all wall types, they may vary slightly depending on the panel material used. Some key conditions include the following:

- The standard panel size shall not be less than $4\,ft \times 8\,ft$ expect at edges or change in framing.
- The minimum nail edge distance should at least be $3/8\,in.$
- Nails should be spaced $< 6\,in\,O.C.$ at edges or at intermediate framing members.
- Studs maximum spacing $24\,in$ (for fireboard panel $16\,in$).

In certain cases, adjustments to shear capacities may be necessary. These adjustments along with other adjustment coefficients are summarized in Table 6-4 below.

Table 6-4 Adjustment Coefficients for Shear Wall Unit Capacities

Condition of Use	Equation	SDPWS 2021 Reference
Aspect ratio $_{wood\,panel} > 2:1$ Aspect ratio $_{fireboard} > 1:1$	$WSP_{wood\,panel} = 1.25 - 0.125\frac{h}{b}$ $WSP_{fireboard\,panel} = 1.09 - 0.09\frac{h}{b}$ $v_{n(adjusted)} = WSP \times v_n$	Section 4.3.3
Unblocked wood structural panel	$v_{n,(ub)} = v_{n(b)} \times C_{(ub)}$ *ub: unblocked, b: blocked*	Section 4.3.5 Table 4.3.5.3 for $C_{(ub)}$ Table 4.3A for $V_{n(b)}$
Two sides sheathing: Dissimilar sheathing material	The greatest of: (1) Twice the smallest side nominal shear, or (2) The Nominal shear of the largest side.	Section 4.3.5.4.2
Two sides sheathing: Similar sheathing material	$v_{n(c)} = K_{min}G_{a(c)}$ $K_{min} = \min\left(\frac{v_{n(1)}}{G_{a(1)}} \text{ or } \frac{v_{n(2)}}{G_{a(2)}}\right)$ $G_{a(c)} = G_{a(1)} + G_{a(2)}$ *1: side 1, 2: side 2, c: combined, G_a: apparent shear stiffness*	Section 4.3.5.4.1
Perforated shear walls	$V_n = v_nC_o\sum b_i$ *b_i: length of shear wall segment, V_n: total nominal shear capacity*	Section 4.3.5.6 Table 4.3.5.6 for C_o

6.3.5.   Wood Sheathing Material

In addition to common sheathing materials for wood diaphragms and shear walls, less common options include lumber sheathing per SDPWS Table 4.2D and Cross-Laminated Timber (CLT) per SDPWS Sections 4.5 and 4.6. Note however that the California Building Code 2022, Section 2301.1.4, prohibits certain methods and materials for DSA and OSHPD, including straight-sheathed horizontal lumber diaphragms, gypsum-based sheathing shear walls, staples in wood structural panels, and unblocked shear walls.

6.3.6.   Wood Structures Seismic Requirements in CBC 2022

Table 6-5 below summarizes some of the key seismic design requirements from CBC Chapter 23. You are encouraged to explore the full set of requirements at your own pace for a comprehensive understanding directly from the code.

Table 6-5 CBC Chapter 23 Specific Wood Seismic Design Requirements

Element	SDC	Requirement	CBC Reference
Fastened with <u>staples rather than nails</u>	-	Diaphragm and shear wall deflection/ blocked, uniformly fastened: Use Equation 23-1 & 2	Section 2305
		Allowable (*not nominal*) shear capacities for diaphragms: Table 2306.2 (1) & (2).	Section 2306
		Allowable (*not nominal*) shear capacities for shear walls: Table 2306.3 (1), (2) & (3).	Section 2306
Permitted stories for light frames	A & B	Three stories.	Table 2308.2.1
	C	Two stories.	
	D & E	One story.	
Risk category limitation	B, C, D or F	Light framed **Not Permitted** for risk category IV.	Section 2308.2.6
Openings in diaphragms	B to E	When dimension > 4 ft, blocking left and right of opening equivalent to the dimension of opening with metal straps imbedded around openings for the provided length of blocking.	Section 2308.4.4.1
Diaphragms vertical offset	D & E	This applies to two adjacent diaphragms, each at a slightly different level (i.e., vertical offset), with a wall in between. If they cannot be tied together, then this arrangement is **Not Permitted**.	Section 2308.4.4.2
Braced Wall Lines (BWL)	D & E	Braced wall lines (BWL) shall intersect perpendicular to each other.	Section 2308.6.1
Cripple walls	*Definition*	*A cripple wall is a short wall between the foundation and the first-floor building joists.*	Section 2308.6.6.1
	D & E	Studs should be < 14 in, and they shall be solid and blocked for the full perimeter.	
Floor & roof diaphragm support	D & E	Shall be laterally supported by braced wall lines on all edges.	Section 2308.6.8.2

6.4. Seismic Design and Detailing Requirements for Masonry Structures
6.4.1.   Shear Wall Types
While various masonry design and detailing requirements have been discussed earlier in this chapter, this section focuses on key concepts and specific detailing requirements for masonry structures as outlined in TMS 402/2016 and the California Building Code 2022.

Based on this, below are the different types of masonry shear walls, following that is their detailing requirements and design methods.

1. Empirical method: This method is based on simplified calculations and is typically used for low-rise buildings and low seismic regions. It relies on prescriptive rules and does not include reinforcement.

2. Ordinary plain: Shear walls are unreinforced and designed to resist in-plane shear forces through the masonry's inherent strength. These are suitable for areas with low seismic activity.

3. Detailed plain: Similar to ordinary plain, but shear walls have additional detailing to improve performance under lateral loads. These walls are also unreinforced but with minimum requirements as specified in Figure 6-5.

4. Ordinary reinforced: These walls include reinforcement to enhance their strength and ductility. They are suitable for moderate seismic regions. The reinforcement typically includes vertical and horizontal steel rebars.

5. Intermediate reinforced: These walls have more reinforcement and stringent detailing requirements compared to ordinary reinforced walls. They are designed for higher SDCs.

6. Special reinforced: These walls have the highest level of reinforcement and detailing requirements. They are designed for the most severe seismic conditions and provide the best performance and lateral load resistance, and they require extensive horizontal and vertical reinforcements.

The following table provides a summary of the permitted shear wall types for the Seismic Design Categories (SDCs). It is important to note that the requirements listed for each SDC are cumulative. This means that a masonry type assigned to SDC B must meet the requirements for both SDC A and SDC B, and so on.

Table 6-6 Permitted Masonry Type for Design Seismic Category

SDC	Empirical	Ordinary Plain	Detailed Plain	Ordinary Reinforced	Intermediate Reinforced	Special Reinforced
A	X	X	X	X	X	X
B		X	X	X	X	X
C				X	X	X
D						X
E						X
F						X

6.4.2.   Nonparticipating Walls

In masonry structures, shear walls can be classified as either participating or non-participating. Participating shear walls are designed to be part of the Seismic Force Resisting System (SFRS) and are intended to withstand lateral loads during seismic events. Non-participating shear walls, on the other hand, are not designed to resist these lateral loads and should be isolated from the SFRS. This isolation ensures that nonparticipating walls do not inadvertently take on loads they are not designed to handle.

Nonparticipating walls should also include minimum reinforcement to ensure they are protected against any unaccounted for in-plane or out-of-plane loads. These minimum reinforcement requirements are outlined in TMS 402/2016 Section 7.4, with key elements summarized in Table 6-2 in Section 6.2.5 of this book.

6.4.3.   Participating Walls

Section 7.3.2 of the TMS 402 outlines the design requirements for each masonry wall type listed in Table 6-6, referencing the appropriate sections for their specific design guidelines in TMS. This section also specifies the minimum reinforcement requirements, which must be met.

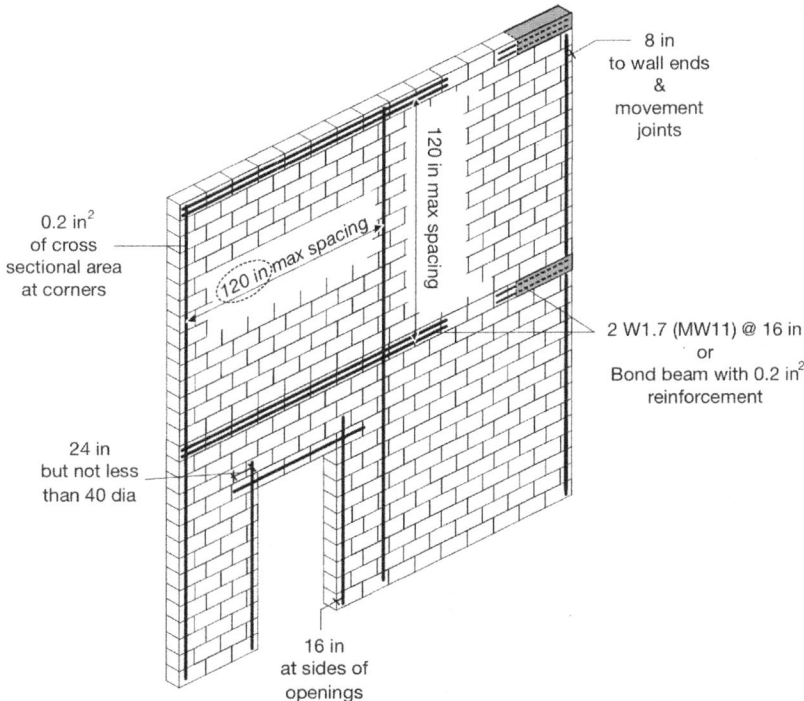

Figure 6-5 Minimum Reinforcement for Detailed Plain & Ordinary Reinforced Masonry Walls

To simplify understanding, we have provided sketches of the key minimum reinforcement requirements for the wall types mentioned in Table 6-6 in Figures 6-5, 6-6, and 6-7. Also note that Autoclaved Aerated Concrete (AAC) walls are not included in these sketches. For more information on the minimum reinforcement requirements for AAC walls, refer to Sections 7.3.2.7 through 7.3.2.9 of the TMS 402. For prestressed masonry, refer to Sections 7.3.2.10 through 7.3.2.12 of TMS 402.

Design Codes, Detailing and Requirements

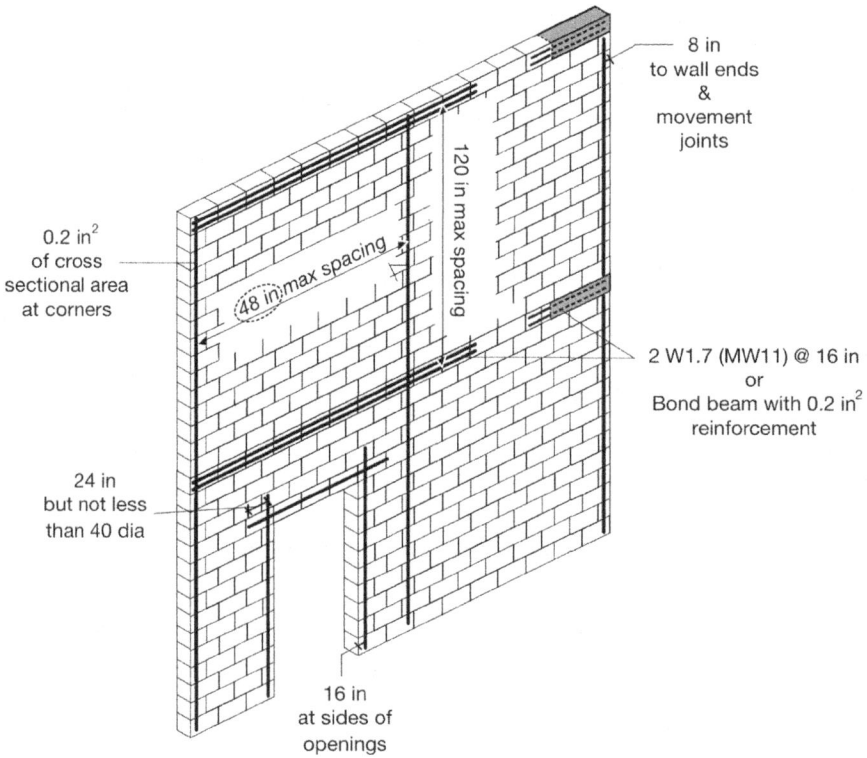

Figure 6-6 Minimum Reinforcement for Intermediate Reinforced Masonry Walls

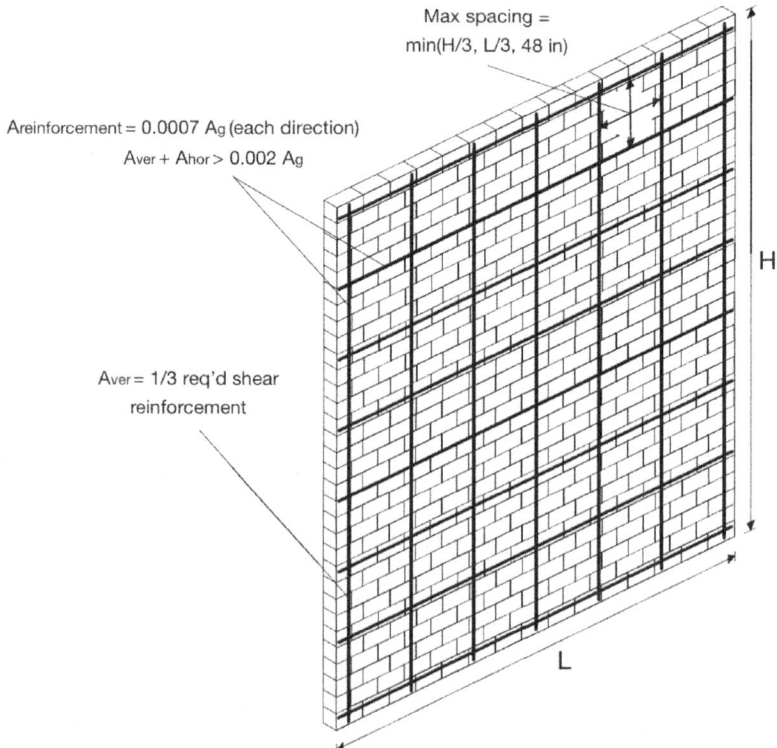

Figure 6-7 Minimum Reinforcement for Special Reinforced Masonry Walls

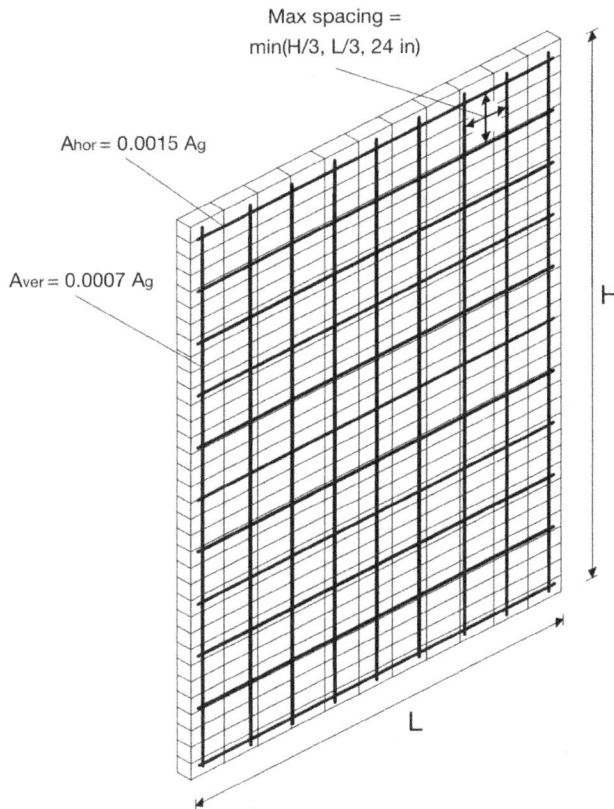

Figure 6-8 Minimum Reinforcement for Special Reinforced Masonry Walls Not Laid in Running Bond

6.4.4.   Other Seismic Requirements for Masonry Construction

In addition to the seismic design requirements for masonry structures outlined here and in TMS, the California Building Code and the ASCE 7 Standard both have other requirements as well.

Office of Statewide Health Planning & Development and Community Colleges

It is crucial to be aware of some of the specific requirements and prohibitions enforced by the Office of Statewide Health Planning and Development (OSHPD), and Community Colleges (DSA-SS/CC). Below is a key summary, however, you are encouraged to visit the relevant sections in the CBC for more information:

- According to CBC Section 2101.2.2, most masonry wall types are prohibited for OSHPD 1R, 2B & 5, and for Community Colleges, with a few exceptions, such as the special reinforced masonry shear walls.

- According to CBC Sections 2106.1.3 and 2115.9.2, minimum reinforcements used in masonry walls for SDC D for OSHPD and DSA-SS/CC should have a minimum total reinforcement of no less than 0.003 of the cross-sectional area with neither horizonal nor vertical reinforcement less than one third of the total. Also, a minimum of No.4 bar should be used at a spacing of no less than 24 *in*.

- Nonparticipating walls that are not laid in running bond in SDC E & F are no longer recommended in CBC for OSHPD and DSA-SS/CC.

Masonry Fireplaces and Heaters

When the seismic design category exceeds SDC C, masonry fireplaces and heaters are subject to specific seismic design requirements outlined in CBC Sections 2111 and 2112 (for detailing see CBC Section 2113.3 for both). These include minimum reinforcement detailing, such as the requirement for continuous vertical and horizontal reinforcement to resist lateral forces. Additionally, anchorage to the foundation must meet strict interval spacing to ensure proper stability during seismic events. The fireplace structure should be securely tied to the building frame to prevent separation, and anchorage to the building's foundation must be designed to resist both lateral and vertical seismic forces.

Concrete or Masonry Veneers

Veneers attached to masonry or concrete walls, including those made from concrete, stone or masonry, are not permitted to extend above the basement level in Seismic Design Categories SDC B through E except in certain cases as specified in CBC Section 2308.6.10.1 & 2. Under these exceptions, veneers may be used provided that specific requirements are met, such as the type of bracing, the location of braces, and the installation of hold-down connectors.

ASCE 7 Standard Requirements and Adjustments to TMS 402

Chapter 14 of the ASCE 7 Standard provides recommendations that complement the seismic requirements outlined in TMS 402. Below are some of those key requirements as summarized in ASCE 7 Section 14.4, and for further details, you are encouraged to refer directly to ASCE:

- Minimum rebars sizes, nominal diameters and splices requirements for masonry construction adjusted per ASCE 14.4.4.1.

- Anchoring bolts embedded in masonry should be designed to resist no less than 2 two times the factored load transmitted by the assembly per ASCE 14.4.5.

- Shear key specifications and dimensions for special reinforced shear walls have been proposed in ASCE 14.4.5.4

- Requirements for veneer anchoring are covered in TMS 402 Section 12.2.2.11. ASCE adds one more requirement in ASCE Section 14.4.6.1 for SDC C and above as presented in Figure 6-9 below. (Other requirements for higher SDCs may apply).

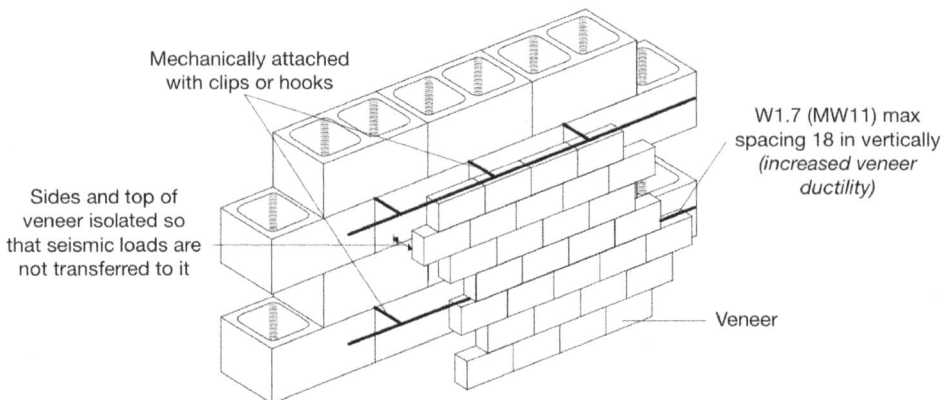

Figure 6-9 Anchored Veneers to SDC C and above

6.5. Seismic Design and Detailing Requirements for Concrete Structures

6.5.1.   Overview

Seismic detailing for concrete structures can be quite involved, with many requirements outlined in Chapter 18 of ACI 318 along with other chapters of the ACI code. Additionally, Chapter 14 of the ASCE 7 Standard along with Chapter 19 of the 2022 California Building Code provide modifications to these requirements. Some key requirements are addressed here.

While ACI 318 is the primary source for concrete seismic and detailing requirements, this reference is not required for the California seismic exam, likely because it demands a deep understanding of structural engineering and concrete design. Therefore, in this section, we will present high-level information for some of the key seismic requirements to help you tie together some of the concepts presented in Chapter 2 of this book and elsewhere, particularly regarding the various Seismic Force Resisting Systems (SFRS). This includes distinctions between concrete structural walls (ordinary, intermediate, and special) and concrete moment frames (ordinary, intermediate, and special), clarifying their differences and appropriate uses.

6.5.2.   Concrete Shear Walls

Seismic Force Resisting System (SFRS)

We will focus on discussing the special reinforced concrete shear walls in the following section, acknowledging that any alternative wall type would be considered a downgrade in comparison. This is followed by a brief discussion on *ordinary plain* and *detailed plain* concrete walls.

Building on the information from Chapter 2 of this book, where we first defined *special reinforced concrete walls* as outlined in Table 12.2-1 of the ASCE 7 Standard, it is important to observe that these special walls are included in four distinct SFRS categories:

(1) in A.1 for the **Bearing Wall System**,
(2) in B.4 for the **Building Frames System**, and
(3) & (4) in D.3 and E.2 for the **Dual Systems**.

Each of these has difference height requirements and difference response modification factors. The response modification factor (R) for those same walls differs under the different SFRS category that they belong to. For example, and per Table 12.2-1, for A.1, $R = 5$, while for B.4, $R = 6$. This is mainly attributed to the use of the SFRS. A common way to distinguish between a **Bearing System** and a **Building Frame System** is that a Bearing System supports more than 5% of the total building floor and roof loads, in addition to its self-weight, which explains the lower (R) for the special concrete walls.

Additionally, when comparing special walls to ordinary walls, particularly within the Bearing Wall System, it is evident that a special wall system is permitted in Seismic Design Categories D, E, and F at heights of 160 ft, 160 ft, and 100 ft, respectively. This height limit can be increased to 240 ft, 240 ft, and 160 ft if no extreme torsional irregularities exist when no shear wall line resists more than 60% of the story shear. While this type of wall can be used in lower SDCs, it is generally not economical or advisable in such cases and hence we stick with the ordinary ones.

Ordinary, Intermediate and Special Reinforced Concrete Walls

A boundary element in reinforced concrete walls refers to the edges of the wall, including the areas around openings such as doors and windows, where additional reinforcement may be required to resist excessive shear and moment forces. Special concrete walls, in particular, have higher demands for resisting these forces, making boundary elements more critical. This is one of the key distinguishing features between ordinary and special walls. For more details, see Figure 6-10.

As shown in Figure 6-10, the boundary element of a special wall is reinforced with hoops as transverse reinforcement, which serve to secure the longitudinal reinforcement and prevent it from spalling or buckling, especially during a severe seismic event. This significantly increases the capacity of the wall. The hoops have seismic hooks, either 90-degree or 135-degree, as seen in the referenced Figure, which help to keep the hoops, reinforcement, and the concrete within, confined in place. Additional discussion on the use of hoops can be found in Section 6.5.3.

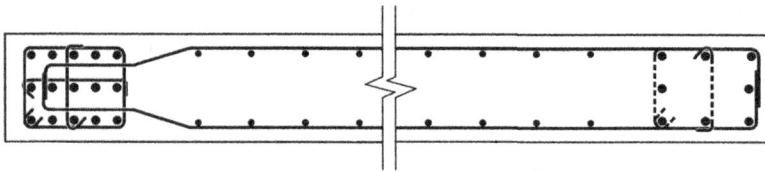

Figure 6-10 Plan View for Special Concrete Wall (left) and Ordinary/Intermediate Concrete wall (right)

Note that we have not provided specific diameters or detailed reinforcement requirements here. If you would like to learn more, please refer to the relevant list of references provided at the end of this chapter.

Ordinary Plain Concrete Walls

Plain concrete walls are commonly used in the construction of basement walls for residential and light commercial buildings in areas with low seismic risk. However, they are not permitted as part of the Seismic Force Resisting System for buildings larger than SDC B. While the Code does not impose a strict height limitation on the use of plain concrete walls for SDC B, as outlined in ASCE 7 Table 12.2-1, the use of plain concrete in relatively minor structures should not be directly applied to multistory construction without careful consideration.

The minimum thickness for plain concrete walls is 5.5 *in* or the lesser of unsupported length or height divided by 24, whichever is greater.

Reinforcement is provided around openings, such as windows and doors, where at least two *No.*5 bars should be used. Reinforcement should extend at least 24 *in* beyond the corners of these openings.

Detailed Plain Concrete Walls

Detailed plain concrete walls are not entirely plain; they include reinforcement as outlined in CBC 2022 Section 1905.1.6. Vertical reinforcement shall have a minimum area of 0.2 *in*² in the cross-section. Near openings, when two *No.*5 bars are required as specified for plain concrete, the vertical reinforcement can substitute for one of the *No.*5 bars.

For horizontal reinforcement, a minimum of 0.2 *in*² should also be provided in the cross-sectional area, with a maximum spacing of 120 *in*. Additionally, the reinforcement at the top and bottom of openings must be continuous along the wall.

6.5.3.   Special Concrete Moment Frames

Overview

As discussed in the previous section on walls, and similarly to moment frames, this section presents the requirements, definitions, and some key specifications for special concrete moment frames. The requirements for intermediate and ordinary frames are considered a downgrade in comparison.

Reinforced concrete special moment frames are made up of beams, columns, and beam-column joints. The frames are proportioned to resist flexural, axial and shear forces that result from buildings swaying during seismic activity. Special proportioning can result in higher stiffness for those frames, and due to this, it improves the seismic resistance in comparison to ordinary or intermediate frames.

Seismic Force Resisting System (SFRS)

You will notice in ASCE 7, Table 12.2-1, that special frames are also included in the dual system, where they are required to resist 25% of the forces, while shear walls (or other rigid SFRS components) resist the remainder. The purpose of this dual system is to allow the use of a larger response modification factor for the more rigid SFRS component in that system, thereby, reducing the overall seismic strength requirements. To illustrate this, refer to item C.5 of ASCE Table 12.2-1, which lists special concrete frames with $R = 8$. When combined with a special shear wall with $R = 5$, as shown in item D.3, the resultant response modification factor for the final SFRS is $R = 7$. Additionally, like special concrete walls, special moment frames are required for Seismic Design Categories D, E, and F. However, unlike special walls, there is no height limitation for special moment frames.

Beam-Column Joint

A typical span length for special moment frames ranges from $20\,ft$ to $30\,ft$, which results in a deeper beam (approximately one-quarter of the clear span). However, caution is needed, as the goal is to design weaker beams and stronger columns. The effective depth of the beams should not exceed twice the depth of the columns in the framing direction, as this could impact the force transfer at the joint – see Figure 6-11 below. These joints must be effective in force transfer and should be stronger than both the beam and the column.

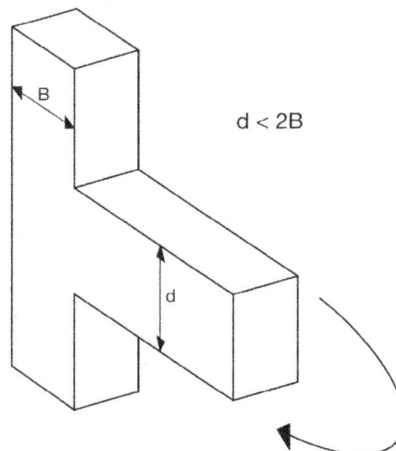

Figure 6-11 Special Concrete Moment Frame Proportions for Beam/Column Connection

Flexural and Longitudinal Reinforcement Splicing Requirements

Due to flexural forces, spalling or loss of concrete cover poses a risk during severe earthquakes. The loss of concrete cover reduces the bond strength, and if this occurs at a lap splice location of longitudinal reinforcement, the consequences could be catastrophic. Therefore, lap splices, if used, should be placed away from sections experiencing maximum moments (typically occurring at the ends of beams in a frame structure). Additionally, splice locations should be provided with closely spaced hoops to confine the reinforcement and concrete in the event of cover loss – see Figure 6-12. As a general rule, splice locations should cover at least two times the distance from critical sections where flexural yielding is likely to occur.

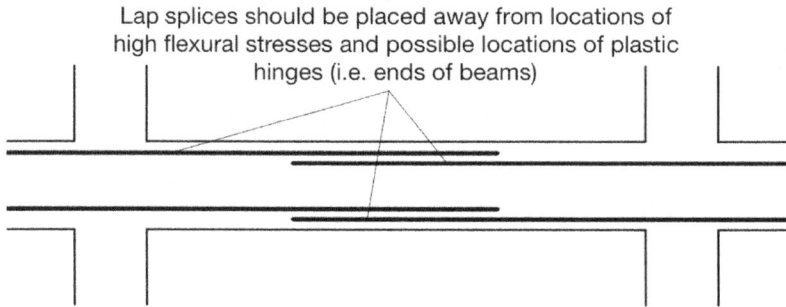

Figure 6-12 Special Concrete Frame Beam Lap Splice Recommended Locations

Beam Shear Requirements

As you may recall from basic structural analysis, shear in a typical beam tends to decrease to zero around its midspan. However, in special moment frames, and due to building swaying, lateral loads, along with reverse action, the resulting shear in the beam does not go to zero at or near midspan as it would in a beam subject only to gravity loads (see Figure 6-13 below). Additionally, in typical beam design, the critical shear section is located a distance (d – *effective depth*) from the face of the support. In special moment frames, however, the critical shear section is taken at the support.

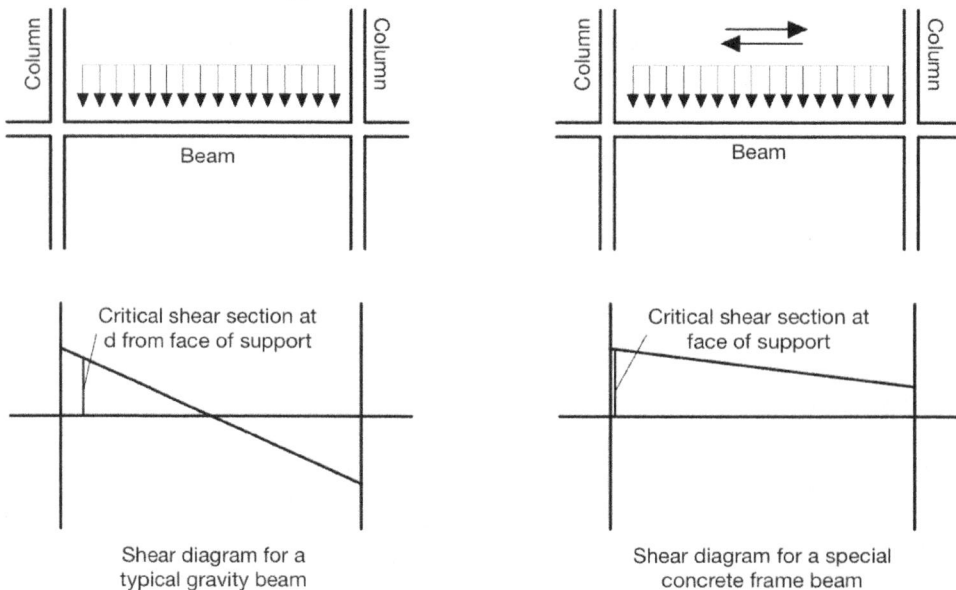

Figure 6-13 Shear Diagrams for a Gravity Beam and a Special Concrete Frame Beam

As a result, the beam is expected to be provided with transverse reinforcement along its entire length. It is important to also consider the requirements for hoops in locations where plastic hinges are likely to form due to flexural actions, which can lead to concrete spalling. Hoops fully enclose the beam, confining the concrete, preventing reinforcement buckling, and improving bond strength, especially when lap splices are present. They also resist shear forces. In contrast, stirrups[17] are not closed and are used in locations where shear resistance alone is required – see Figure 6-14. In most of the cases, it may be more practical to extend the hoops along the entire length of the beam.

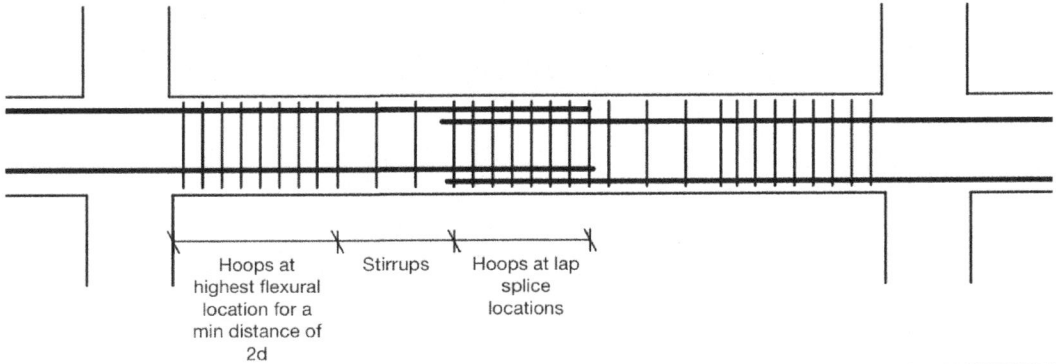

Figure 6-14 Proposed Locations of Hoops and Stirrups in Special Concrete Frame Beams

Columns in Special Concrete Frames

Columns that are part of the seismic force resisting system must satisfy the strong-column/weak-beam requirement for all load combinations.

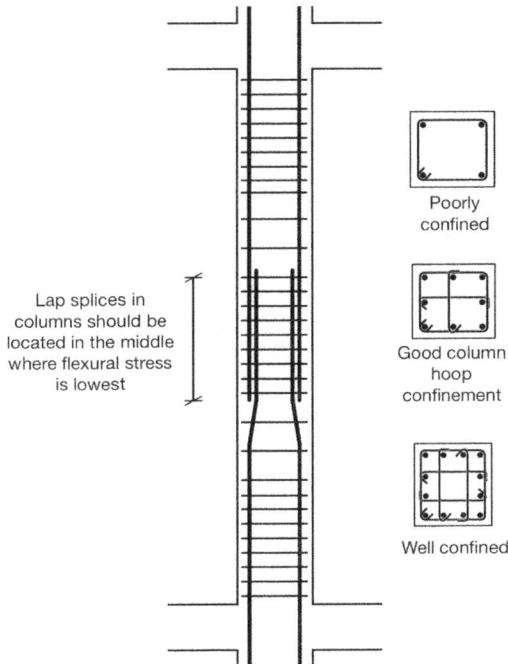

Figure 6-15 Proposed Locations for Laps Splices and Hoops in Columns of Special Moment Frames

---

[17] Hoops can be constructed using stirrups with seismic hooks of either 90- or 135-degree bends.

Also, and similar to beams, the transverse reinforcements in columns serve the same functions: confining the concrete, preventing reinforcement buckling, improving bond strength, and providing shear strength. These reinforcements should be spaced variably along the column length; however, for confinement purposes, they should be closely spaced at columns ends, where flexural forces are expected to be highest. Additionally, to avoid the high flexural force areas near the ends and potential plastic hinge locations, lap splices should be restricted to the center half of the column length. See Figure 6-15 for better understanding.

6.5.4.   Concrete Structures Seismic Requirements in CBC 2022

This section does not aim to summarize the requirements for concrete structures in the California Building Code. However, most of these requirements were covered in the previous sections. It is advisable however to familiarize yourself with Chapter 19 of CBC 2022, which covers some more seismic requirements which were not discussed here. Note the location of key requirements not covered here, such as diaphragm seismic requirements (CBC Section 1910.3.1), construction drawings (CBC Section 1901.5), and few other sections. This will help you remember them and refer to them effortlessly during the exam.

6.6. Seismic Design and Detailing Requirements for Steel Structures
6.6.1.   Steel Special Moment Fames (SMFs)

Overview

In this section, and similar to the introduction made for the concrete moment frames section, we will highlight some of the key distinguishing factors for a better understanding of what sets the special steel moment frames apart. There are several standards used to define and design these frames, such as the AISC 341 for the seismic design of steel structures, and the AISC 356 for the design and selection of prequalified connections. There are also other references, but neither the AISC, nor those other references, are required for the seismic exam, as they tend to be quite detailed, and are mostly needed by experienced structural engineers, which is why this section will remain at a higher level only.

Steel Moment Frames Performance

Steel moment frames have superior performance compared to other framing structures such as reinforced concrete frames. What makes them more superior is not only their ductility, but also the composite action for both steel and concrete as, in most cases, steel structures are covered by concrete for fire rating purposes. This improves the stiffness of the structure.

In light of this, it is important to also consider observations from past earthquakes, such as the Northridge earthquake, and other common failures in steel moment frames, which inform us about improvements needed for these frames. The most famous element failure in steel moment frames is the beam-column connection's brittle failure, especially the bottom flange of the beam when welded to the column flanges – see Figure 6-16, which is self-explanatory.

It is due to this, and the significant importance of these connections, that qualification testing, which is typically expensive, must be conducted specifically for steel Special Moment Frames. To avoid these failures, or the expenses that comes with testing, AISC 356 provides prequalified connections in its standard (Figure 2-8 of Chapter 2, and Figure 6-17 here, present one of these connections referred to as the Reduced Beam Section RBS). Testing should be conducted if the designer chooses to deviate from these prequalified connections.

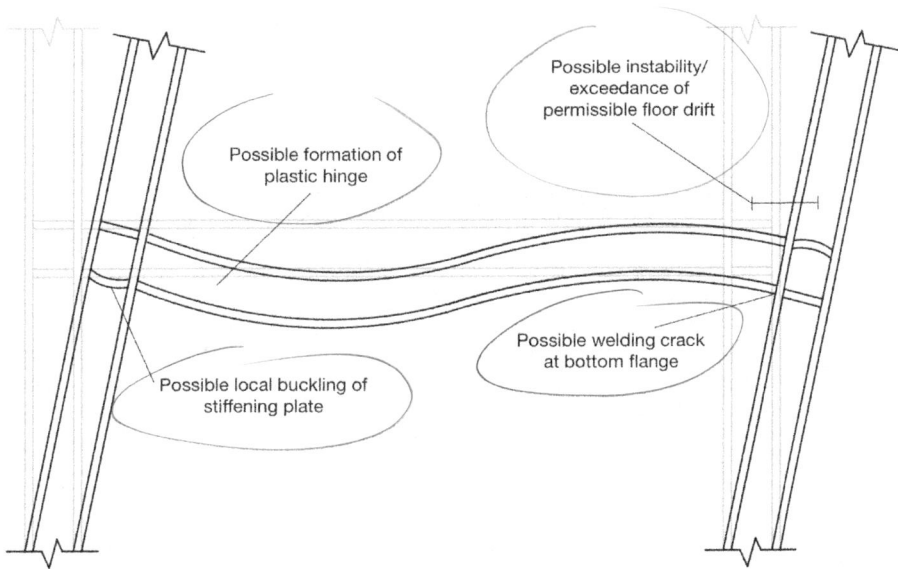

Figure 6-16 Possible Failure Locations in Steel Special Moment Frames

The Reduced Beam Section (RBS)

The Reduced Beam Section (RBS) is one of the key features of seismic design for special steel moment frames, as it offers several important benefits. It is designed to direct the plastic deformations to occur at the beam-end rather than the beam-column joint or the column itself, which enhances energy dissipation during an earthquake. This controlled plastic hinging helps reduce the forces transferred to the column and limits damages that could affect the joint. See Figure 2-8 of Chapter 2, and Figure 6-17 below.

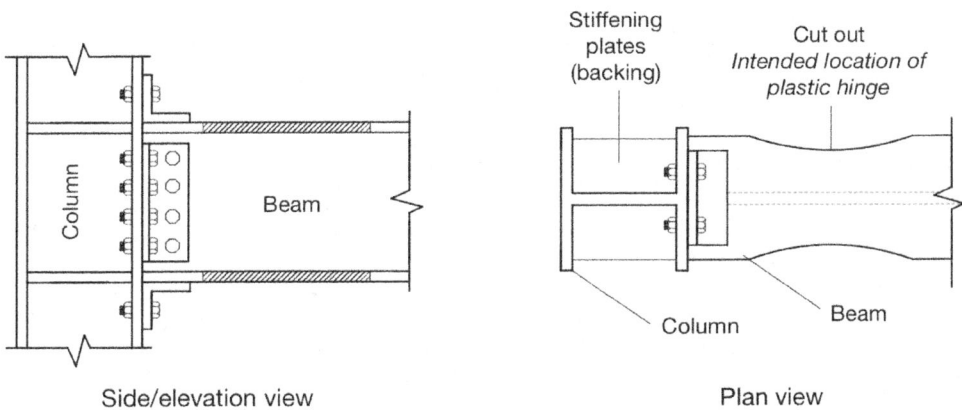

Figure 6-17 Reduced Beam Section in Steel Moment Frames

By reducing the section of the beam near the column face, the RBS prevents premature brittle failure at the beam-column connection, promoting a more ductile response and allowing for a gradual and predictable failure mechanism.

Inelastic Behavior in Steel Special Moment Frames (SMFs)

Table 6-7 summarizes some of the key behaviors and failure modes for the different parts of the steel moment frames. Understanding those should help us understand were to target plastic hinges and generally design better SMFs.

Design Codes, Detailing and Requirements

Table 6-7 Steel SMFs Failure Modes

Steel SMF Component	Failure mode
Beam	Beams could fail due to local buckling of the flanges near the columns, or midspan if large gravity loads are present. This is why the RBS beam section, shown in Figure 2-8 of Chapter 2 and Figure 6-17 of this chapter, is sometimes recommended.
Beam/Column connection	Failure can occur due to fractures around the welds or base material, access holes, or fractures at the net section near bolt holes or bolt bearing.
Joint panel zone	This is the (protected zone) at the beam/column connection that is subject to high compressive and shear forces. Failure at this zone causes column flange bending, web crippling, and/or buckling.
Column	Excessive local buckling and lateral-torsional buckling. Therefore, column design should be based on the strong column/weak beam principle.
Column base	Anchorage stretching or pullout, fractures in the base plate, or local and lateral torsional buckling near the base.
P-Delta effect	Failure due to gravity-amplified load generated from a P-delta effect as explained in Section 2.4.2.
Sideway collapse	This occurs when the story shear exceeds the story shear resistance.

Steel Special Moment Frames

The primary focus in the design of steel Special Moment Frames (SMFs) is stability – i.e., the control of drift and P-delta effect. To control drift[18] and minimize the potential P-delta effect, careful selection of frame sizes is crucial. The design should adhere to the strong column/weak beam principle, where the beam is hinged on the column. In some cases, supplemental lateral bracing may be required by the design standards. The typical span for these systems ranges from 20 ft to 40 ft.

Generally, moment frames do not have shear walls or bracing, but they may include these elements as part of a dual system, as specified in ASCE 7 Table 12-2.1. This lack of shear walls or bracing is an advantage, as it provides greater architectural flexibility by allowing open spaces without obstructions from walls or braces.

For optimal seismic performance, moment frames should be ductile to accommodate inelastic deformations. As discussed earlier, designers should target areas where plastic hinges are intended to form, such as at the ends of beams or at column bases. In many cases, supplemental bracing will help limit large deformations and keep the frame within the allowable drift limits to prevent P-delta failures. The use of compact steel sections is also essential for ensuring ductility. A compact section, where the flange is continuously welded or connected to the web with adequate web and flange thickness, ensures that the section reaches its full plastic capacity, which is vital for effective seismic performance.

---

[18] The design for drift is an iterative process because lateral forces are influenced by the fundamental period of the structure. As the structure is resized to control drift, this fundamental period changes, which in turn affects the lateral forces. This creates a feedback loop that requires multiple adjustments to ensure that the drift remains within acceptable limits.

6.6.2.   Steel Bracing Frame Systems

Overview

Up until this point, we have briefly mentioned building steel bracing systems without going into much detail. In the following section, we will explore these systems in more depth. These systems are listed under Item B (Building Frame System) in Table 12.2-1 of the ASCE 7 standard, and include: steel eccentrically braced frames, steel special concentrically braced frames, ordinary steel concentrically braced frames, and buckling-restrained braced frames. Composite systems with steel are also considered for those items.

This section provides an overview of these bracing systems. While it does not cover the specifics of their design, it is important to understand their key differences, particularly in terms of ductility, to determine which systems are more suitable for different levels of seismic risk.

Concentrically Braced Frames

Two systems fall under this category: the ordinary and the special, with the special system offering more ductility than the ordinary one.

As the name suggests, these systems consist of diagonal braces that converge at a central point in the structure – see Figure 6-18. These braces are designed to resist forces through axial tension or compression. Failure of such a system typically occurs when the tension chord yields or the compression chord buckles, with the latter being more common – see Figure 6-23. Because of this, these systems tend to be less effective in highly dynamic seismic events, as they are prone to buckling under extreme forces. Consequently, in Table 12.2-1 of ASCE 7 Standard, the response modification factor for the ordinary system is 3¼, while for the special system, it is 6. Both systems exhibit lower ductility compared to the eccentrically braced and the buckling-restrained systems.

Figure 6-18 Examples of Steel Concentrically Braced Frames

Eccentrically Braced Frames

One might assume that an eccentric system is less ductile than a concentric system due to its inherent eccentricity, but this is a misconception.

An eccentric bracing system typically consists of a beam, one or two braces, and columns – see Figure 6-19. Its configuration is similar to a concentric system, with one key difference: at least one end of each brace is eccentrically connected to the frame. This eccentric connection introduces shear and bending forces to the small link that has formed (see Figure 6-20 with the link circled). This makes it more resistant to lateral forces compared to a concentric brace. This short link bears the majority of the bending and shear forces, thus providing significant stiffening.

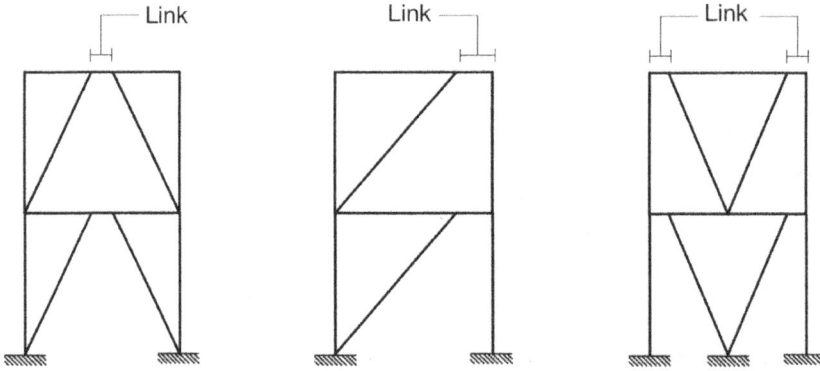

Figure 6-19 Different Eccentric Bracing Systems Showing the Link Beam

Because of this design, eccentric braced frames are considered more ductile, as reflected by their response modification factor in ASCE 7 Table 12.2-1 with $R = 8$. Eccentric braces enhance the system's flexibility and energy-dissipating capacity, enabling them to absorb a greater portion of seismic energy through inelastic deformation. These systems provide better control over displacement. For failure modes for all systems – see Figure 6-23.

Figure 6-20 Example of a Steel Eccentrically Braced Frame

Buckling-Restrained Braced Frames

Buckling-restrained braced frames represent a significant innovation in seismic resistance and lateral force mitigation, making them an important part of modern building design. An example of this type of frame is shown in Figure 6-21 below.

Figure 6-21 An Example of a Buckling-Restrained Braced Frame

These frames share a similar configuration with concentric braced frames, with one crucial difference: the bracing members do not buckle under compression. Instead, they rely on the yielding of the steel members in compression or tension to determine their capacity.

The absence of buckling failure is achieved by introducing continuous restraining for the core steel member which is primarily the one under either tension or compression. Figure 6-22 illustrates the internal structure of these members providing a look inside cross section. The steel core for this member consists of a steel cross section element, which can take various shapes – in this case, an "X" shape. The steel core is surrounded by continuous mortar filling, and the entire assembly is encased in a steel casing. A layer of unbonding material separates the steel core from the mortar, preventing the two from adhering. The end of the steel core is normally the one connected to the frame.

This innovative design offers a higher response modification factor for the system compared to traditional frames, as specified in ASCE 7 Table 12-2.1, Item B.25 for Steel Buckling-Restrained Braced Frames, which has a response modification factor of $R = 8$. This is greater than the response modification factors of special concentrically braced frames ($R = 6$) and ordinary concentrically braced frames ($R = 3\frac{1}{4}$), offering enhanced performance.

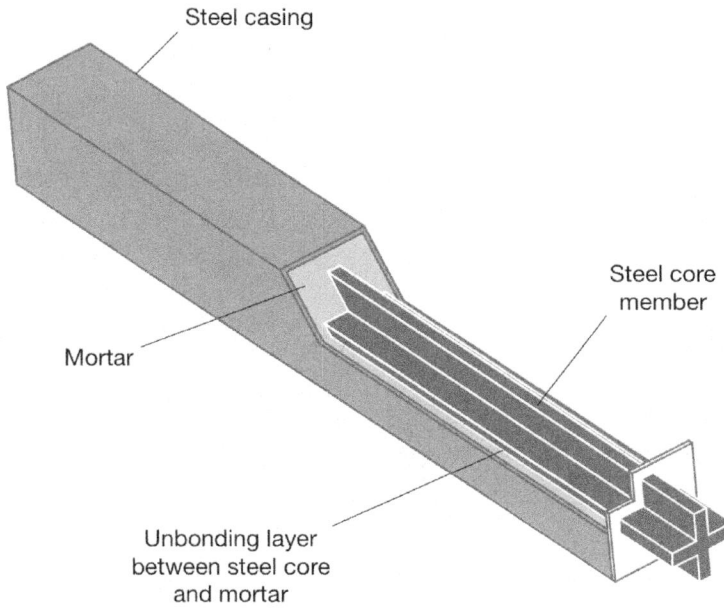

Figure 6-22 A Look Inside a Buckling-Restrained Brace Member

As discussed in this section, the failure of this bracing system occurs due to yielding in either compression or tension, as illustrated in Figure 6-23. This figure also highlights the failure modes of other bracing frame systems for comparison.

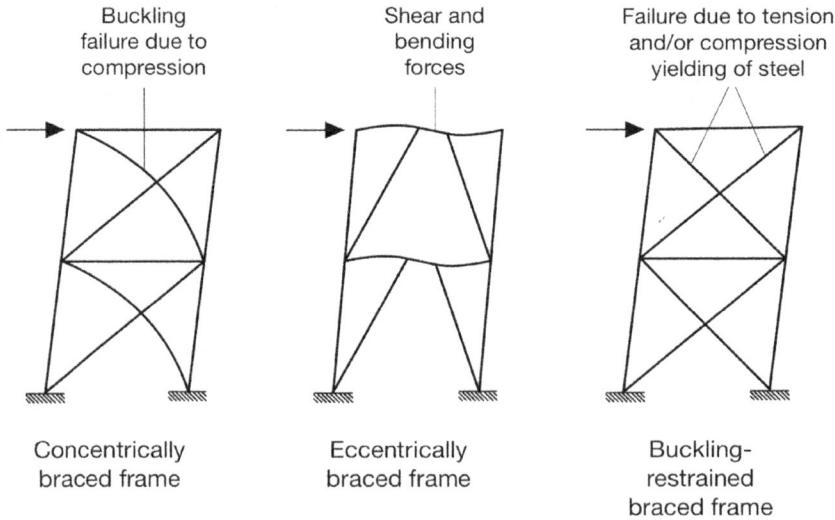

Figure 6-23 Failure Modes for Concentric, Eccentric and Buckling-Restrained Braces

Moreover, and due to their enhanced performance, these systems are particularly suitable for use as dampers in high-seismic areas. See Figure 2-22 in Chapter 2, which showcases dampers within framing systems. The viscoelastic dampers shown in Figure 2-22 can be replaced by those buckling-restrained braces.

6.6.3.   Steel Structures Seismic Requirements in CBC 2022

There are several seismic requirements outlined in California Building Code Sections 2205, 2206, 2211, and 2212, which are summarized in the table below. For more details, feel free to refer to the relevant sections of the CBC.CBC Chapter 22 Steel Structures Seismic Design Requirements.

Table 6-8 CBC Chapter 22 Steel Structures Seismic Design Requirements

Component	SDC	Requirement	CBC Reference
Structural steel	B & C	Use prequalified sections from AISC 341, this applies to beam/columns joints for IMFs and SMFs as well.	2205.2
	D to F	Similar to the above, and… All SFRS systems of ASCE Table 15.4-1 shall be designed in accordance with AISC 341.	2205.2
Cold-formed steel in light frame construction	B to F	If (R) was determined from ASCE Tabe 12.2-1, use AISI S400 for the SFRS design and detailing.	2211
	D to F in OSHPD 1R, 2 & 5	Similar to the above, also, steel stud plates or sills should be bolted to the foundation or foundation walls.	
Composite structural steel and concrete construction	In DSA-SS/CC	- Use Reduced Beam Section (RBS). - Built-up box columns wall thickness > 1.25 *in*, and HSS columns > ½ *in*. - Beam web fillet > 1/4 *in*.	2206 & 2212

6.7. Chapter References

The following references were used in this chapter. The reference(s) in italics is/are required for the exam, as noted in the preface of this book.

[1] ACI 318. Building Code Requirements for Structural Concrete and Commentary, 2019.

[2] AISC 341-22. Seismic Provisions for Structural Steel Buildings. An American National Standard.

[3] AISC 358-22. Prequalified Connections for Special and Intermediate Steel Moment Frames for Seismic Applications. An American National Standard.

[4] *American Society of Civil Engineers, ASCE/SEI 7-16 Minimum Design Loads for Buildings and Other Structures.*

[5] *AWC SDPWS-2021 Special Design Provisions for Wind and Seismic. Accessed through https://awc.org/publications/2021-sdpws in January 2025.*

[6] *Breyer, D., Cobeen, K. & Martin, Z. (2019). Design of Wood Structures – ASD/LRFD, 8th Edition. McGraw Hill.*

[7] *California Building Code (CBC 2022) California Code of Regulations, Title 24, Part 2, Volumes 1 and 2. Accessed through https://codes.iccsafe.org in January 2025.*

[8] EERI & IAEE (2011). Seismic Design Guide for Low Rise Confined Masonry Buildings. Confined Masonry Network, A Project of the World Housing Encyclopedia, with funding support from Risk Management Solutions.

[9] Hamburger, R. Krawinkler, H., Malley, J. & Adan, S. (2009). Seismic Design of Steel Special Moment Frames: A Guide for Practicing Engineers. NEHRP Seismic Design Technical Brief No.2. National Institute of Standards and Technology. U.S. Department of Commerce.

[10] Kersting, R.A., Fahnestock, L.A. & Lopez, W.A. (2015). Seismic Design of Steel Buckling-Restrained Braced Frames: A Guide for Practicing Engineers. NEHRP Seismic Design Technical Brief No.11. National Institute of Standards and Technology. U.S. Department of Commerce.

[11] Kingsley, G.R., Shing, P. & Gangel, T. (2014). Seismic Design of Special Reinforced Masonry Shear Walls. A Guide for Practicing Engineers. NEHRP Seismic Design Technical Brief No.9. National Institute of Standards and Technology. U.S. Department of Commerce.

[12] Moehle, J., Hooper, J. & Lubke, C. (2008). Seismic Design of Reinforced Concrete Special Moment Frames: A Guide for Practicing Engineers. NEHRP Seismic Design Technical Brief No.1. National Institute of Standards and Technology. U.S. Department of Commerce.

[13] Moehle, J., et al (2011). Seismic Design of Cast-in-Place Concrete Special Structural Walls and Coupling Beams: A Guide for Practicing Engineers. NEHRP Seismic Design Technical Brief No.6. National Institute of Standards and Technology. U.S. Department of Commerce.

[14] Sabelli, R., Roeder, C.W. & Hajjar, J.F. (2013). Seismic Design of Steel Special Concentrically Braced Frame Systems: A Guide for Practicing Engineers. NEHRP Seismic Design Technical Brief No.8. National Institute of Standards and Technology. U.S. Department of Commerce.

[15] TEK 05-01B. Concrete Masonry Veneer Details. Concrete Masonry & Hardscapes Association.

[16] TEK 03-06C. Concrete Masonry Veneers. Concrete Masonry & Hardscapes Association.

[17] TEK 14-18B. Seismic Design and Detailing Requirements for Masonry Structures. Concrete Masonry & Hardscapes Association.

[18] *TMS 402/602-16 Building Code Requirements and Specifications for Masonry Structures.*

Chapter Appendix

This appendix contains the following information:

➢ California Building Code 2022 Matrix Adoption Table Abbreviations

California Building Code 2022 Matrix Adoption Table Abbreviations

Abbreviation	Adopting Agency (*) (**)
AGR	Department of Food and Agriculture
BSC	California Building Standards Commission
BSC-CG	California Building Standards Commission - CALGreen
BSCC	Board of State and Community Corrections
CEC	California Energy Commission
CALTRAN	California Department of Transportation
DPH	Department of Public Health
DWR	Department of Water Resources
DSA AC	Division of the State Architect - Access Compliance
DSA SS	Division of the State Architect - Structural Safety
DSA SS/CC	Division of the State Architect - Structural Safety/Community Colleges
HCD	Department of Housing and Community Development
OSHPD 1	Office of Statewide Health Planning and Development (General acute care hospitals)
OSHPD 1R	Office of Statewide Health Planning and Development (Removed acute care hospitals)
OSHPD 2	Office of Statewide Health Planning and Development (Skilled nursing facilities)
OSHPD 3	Office of Statewide Health Planning and Development (General licensed health facilities such as licensed clinics)
OSHPD 4	Office of Statewide Health Planning and Development (Correctional treatment centers)
OSHPD 5	Office of Statewide Health Planning and Development (Acute psychiatric hospitals)
SFM	Office of the State Fire Marshal
SL	State Library
SLC	State Lands Commission

(*) If you were an agency and wish to have your specific requirements incorporated into the California Building Code (CBC), you can apply to have them included. The process begins with submitting a formal request to the Building Standards Commission (BSC), which will review and evaluate your proposed standards to ensure they align with California's building safety and regulatory needs. If approved, your agency's provisions will be added to the CBC's adoption matrix table, specifying which requirements apply to particular types of buildings or construction methods. Once included, these provisions will be applicable to relevant construction projects, and builders and designers will be required to follow them as outlined in the CBC.

(**) Detailed information for all the above, along with definitions of state-regulated and nonregulated buildings can be found in Chapter 1 of the CBC 2022.

This page is intentionally left blank

Problems & Solutions

The following questions have been carefully designed to provide comprehensive coverage of the material presented in this book, while also being educational in nature, as you will notice. Additionally, you may come across a few questions not directly covered in this chapter; these have been intentionally included to ensure broader coverage of the content.

➢ Problem 6.1: SDC for a Health Care Facility
➢ Problem 6.2: Distance from Active Earthquake Fault
➢ Problem 6.3: Day Care Occupant Load
➢ Problem 6.4: Structural Observations
➢ Problem 6.5: Inspection of Construction Activities
➢ Problem 6.6: Masonry Construction Inspection
➢ Problem 6.7: Cast-in-Place Deep Foundation Elements
➢ Problem 6.8: Continuous Special Inspection for a Wood Structure
➢ Problem 6.9: SFRS for a State Regulated Acute Health Facility
➢ Problem 6.10: Masonry Seismic Requirements
➢ Problem 6.11: Foundation Ties
➢ Problem 6.12: Buildings Seismic Instrumentation
➢ Problem 6.13: Cantilevered Open Diaphragms
➢ Problem 6.14: Wood Diaphragm Nominal Shear
➢ Problem 6.15: Wood Diaphragm Shear Strength Safety Factor
➢ Problem 6.16: Wall Chord Forces
➢ Problem 6.17: Unblocked Shear Wall Capacity
➢ Problem 6.18: Perforated Wall Total Shear Capacity
➢ Problem 6.19: Openings in Wooden Sheathed Diaphragms
➢ Problem 6.20: Type of Masonry Shear Wall Construction
➢ Problem 6.21: Detailed Plain Masonry Shear Walls
➢ Problem 6.22: Bolts Anchored at Concrete Columns or Pedestals
➢ Problem 6.23: Detailed Plain Concrete Walls
➢ Problem 6.24: The Main Function of Hoops in Special Moment Frames
➢ Problem 6.25: Best System for Energy Dissipation

PROBLEM 6.1 *SDC for a Health Care Facility*

A health care facility planned to have more than 50 care recipients with no emergency surgery is located in San Francisco (*). The building is founded on soil type C.

The Seismic Design Category SDC for this building is:

(A) SDC C

(B) SDC D

(C) SDC E

(D) SDC F

(*) Use the following spectral parameters for this facility in San Francisco:

$$S_S = 1.5$$
$$S_1 = 0.95$$

PROBLEM 6.2 *Distance from Active Earthquake Fault*

According to the California Building Code (CBC), how is the distance from an active earthquake fault measured?

(A) The horizontal distance from the building to the surface trace of the fault line.

(B) The distance from the nearest point of the building to the closest edge of an Alquist-Priolo fault zone for an active fault.

(C) The measurement taken from the fault's rupture zone to the closest building edge.

(D) The distance from the building to the fault trace, excluding buffer zones.

PROBLEM 6.3 *Day Care Occupant Load*

What is the Occupant Load for a day care facility with a gross area of $4,375\,ft^2$ and a net area of $3,500\,ft^2$?

(A) 100

(B) 125

(C) 150

(D) 175

PROBLEM 6.4 *Structural Observations*

Which of the following conditions require structural observations?

(A) The structure is classified as Risk Category II and above.

(B) The structure is a high-rise building.

(C) The structure is assigned to Seismic Design Category D and is greater than two stories above the grade plane.

(D) B + C.

PROBLEM 6.5 *Inspection of Construction Activities*

The following activities require continuous special inspections for an OSHPD 4 (correctional treatment centers) in the state of California based on CBC 2022 except:

(A) Inspect and test reinforcement in special moment frames.

(B) Confirm sizes and lengths of material for driven deep foundations.

(C) Confirm that the shallow foundation substrates can achieve the design bearing capacity.

(D) Inspect tension adhesion anchors installed in finished concrete elements that are either horizontal or vertical.

PROBLEM 6.6 *Masonry Construction Inspection*

The following constitutes the minimum continuous special inspections for risk category IV masonry structures designed in accordance with Part 3 of the TMS 402/602-16:

    (A) Checking the size of reinforcements along with prestressing steel, bolts, and anchorage.

    (B) Observing the preparation of grout and mortar specimens.

    (C) Inspecting the building of masonry units and mortar joints.

    (D) The inspection of the location and the sizing of structural elements.

PROBLEM 6.7 *Cast-in-Place Deep Foundation Elements*

What is the minimum reinforcement required for cast-in-place deep foundation elements for an OSHPD 1 (acute care hospital) assigned to Seismic Design Category E per the California Building Code CBC 2022?

    (A) A minimum of two longitudinal bars with ratio ≥ 0.005.

    (B) A minimum of three longitudinal bars with ratio ≥ 0.005.

    (C) A minimum of four longitudinal bars with ratio ≥ 0.005.

    (D) No specific reinforcement requirements.

PROBLEM 6.8 *Continuous Special Inspection for a Wood Structure*

For structural wood used in Seismic Force Resisting Systems (SFRS) for an OSHPD 4 (correctional treatment centers) in the state of California based on CBC 2022, continuous inspection is required during field gluing of SFRS elements in which Seismic Design Categories?

    (A) SDC B

    (B) SDC C

    (C) SDC D

    (D) Options (B) + (C)

PROBLEM 6.9 *SFRS for a State Regulated Acute Health Facility*

Which of the following moment resisting frames systems is permitted in a state regulated OSHPD acute health facility per the California Building Code 2022?

    (A) Steel special moment frames.

    (B) Steel special truss moment frames.

    (C) Steel intermediate moment frames.

    (D) Cold-formed steel special bolted moment frames.

PROBLEM 6.10 *Masonry Seismic Requirements*

A retail building located in Los Angeles, West Hollywood (*), soil type B, with a foundation masonry wall $8\,in$ thick that is not part of the SFRS and not laid in a running bond.

The minimum horizontal reinforcements used in this wall should be:

    (A) Two longitudinal wires of $W1.7$ spaced no more than $16\,in$.

    (B) $No.4$ bars spaced no more than $48\,in$.

    (C) 0.0015 of the gross cross section with maximum spacing of $24\,in$.

    (D) $No.6$ bars spaced no more than $48\,in$.

(*) Use the following spectral parameters:

$S_S = 2.38$

$S_1 = 0.86$

PROBLEM 6.11 Foundation Ties

For a building with shallow footings on Soil Class E assigned to a Seismic Design Category D, and $S_{DS} = 0.5$, foundation ties are required.

Given that the larger footings carry a gravity load of 25 *kip* and the smaller footings carry a gravity load of 15 *kip*, what should be the design force for the ties connecting these footings per the California Building Code?

(A) 1.25 *kip*

(B) 2.5 *kip*

(C) 3.75 *kip*

(D) 4.5 *kip*

PROBLEM 6.12 *Buildings Seismic Instrumentation*

According to the 2022 California Building Code, how many accelerographs should be provided for a structure located in a seismic zone with an $S_1 > 0.4$ and exceeds ten stories in height?

(A) ≥ 1

(B) ≥ 2

(C) ≥ 3

(D) More information needed to decide.

PROBLEM 6.13 *Cantilevered Open Wooden Diaphragms*

Which of the following requirements apply to open structures cantilevered diaphragms made of wood sheathing?

(A) The maximum panel sizes for these diaphragms 8 *ft* × 8*ft*.

(B) Aspect ratio for nonhigh loads blocked diaphragms < 1.5: 1.

(C) Nail edge distance should be 0.5 *in*.

(D) Diaphragms with torsional irregularity aspect ratio < 1.5: 1.

PROBLEM 6.14 *Wood Diaphragm Nominal Shear*

What is the nominal unit shear capacity for the following Structural I that is ½ *in* thick sheathing grade wood diaphragm when the loading is applied in the direction shown?

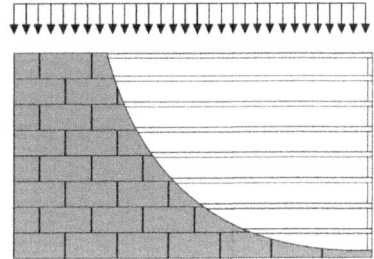

Given the following:

o   The framing is Southern Pine.

o   6 *in. O. C. 6d* nail spacing.

o   Width of nailed framing face at adjoining panels is 2 *in*.

(A) 600 *plf*

(B) 640 *plf*

(C) 670 *plf*

(D) 800 *plf*

PROBLEM 6.15 *Wood Diaphragm Shear Strength Safety Factor*

The below is a 30 *ft* wide blocked wooden diaphragm, case 5 loading per the load direction shown below, framing is a 2 *in* wide Southern Pine, 8*d* 6 *in O. C.* nails are used, sheathing grade is a normal sheathing 15/32 *in* thick.

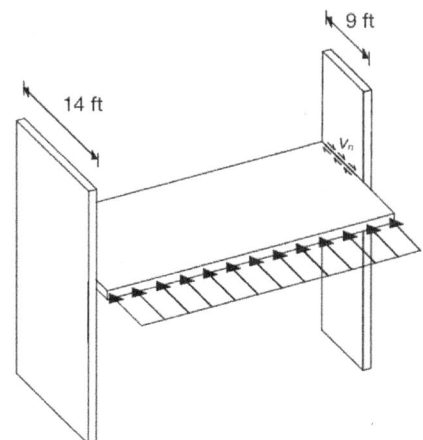

The total seismic force applied to the diaphragm at this floor level in the specified direction is 3,600 lb, calculated using the ASD load combination method.

What is the safety factor for shear strength of the diaphragm based on this seismic force calculated against the allowable shear capacity?

    (A) 1.4

    (B) 2.4

    (C) 3.3

    (D) 4.1

PROBLEM 6.16 *Wall Chord Forces*
The below segmented shear wall structure is subject to a lateral load of 20 kip in the direction shown. The walls are made of the same material.

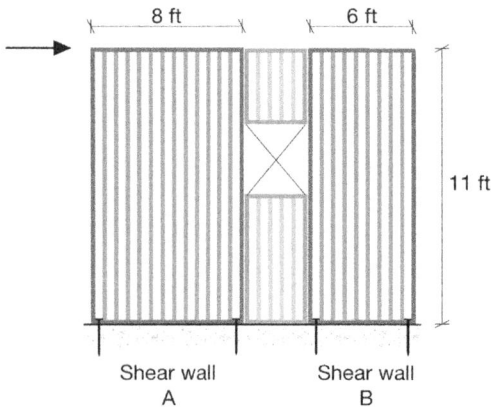

What is the tension force generated in the chord of shear wall A?

    (A) 32.6 kip

    (B) 29.8 kip

    (C) 25.0 kip

    (D) 15.7 kip

PROBLEM 6.17 *Unblocked Shear Wall Capacity*
The following is a 11 ft tall, 8 ft wide, unblocked wood structural panel shear wall. The sheathing is a normal 3/8 in thick and

is attached to the studs with 8d nails spaced 6 in O.C.

The studs are spaced 24 in apart, and the panels used are 4 ft × 8 ft in size.

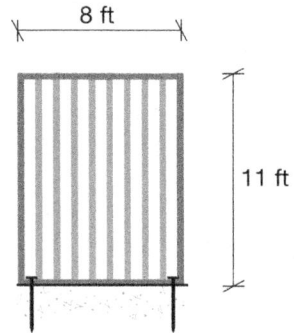

Based on this information, what is the nominal unit shear capacity for this wall?

    (A) 310 plf

    (B) 350 plf

    (C) 615 plf

    (D) 635 plf

(*) Structural I per shear capacity table is a superior and more expensive grade, used only when required. Thus, a "normal" grade implies Structural I is not selected.

PROBLEM 6.18 *Perforated Wall Total Shear Capacity*
The unit shear capacity for the below perforated wood/sheath paneled shear wall has been determined as 1,000 plf without taking into account the provided openings.

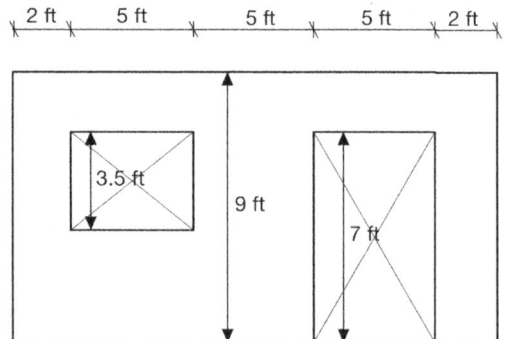

Knowing that sheathing is provided to this wall fully from top to bottom (excluding openings), what would be the adjusted total shear capacity for this wall?

    (A) 3,150 *lb*

    (B) 3,600 *lb*

    (C) 4,250 *lb*

    (D) 4755 *lb*

PROBLEM 6.19 *Openings in Wooden Sheathed Diaphragms*

In accordance with the 2022 California Building Code, for Seismic Design Category E, what is the requirement for openings larger than 4 *ft* in a diaphragm wood-sheathed floor for a light-frame structure?

    (A) Provide blocking for a distance of 4 *ft* at the sides of this opening.

    (B) Provide blocking for a distance equivalent to the opening dimension at the sides.

    (C) Provide blocking for a distance equivalent to 1.5 times the opening dimension at the sides.

    (D) Openings of this size are not permitted in SDC E.

PROBLEM 6.20 *Type of Masonry Shear Wall Construction*

Which Seismic Design Categor(ies) is/are most suitable for the use of an intermediate masonry shear wall?

(Choose the best answer)

    (A) SDC A

    (B) SDC A & SDC B

    (C) SDC A, SDC B & SDC C

    (D) SDC A, SDC B, SDC C & SDC D

PROBLEM 6.21 *Detailed Plain Masonry Shear Walls*

What is the minimum requirement for horizontal reinforcement for a detailed unreinforced masonry shear wall?

    (A) Two longitudinal wires of *W*1.7 spaced no more than 16 *in* every 120 *in*.

    (B) Two longitudinal wires of *W*1.7 spaced no more than 16 *in* every 48 *in*.

    (C) 0.0015 of the gross cross section with maximum spacing of 24 *in*.

    (D) No reinforcement is required.

PROBLEM 6.22 *Bolts Anchored at Concrete Columns or Pedestals*

Per ASCE 7 Standard, for structures assigned to Seismic Design Categories C, D, E, or F, what is the required transverse reinforcement arrangement when anchor bolts are placed at the top of a column or pedestal?

    (A) Surround anchor bolts by at least four longitudinal bars within 5 *in* of the top with at least two *No.* 4. Ties must have hooks.

    (B) Surround anchor bolts by at least four longitudinal bars within 5 *in* of the top with at least two *No.* 3. Ties must have hooks.

    (C) Surround anchor bolts by at least four longitudinal bars within 10 *in* of the top with at least two *No.* 4. Ties must have hooks.

    (D) Surround anchor bolts by at least four longitudinal bars within 10 *in* of the top with at least two *No.* 3. Ties must have hooks.

PROBLEM 6.23 *Detailed Plain Concrete Walls*

What is the minimum requirement for horizontal reinforcement for a detailed plain concrete wall?

(A) A minimum of 0.2 in^2 of the cross section with a maximum spacing of 120 in.

(B) *No.*3 rebars placed every 180 in.

(C) Openings are only required to be reinforced using continuous two *No.*5 at all edges.

(D) No reinforcement is required.

PROBLEM 6.24 *The Main Function of Hoops in Special Moment Frames*

Which of the following is the primary function of hoops in special concrete moment frames?

(A) Provide transverse reinforcement for shear resistance..

(B) Confine concrete, prevent rebars buckling, improve bond strength, in locations of possible plastic hinges.

(C) Prevent plastic hinge formation and reduce the likelihood of concrete spalling.

(D) To act as lap splices for reinforcing bars within beams.

PROBLEM 6.25 *Best System for Energy Dissipation*

Which of the following structural systems is generally considered the most ductile based on their response energy dissipation characteristics?

(A) Steel special concentrically braced frames.

(B) Steel ordinary concentrically braced frames.

(C) Steel eccentrically braced frames.

(D) Steel special truss moment frames.

SOLUTION 6.1

The Seismic Design Category is determined using Section ASCE 7 11.6 when $S_1 > 0.75$. This information is also summarized in Table 1-9 in Chapter 1 of this book.

The risk category for a health care facility with more than 50 care recipients with no emergency surgery provided is given in CBC Table 1604.5, also mentioned in this book in Section 6.2.2, and shall be assigned a risk category III.

Based on Table 1-9 of Chapter 1 of this book, the Seismic Design Category is E.

Correct Answer is (C)

SOLUTION 6.2

In accordance with the definitions chapter of the CBC code, it states that the required distance is measured from the nearest point of the building to the closet edge of an Alquist-Priolo Earthquake Fault Zone for an active fault. Alquist-Priolo fault zones has been defined in Section 1.2 of Chapter 1 in this book.

Correct Answer is (B)

SOLUTION 6.3

In accordance with Table 1004.5 of the CBC code, the Occupant Load Factor OLF for a day care function or space is a net of $35\ ft^2/person$.

$$OL = \frac{3,500\ ft^2}{35\ ft^2/person} = 100\ (*)$$

Correct Answer is (A)

(*) CBC Table 1006.3.3 requires two number of access or exits for this facility in this case.

SOLUTION 6.4

In accordance with Section 1704.6.1, only one of the three items mentioned in this question requires structural observation, which is when the structure is a high-rise building.

Correct Answer is (B)

SOLUTION 6.5

Check CBC Chapter 17A Special Inspections and Tests for OSHPD 4 (correctional treatment centers) and the following tables:

- Table 1705A.3 Concrete Construction.
- Table 1705A.6 Tests of Soils.
- Table 1705A.7 Driven Deep Foundation Elements.

The below items from the abovementioned tables correspond to the statements in the question and are the only ones requiring continuous special inspections:

Item	Continuous Special Inspection
1705A.3 Item 4.a	Inspect anchors post-installed in hardened concrete members: (a) Adhesive anchors installed in horizontally or upwardly inclined orientations to resist sustained tension loads.
1705A.3 Item 1.a	Inspect and test reinforcement, including prestressing tendons, and verify placement: (a) Reinforcement in special moment frames, boundary elements of special structural walls and coupling beams.
1705A.7 Item 1	Verify element materials, sizes and lengths comply with the requirements.

The only activity from the statements mentioned in the body of the question that does not require continuous special inspections is activity (C).

The following activity requires periodic special inspection only:

Item	Periodic Special Inspection
1705A.6 Item 1	Verify materials below shallow foundations are adequate to achieve the design bearing capacity.

Correct Answer is (C)

SOLUTION 6.6
The question refers to the TMS 402 and TMS 602 Building Code Requirements and Specifications for Masonry Structures version 2016.

It is important to observe the differences between this version and the 2013 version. Chapter 3 of TMS 402/602-2013 presented three quality assurance tables (Quality Assurance Level A, B, and C). However, these tables were removed from TMS 402 and TMS 402 now references TMS 602. The tables were also modified into two tables with levels of compliance in pages S24-26.

In TMS 402/602-2016, Chapter 3, Table 3.1 assigns a level 3 for risk category IV structures that are designed in accordance with Part 3 of the same code. Quality assurance level 3 minimum requirements are listed in Table 4 of TMS 602-2016, page S26. This table identifies the minimum special inspections (periodic or continuous) for this level.

When all options in the question belong to the periodic requirements of inspection of this program, only option B, which specifies observing grout specimens' preparation, was mentioned under the continuous inspection requirement.

Correct Answer is (B)

SOLUTION 6.7
Reference is made to the California Building Code CBC 2022 Section 1810A.3.9.4.2 for OSHPD 1. Also, Table 6-3 provided in this chapter. A minimum of four longitudinal bars with ratio ≥ 0.005 is what is required for a cast-in-place deep foundation elements in Seismic Design Category E.

Correct Answer is (C)

SOLUTION 6.8
Reference is made to the California Building Code CBC 2022 Section 1705A.13.2 for OSHPD 4 (correctional treatment centers).

Continuous inspection is required during field gluing of SFRS elements in Seismic Design Category D.

These are also summarized in Section 6.2.3 of this book.

Correct Answer is (C)

SOLUTION 6.9
Reference is made to the California Building Code 2022 Section 1617A.1.4 item (C) – you can also refer to the Appendix of Chapter 2 in this book were the NOT PERMITTED systems per CBC 2022 are shaded for ease of reference.

Based on the above, the following systems are not permitted:

2. Special steel truss moment frames – Not permitted by OSHPD.

3. Intermediate steel moment frames – Not permitted by OSHPD.

12. Cold-formed steel special bolted moment frame – Not permitted by DSA-SS and OSHPD.

Hence, the only OSHPD permitted system with the use of moment resisting frames systems a "steel special moment frames".

Correct Answer is (A)

SOLUTION 6.10

Based on the spectral response acceleration parameter of one second for this location, $(S_1 = 0.86) > 0.75$, and, risk category II (retail), the Seismic Design Category for this building is E – See Table 1-9 here.

Given this is a general retail building, CBC 2022 Section 1807.1.6.3.2 for masonry foundation walls – which refers to TMS 402-16 Section 7.4.5.1 (summarized in Table 6-1 and 6-2 in this book) – shall be used.

The above code section (or Table 6-2 of this book) specifies for Seismic Design Category E, with nonparticipating masonry not laid in running bond, an area of horizontal reinforcement of at least 0.0015 multiplied by the gross area of masonry with a maximum spacing of 24 in.

Correct Answer is (C)

SOLUTION 6.11

Reference is made to the CBC 2022 Chapter 18 Section 1809.13 or 1809A.13. This information is also summarized and provided in Tables 6-1 or 6-3 of this chapter, and is calculated as follows:

$$= Min\left\{\left(\frac{25\ kip \times 0.5}{10}\right) or\ (25\% \times 15\ kip)\right\} = 1.25\ kip$$

Correct Answer is (A)

SOLUTION 6.12

Reference is made to CBC 2022 Appendix L, or Section 6.2.7 of this book.

Structures to be equipped with more than three approved recording accelerographs, (and at least shall include one tri-axial free-field instrument or equivalent for OSHPD 1 & 4), when exceeding ten stories in height regardless of floor area, or, when exceeding six stories with aggregate floor area > 60,000 ft^2.

Correct Answer is (C)

SOLUTION 6.13

Referring to SDPWS 2021 Section 4.2.6, the most applicable requirement is (B), which states that the aspect ratio for a blocked diaphragm (non-high) must be less than 1.5:1, as outlined in Section 4.2.8.1 of the same document – also check page 223 here.

Correct Answer is (B)

SOLUTION 6.14

Reference is made to Section 6.3 in this book, also to SDPWS 2021 Table 4.2C for unblocked sheathed wood diaphragms.

This floor is made of non-staggered panels. The long continuous edge is parallel to the framing system, and the load applied is in the short direction. This corresponds to load case 2.

Using Structural I sheathing grade, $6d$ nails at $6\ in\ O.C.$ with a 2 in width of nailed face Southern Pine frame generates the following unit capacity from using Table 4.2C:

$$v = 600\ plf\ (*)$$

Correct Answer is (A)
(*) <u>This is a trick question</u>:

The thickness of ½ in is not listed in the SDPWS table. However, by consulting the required references for this exam (refer to reference [6] in the chapter reference list), page 467 of this reference specifies that if a panel thicker than those provided in the SDPWS tables is used, the unit shear should be based on the nailing configuration.

The same principle applies to nails when a nailing configuration is not provided in those tables, where the minimum value from all applicable cases should be used.

In other words, no increase in unit shear is permitted due to increase in thickness or nails configuration beyond what is provided in those tables.

SOLUTION 6.15

Reference is made to SDPWS 2021, Table 4.2A for blocked diaphragms.

Given the question data, the nominal unit shear capacity for this diaphragm is:

$$v_n = 755\,plf$$

The allowable unit shear using SDPWS Section 4.1.4 is:

$$v_a = \frac{755\,plf}{2.8} \cong 270\,plf$$

Calculate the maximum unit shear based on the allowable load applied to it, assuming the diaphragm acts as a simple beam using the assumptions of Figure 4-2 of Chapter 4 of this book:

$$R_{a,shear\,wall} = 0.5 \times 3{,}600 = 1{,}800\,lb$$

The above is the reaction on any of the two shown shear walls.

To convert this value into unit shear, divide it by the shortest wall length (i.e., $9\,ft$) to generate the maximum applied unit shear:

$$= \frac{1{,}800\,lb}{9\,ft} = 200\,plf$$

The safety factor is therefore calculated as follows:

$$= \frac{270\,plf}{200\,plf} \cong 1.4$$

Correct Answer is (A)

SOLUTION 6.16

In reference to Figure 4-1 in Chapter 4 of this book, these walls are made of similar material, wood, and they are placed in a Single Line of Action, this means that the lateral load is distributed to each wall based on length.

Wall A lateral load portion is therefore calculated as follows:

$$V = \frac{8}{8+6} \times 20 = 11.4\,kip$$

Footnote of Section 6.3.4 of this book states that chord forces are determined using the overturning moment. This is also per SDPWS Equation 4.3-7.

$$T = \frac{Vh}{b} = \frac{11.4 \times 11}{8} = 15.7\,kip$$

Correct Answer is (D)

SOLUTION 6.17

In this question you can refer to Table 6-4 of this book which refers you to SDWPS Equation 4.3-2 and Table 4.3.5.3.

$$v_{n,(ub)} = V_{n(b)} \times C_{n(ub)}$$

Per SDPWS Table 4.3A:

$$v_{n(b)} = 615\,plf$$

Aspect ratio is $11/8 = 1.38 : 1 < 2 : 1$ hence no need for any modification to account for aspect ratio.

Per SDPWS Table 4.3.5.3, and the information provided in the question, use a value of $C_{n(ub)} = 0.5$.

$$v_{n,(ub)} = 615 \times 0.5 = 307.5 \approx 310\,plf$$

Correct Answer is (A)

SOLUTION 6.18

In this question you can refer to Table 6-4 of this book which refers you to SDPWS 2021 Section 4.3.5.6, and Table 4.3.5.6 for determining (C_o).

$$V_n = v_n C_o \sum b_i$$

In order to determine C_o from the SDPWS Table 4.3.5.6, we need to determine few factors first:

$$\frac{A_o}{A_{wall}} = \frac{5 \times 3.5 + 5 \times 7}{19 \times 9} \cong 30\%$$

$$\frac{A_{fhs}}{A_{wall}} = \frac{2 \times 9 + 5 \times 9 + 2 \times 9}{19 \times 9} \cong 47\%$$

Where in this case A_{fhs} is defined in Section 4.3.5.6 of the SDPWS 2021 as the area of the full-height sheathing in the wood-frame shear walls. This area refers to the portion of the wall that is continuously sheathed from top to bottom without interruptions by openings. See below sketch.

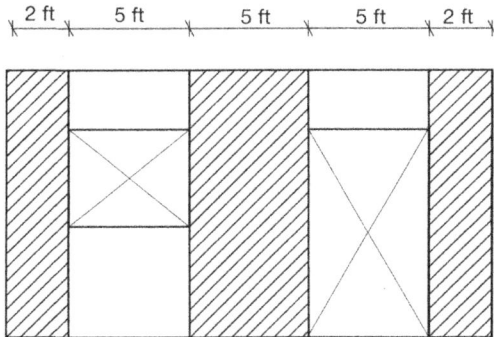

Based on this, and using interpolation, SDPWS Table 4.3.5.6 (or with the use of Equation 4.3-6) provides the following result:

$$C_o = 0.72$$

$\sum b_i$ is the sum of the remaining shear wall segments lengths, provided their aspect ratio $< 3.5 : 1$, and if $> 2 : 1$ (which is not the case here), should be adjusted using SDPWS Section 4.3.3.4.

$$V_n = v_n C_o \sum b_i$$
$$= 1{,}000 \times 0.72 \times (0 + 5 + 0)$$
$$= 3{,}600 \ lb$$

Correct Answer is (B)

SOLUTION 6.19
In reference to Table 6-5 of this book, which refers to the requirements for openings larger than **4 ft** for SDC B to E in Section 2308.4.4.1 of the CBC 2022. The latter specifies that blocking should be provided for a distance equivalent to the opening's dimension with a metal strap or tie provided along this distance as well.

Correct Answer is (B)

SOLUTION 6.20
Reference is made to Table 6-6 of this chapter, or TMS 402 Table CC-7.3.2-1.

Based on those references, intermediate reinforced sear walls can be used in SDC A, B & C.

Correct Answer is (C)

SOLUTION 6.21
Reference is made to Figure 6-5 of this chapter, or TMS Section 7.3.2.3.1.

Based on those references, detailed plain (unreinforced) masonry shear wall minimum horizontal reinforcement should be Two longitudinal wires of $W1.7$ spaced no more than 16 *in* every 120 *in*.

Correct Answer is (A)

SOLUTION 6.22
Reference is made to ASCE 7 Standard Section 14.2.2.2.

In ASCE, it is stated that bolts placed in the end of columns or pedestals shall be surrounded with at least four longitudinal bars within 5 *in* of the top with at least two *No.4* (or at least three *No.3*) and ties must have hooks.

Correct Answer is (A)

SOLUTION 6.23
Reference is made to CBC 2022 Section 1905.1.6 – also see Section 6.5.4 here.

Per the above, horizontal reinforcement should be a minimum of $0.2 \ in^2$ provided in the cross-sectional area with a maximum spacing of 120 *in*.

Correct Answer is (A)

SOLUTION 6.24
Reference is made to Section 6.5.3 in this book, where the role of hoops in special concrete moment frames is discussed.

Hoops are primarily used to confine concrete, prevent reinforcement buckling to improve bond strength, especially in locations where lap splices are present. They also resist shear forces and are crucial in areas where plastic hinges are likely to form due to flexural actions, which can lead to concrete spalling. This makes them distinct from stirrups, which are not closed and are used primarily for shear resistance in locations where confinement is not a concern.

Correct Answer is (B)

SOLUTION 6.25
Steel eccentric braced frames are generally the most ductile among the listed systems. They are designed to absorb and dissipate seismic energy through controlled inelastic deformations in the short link(s) between the bracing connections as shown in Figure 6-23 of this chapter. This results in a higher Response Modification Factor $(R = 8)$, compared to the response modification factor for the other listed system. This reflects the system's ability to undergo large deformations without failure, indicating high ductility and better energy dissipation.

Correct Answer is (C)

CHAPTER
7

Seismic Vulnerability and Improvements

7.1. Purpose of this Chapter

By now, if you completed all the chapters leading up to this one, you should have a solid understanding of the lateral seismic system – how it resists lateral forces, the critical points that could severely affect its performance, and its interaction with the gravity-resisting system. With this foundation, you should now be well-equipped to assess a lateral system's vulnerability after ground shaking, identifying damages, potential weaknesses, and areas for improvement. This chapter will help you consolidate that knowledge and guide you through a process to identify clear signs of damage in a post-earthquake system that may be hazardous and require immediate attention or further improvement.

This chapter begins by defining the assessment process, outlining the role and qualifications of inspectors, and describing the steps they take to evaluate buildings after ground shaking. It then highlights key observations in buildings (or the surrounding ground conditions) that may lead to critical decisions, such as declaring a building, or an area, as unsafe, identifying the need for improvements, or determining the need for further assessment.

Throughout this chapter, we reference the ATC 20 Procedures for Postearthquake Safety Evaluation of Buildings, along with its addendum and manual. These references provide additional information and include helpful figures that are easy to read and follow. It is highly recommended to obtain a copy of these documents if you wish to gain a comprehensive understanding of this topic and to refer to them for further details.

7.2. Buildings Evaluation Procedure
7.2.1. Overview

The procedure consists of three stages, the Rapid Evaluation (10 to 20 minutes per structure), the Detailed Evaluation (1 to 4 hours per structure), and the Engineering Evaluation (1 to 7 days or more per structure).

The rapid evaluation quickly assesses buildings' apparent safety. Structures that are obviously unsafe are designated as such using a red placard, while those deemed safe are marked as *Inspected* with a green placard. Buildings which are undecided and requiring further inspection, are marked as *Limited Entry* or *Restricted Use* using a yellow placard and are flagged for a more detailed assessment by a team of structural engineers, which marks the start of the second stage for those specific buildings. The rapid evaluation may not involve structural engineers due to resource limitations in the aftermath of a devastating event.

The second phase, conducted by structural engineers (or geotechnical engineers if needed), is more thorough. This phase classifies the undecided (grey area) structures as either safe and inspected using a green placard, potentially dangerous with limited entry using yellow, or unsafe with the use if a red placard. Hospitals, police stations, and fire stations should be prioritized for detailed evaluation.

Any further evaluation beyond this stage should be performed by a structural engineering consultant hired by the owner. This phase may involve calculations and modeling, and the outcome could be suggested improvements, or the removal of structure(s).

For posting, the following color codes are used, regardless of the phase of assessment:

- Green: Inspected and generally safe.
- Yellow: Limited entry is allowed; the building is either dangerous or undecided. The owner may enter for emergency purposes at their own risk. The building cannot be used continuously, and the public is not permitted.
- Red: Unsafe, extreme hazard, potential collapse. Entry is prohibited except for authorities.
- Barricade or Yellow Tape (without placards): The area is unsafe due to potential falling hazards, such as a parapet, chimney, or other structural elements.

Re-inspections are normally conducted to update placards based on new findings or to correct previous mistakes.

7.2.2.   Rapid Evaluation

Due to the shortage of structural engineers post a damaging earthquake, the rapid evaluation depends on inspectors and volunteers who should have at least five years of experience in general building design and construction. Ideally, structural engineers familiar with the structural aspects of building construction would conduct the evaluation, but this is not always feasible.

As outlined in the overview section, the rapid evaluation takes 10 to 20 minutes of exterior inspection with no entry into the building, unless specific issues are reported. If a building is deemed unsafe, an *Unsafe* placard should be placed immediate. These placards, along with others as mentioned in the overview section, must be displayed at each entrance of the building.

During the evaluation, inspectors fill out a rapid assessment form that addresses six key questions. A "yes" answer to questions 1, 2, 3, or 5 below requires posting an *Unsafe* placard. A "yes" answer to question 4, or certain aspects of question 6, such as hazardous material spills, marks an area as *Area Unsafe*. In cases where the evaluation is inconclusive, a *Limited Entry* or *Restricted Use* placard is used. The questions are as follows:

1.   Is there collapse, partial collapse, or is the building off its foundation?
2.   Is the building or a story noticeably leaning?
3.   Is there severe racking of walls, or obvious severe damage and distress?
4.   Are there falling hazards such as chimneys, parapets, or other structural elements?
5.   Is there severe ground or slope movement present?
6.   Are there other hazards present?

7.2.3.   Detailed Evaluation

Essential facilities are given priority for detailed evaluation, such as: hospitals, health care facilities, police and fire stations, jails and detention centers, communication centers and emergency operations centers. Two structural engineers should conduct this evaluation and for buildings designated as *Restricted Use* (or *Limited Entry*). Due to shortage of engineers, one structural engineer and one inspector can be deemed acceptable. Volunteers from the California Office of Emergency Services (OES) Engineering Program are also considered for participation.

This evaluation does not involve destructive exploration, such as the removal of plaster. If necessary, the building owner will be requested to remove plaster walls or other architectural features for inspection. In some cases, specialists may be called upon for more complex structures, such as high-rise buildings or other unique structures. The evaluation focuses on assessing both the lateral and vertical load-carrying systems, ensuring they are fully functional. The lateral system, in particular, must be capable of handling aftershocks.

There are general guidelines for detailed evaluations of common building types, including two-story apartments, office buildings, multi-story structures, and tilt-up industrial buildings. These guidelines require inspectors to examine specific structural elements to assess the safety of the building and its surroundings. See Table 7-1 for those guidelines.

Table 7-1 General Guidelines for Unsafe Conditions

Overall Damage	P-Delta Effect	Falling Hazards
Collapse or partial collapse	Multistory frame building with residual drift	Such as parapets or chimneys or others
The entire building or one single story leaning	-	Veneer separation
Fractured foundation	-	Other failures
Vertical Load System	**Lateral Load System**	**Slope/Foundation Distress**
Columns out of plumb	Broken or leaning moment frames	Building base pulled apart or deferential settlement occurred
Buckled or failed columns	Severely cracked shear walls	Suspected major slope movement
Roof/floor framing separation from vertical support	Broken or buckled vertical braces	Sliding or falling landslide debris to affect buildings
Bearing wall, pilaster or corbel cracking	Broken or seriously damaged diaphragms or horizontal brace	-

Special attention during the evaluation should be focused on areas exhibiting horizontal torsional or vertical irregularities, as shown in Figures 2-16 and 2-17.

Post the inspection of restricted entry buildings, the placard can be updated to either *Unsafe* or *Inspected*. While entry remains restricted for all buildings with the exception of the *Green* ones, if an owner wishes to access the property to retrieve personal items or other belongings, and for other search and rescue operations, special permission must be obtained. For guidance, refer to guidelines outlined in Appendix D of the ATC 20-1 Field Manual. This includes recommendations related to the mainshock magnitude, the stability of unsafe buildings, and any required waiting periods for buildings deemed unsafe but stable before gaining entry.

7.2.4. Detailed Evaluation for Targeted Buildings Type

This section outlines key areas in affected buildings that should be thoroughly examined, based on observations from major earthquakes. These areas are relevant to various building types and geotechnical hazards, and such assessments may determine whether the building, structure, or its surroundings are unsafe or require adjustments. Table 7-2 below addresses wood, masonry, and concrete structures, while Table 7-3 focuses on steel buildings.

Table 7-2 Suggested Inspection Points for Wood, Masonry and Concrete Structures

	Small Wood Buildings	Larger Wood buildings	Unreinforced Masonry	Reinforced Masonry	Tilt-up Structures	Concrete Cast-in-Place	Precast Concrete	Liftslab Structures
General/Overall Damage								
Collapse or partial collapse	x	x	x	x	x	x		
Building or individual story noticeably leaning		x	x	x	x	x		
Separation between two parts of a building/dwelling	x							
Hillside homes with posts with residual lean, broken braces or displacement	x							
Failure of significant vertical load carrying element or connection	x	x	x	x	x	x		
First floor leaning in an open front apartments or similar structures		x						
Foundations								
House off foundation or signs of movement	x	x						
Fractured foundations	x	x	x	x	x	x		
Bowing of underground concrete walls						x		
Floors, Roofs or Diaphragms								
Failed member or connection in roof truss		x			x			
Roof truss or heavy framing separation from vertical support		x			x			
Roof/floor framing has pulled away from vertical support		x			x			
Roof or floor displaced from walls	x							
Roof or floor framing separation from wall			x	x				
Bowed, broken or seriously damaged diaphragms			x	x	x			
Broken or severely damaged diaphragm chord or collector			x	x	x			
Movement or failure at shear connection between diaphragm and wall			x	x	x			
Broken floor wall connection in concrete or seriously damaged diaphragm						x		
Separated sheets or framing in plywood diaphragms		x						
Plaster cracking in roof or floor framing elements			x					
Columns								
Columns noticeably out of plumb			x	x		x		
Buckled or failed columns			x	x		x	x	
Massive spalling for concrete columns and exposure of reinforcements						x		
Large diagonal crack extending through concrete columns						x		

	Small Wood Buildings	Wood larger buildings	Unreinforced Masonry	Reinforced Masonry	Tilt-up Structures	Concrete Cast-in-Place	Precast Concrete	Liftslab Structures
Walls								
Failed wood shear walls or piers (in-plane) with many opening		X						
Failure or imminent failure of wood walls providing vertical support		X						
Cripple wall failure	X							
Wood walls residual racking > **1 to 2 in** per story	X	X						
Masonry walls out of plumb			X					
Masonry walls diagonal or other large cracking			X					
Reinforced masonry/concrete walls with > **1/8 in** diagonal (shear) crack				X		X		
Concrete walls with > **1/8 in** horizontal crack						X		
Slippage of **1/8 in** in horizontal concrete walls construction joints						X		
Several failed piers at any story level			X			X		
Walls, corbels (for concrete) or pilasters cracking jeopardizing vertical support			X		X		X	
Failure of *infill masonry wall* or large cracks in it extending to concrete frames						X		
Wythe separation and masonry spalling in masonry walls			X					
Adobe building with structural damage			X					
Tilt up walls – Outward leaning wall panel					X			
Tilt up walls – Broken or severely damaged closure strips or broken tie plates					X			
Slabs, Flat Slabs and Beams								
Concrete slabs and beams separation from vertical support						X		
Failed *spandrel beams*						X		
Seriously degraded concrete moment frames						X		
Severe *panel zone* cracking in concrete moment frames						X		
Flat slab with punching shear failure or failure at columns						X		X
Precast Element Structures								
Connection of elements, and joints for beams/columns/slabs							X	
Failed joints of primary members/ spalled concrete /corbel cracking							X	

Below are some definitions related to *italics highlighted* fonts in the table above:
  o   Adobe construction: a building technique using sundried mud bricks for walls.
  o   Cripple wall: a short, often wooden wall between the foundation and the floor framing.
  o   House off foundation: in older wood construction, houses were not anchored to foundations.
  o   Infill masonry walls: non-load bearing, but resist lateral forces, made of brick or concrete blocks.
  o   Liftslab structures: construction method where slabs are poured at ground level and then lifted into place.
  o   Larger wood buildings: this refers to commercial and industrial
  o   Panel zone: the region where beams connect to columns which is critical for resisting lateral forces.
  o   Spandrel beam: supports the edge of a floor slab, located between windows in a building's facade.

Seismic Vulnerability and Improvements

Table 7-3 Suggested Inspection Points for Steel Construction

	Braced Steel Frames	Moment-Resisting Frames	Prefabricated Metal Buildings	Frames with Infill Masonry Walls	Frames with CIP or R/Masonry
Roof and Floor Framing					
Broken or buckled member in truss	X	X		X	X
Failed connection in truss	X	X		X	X
Roof or floor framing separation from vertical support	X	X		X	X
Failure of significant vertical load carrying element or connection	X	X		X	X
Columns					
Shear failure at base	X	X		X	X
Columns out of plumb	X	X		X	X
Buckled or bowed columns	X	X		X	X
Vertical or Horizonal Bracing					
Broken brace or connection	X	X		X	X
Buckled or greatly stretched brace	X	X		X	X
Failed chord or connection	X	X		X	X
Horizontal Diaphragm					
Bowed, broken or seriously damaged diaphragm	X	X		X	X
Broken or seriously damaged chord or collector	X	X		X	X
Movement or failure at shear connection between diaphragm or beam	X	X		X	X
Horizontal cracks > ½ *in* wide in concrete structural slabs	X	X		X	X
Foundations					
Bowing of underground walls	X	X		X	X
New fractures > ½ *in* wide in basement floor slabs	X	X		X	X
Fractured foundations	X	X	X	X	X
Moment Frames					
Broken, leaning or seriously degraded moment frames		X		X	X
Weld or other connection failure at moment joint		X		X	X
Flange buckling near joints		X		X	X
Prefabricated Metal Buildings					
Rod connection failure in wall or roof bracing			X		
Broken Rods or greatly stretched rods in walls bracing			X		
Buckled flange in bent (i.e., a bent is similar to three hinged arches)			X		
Failed bolted moment connection			X		
Base plate with Sheared anchor bolt			X		
Frame out of plumb > **1 *to* 2 *in***			X		

7.2.5.   Geotechnical Evaluation

Several geotechnical hazards can affect structures and must be examined to assess the overall safety of the building. Some key points to consider which marks the associated building as unsafe include the following:

✓ Surface fault rupture which could have caused the damage to the building in question.

✓ Slope failure causing foundation damage or loss of foundation support.

✓ Slope movement continuing under static condition (this designates the area as unsafe).

✓ Building in active slope failure zone.

✓ Building in path of debris, including rock fall, from active slope failure zone.

✓ Retaining wall leaning outward 5° (or $1:12$ slope) or more (area underneath is unsafe).

✓ Building damaged by ground displacement.

✓ Ground fissures and scraps more than 4 *in* wide near buildings.

✓ Large cracks, increased seepage or embankment failure in earth dam.

✓ Overtopping of dam by wave.

7.3. Chapter References

The following references were used in this chapter. The reference(s) in italics is/are required for the exam, as noted in the preface of this book.

[1] *ATC-20. Procedures for Postearthquake Safety Evaluation of Buildings, 1995.*

[2] *ATC-20-1. Field Manual, 2nd Edition: Postearthquake Safety Evaluation of Buildings, 2nd Edition.*

This page is intentionally left blank

Problems & Solutions

The following questions have been carefully designed to provide comprehensive coverage of the material presented in this book, while also being educational in nature, as you will notice. Additionally, you may come across a few questions not directly covered in this chapter; these have been intentionally included to ensure broader coverage of the content.

- ➤ Problem 7.1: Engineering Evaluation After an Earthquake
- ➤ Problem 7.2: Detailed Evaluation After an Earthquake
- ➤ Problem 7.3: Unsafe Buildings Aftermath of an Earthquake
- ➤ Problem 7.4: Entry into Damaged Buildings
- ➤ Problem 7.5: Common Failures in Small Wooden Buildings

PROBLEM 7.1 *Engineering Evaluation After an Earthquake*

In the aftermath of a severe earthquake, certain buildings are designated for engineering evaluation following a detailed evaluation, who is responsible for carrying out these evaluations?

(A) Authority structural engineers.

(B) Private engineering consultant hired by the building owners.

(C) Private engineering consultant hired by the relevant authority.

(D) Authority engineering department.

PROBLEM 7.2 *Detailed Evaluation After an Earthquake*

How long does a typical detailed evaluation of a building take after an earthquake?

(A) Half an hour

(B) 1 to 4 hours

(C) 1 day

(D) More than 2 days

PROBLEM 7.3 *Unsafe Buildings Aftermath of an Earthquake*

Which of the following conditions can deem a building as unsafe?

(A) One story of the building has a residual drift.

(B) A large chimney seems unstable.

(C) The long parapet of the building could fall anytime.

(D) A vertical crack in a wall.

PROBLEM 7.4 *Entry into Damaged Buildings*

After an earthquake that had a magnitude of $M = 5$, the owner of a home that was tagged as unsafe, but stable with no major leaning or partial collapse, wishes to access the property for one hour to retrieve valuable possessions and important paperwork.

When can the owner safely enter the property?

(A) One day after the mainshock.

(B) Two days after the mainshock.

(C) Three days after the mainshock.

(D) Entry to buildings tagged as unsafe is not permitted.

PROBLEM 7.5 *Common Failures in Small Wooden Buildings*

Which of the following is a common failure in a small wooden dwelling for single-family housing following a severe earthquake?

(A) Cripple wall failure.

(B) Wythe separation and masonry spalling.

(C) Punching shear in slabs.

(D) Reinforced masonry wall with diagonal crack $> 1/8$ *in.*

SOLUTION 7.1

According to Section 3.7 of the ATC 20 Procedures for Postearthquake Safety Evaluations of Buildings, as well as Section 7.2.1 of this book, the evaluation is carried out by a private engineering consultant through arrangements done by the building owner.

Correct Answer is (B)

SOLUTION 7.2

According to Table 3.1 of the ATC 20 Procedures for Postearthquake Safety Evaluations of Buildings, as well as Section 7.2.1 of this book, detailed evaluation ranges from 1 to 4 hours.

Correct Answer is (B)

SOLUTION 7.3

A residual drift in one story of the building can compromise overall stability making the building unsafe due to the P-Delta effect (refer to Table 7-1 of this book, or Section 5.5 item 1 of ATC 20).

In contrast, an unstable large chimney or a parapet that could fall poses localized hazards only, making the affected areas unsafe. These areas should be barricaded to prevent harm, but the overall structural integrity of the building may not be affected or may need further investigation.

A vertical crack in a wall typically does not deem the building unsafe unless it severely affects the structure's load-bearing capacity. Such cracks should be inspected and monitored for further damage, but they usually don't require labelling the entire building as unsafe.

Correct Answer is (A)

SOLUTION 7.4

According to Appendix D of the ATC 20-1 Field Manual, entering a building after an earthquake may be necessary for search and rescue operations, or for the owner to retrieve important possessions, and therefore a guideline is presented in this appendix.

Aftershocks few days following a significant earthquake are common. Therefore, only stable buildings – i.e., those that are not leaning, damaged, or showing signs of partial collapse – can be considered for entry while adhering to the guidelines outlined in the table below:

Recommended Days to Wait Before Entry of Buildings Posted Unsafe but Stable (*)			
Magnitude (M)	Enter for 2 hours	Enter for 8 hours	Enter for 24 hours (**)
$M \geq 6.5$	1 day	3 days	8 days
$6 \leq M < 6.5$	1 day	2 days	4 days
$M < 6$	1 day	1 day	2 days

Based on the above, and for an earthquake with a mainshock magnitude of $M = 5$, the owner can gain access for 2 hours after waiting for 1 day post the main earthquake (*).

Correct Answer is (A)

(*) A stable building is not expected to collapse or partially collapse under its own weight or in an aftershock.

(**) For continuous access only by essential personnel and repair workers.

SOLUTION 7.5

According to Table 7-2 of this book, also in reference to the ATC 20 Procedures for Postearthquake Safety Evaluations of Buildings, Section 6.1, item 4, cripple wall is a most common failure for smaller size wooden structures.

A cripple wall is a short, often wooden wall between the foundation and the floor framing.

Correct Answer is (A)

Index

&

Equation Finder

Index

Equation Finder

Notes

&

Common Mistakes

Notes:

Use this section to note down important equations or concepts that you did not

find in this book, or found it in an example that you want to remember:

<u>I will start here with some, and feel free to add yours as you go:</u>

1- You can change C_s per ASCE C11.4.8 (Table 3-4 here) when ground motion

hazard analysis is not required for Site Class D when $S_1 \geq 0.2$ as follows:

A) When $T \leq 1.5T_s$: $C_s = S_{DS}/(R/I_e)$

B) When $T_L \geq T > 1.5T_s$: $C_s = 1.5 \times \{S_{D1}/(T(R/I_e))\}$

This does not apply to seismically isolated or structures with damping system.

2- You can use $F_a = 1.2$ per ASCE C11.4.8 (or Table 1-2 here) when ground

motion hazard analysis is not required for Site Class E when $S_s \geq 1.0$.

This does not apply to seismically isolated or structures with damping system.

3- In load combinations, if vertical acceleration was given or was required use

$E_v = 0.3S_{av}D$ where S_{av} calculation is provided in Section 1.8 here.

4- You can use $E_v =$ zero in load combinations for SDC B.

5- Critical damping C_{cr} is the minimum level of damping required to prevent

oscillation. This means a critically damped structure returns to its original

position with no oscillation, where $C_{cr} = 2\sqrt{(wk/g)}$

6- Damping C is the process by which vibrational energy is dissipated and helps

reduce the amplitude of oscillation. In seismology, damping is expressed as a

%age of critical damping $\zeta = C/C_{cr}$ - typically 5% for concrete structures.

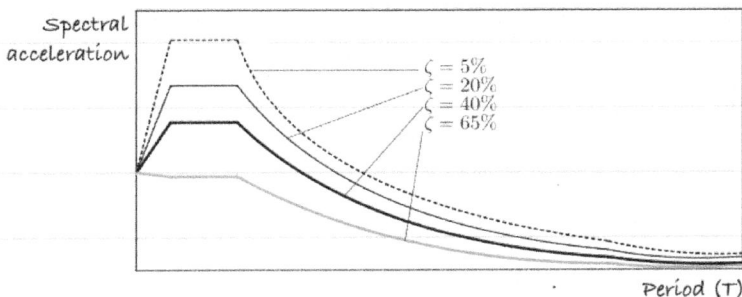

This means if damping increases, response spectrum is pushed down, hence the the spectral acceleration reduces and this generates lesser seismic forces.

7- The natural frequency (f) of a building is the rate which the building tends to vibrate. It is the number of complete oscillations per second, Hertz (Hz), and is calculated as $f = 1/T$, where (T) is the fundamental period.

8- The maximum wood diaphragm aspect ratios L/W:

- Wood structural panel, unblocked (3:1), blocked (4:1)

- Single-layer horizontally-sheathed lumber (2:1), diagonally (3:1)

- Double-layer diagonally-sheathed lumber (4:1)

9- A cantilever wood diaphragm maximum length is 35 ft (SDPWS 4.2.6).

10- The redundancy factor Rho (ρ) should always be taken as 1.3 for SDC D and above, but can be reduced to 1.0 for <u>no extreme torsional irregularity</u>, and:

(1) You have 2 bays or more resisting, or,

(2) Removing a shear wall or a single brace or a loss in a beam in a moment frame will not result in a loss of more than 33% in story strength.

—> More details on this check ASCE 12.3.4.2.

11- Location of building base for the sake of calculating shear forces:

(1) Taken above stiff soil, hence, if you have liquefiable soil, take it below it.

(2) Taken above the basement level at ground, if basement had substantial openings, then new base shifts below those.

(3) If at a slope, normally taken at the foundation level or the lowest level.

—> More details on this check ASCE C11.2

Common Mistakes:

You can learn from your own mistakes and from others as well. Below is the list of the most common mistakes with (1) being the most common. Note any mistakes that you commit during your practice and your exam, and feel free to to share them via the email in this book and I will include them in a future edition for others to learn from.

1- Not reading all the SFRS tables well BEFORE the exam, also during!

2- Not paying attention to the required minimum and maximum forces.

3- Not reading the question in full, jumping into conclusions, fat fingers.

4- Not reading all risk categories in CBC in full PRIOR to the exam.

5- Lack of practice on certain equations, or key tables, PRIOR to the exam.

6- Not using the correct $C_{s,min}$ value when $S_1 > 0.6$.

7- Lack of practice on the horizontal and vertical irregularity equations.

8- Importance factor for sprinklers and egress stairs always 1.5 regardless of the building's risk category.

9- If site class is unknown you can assume class D but $F_a \geq 1.2$.

10- Deflection and shear equations for wood diaphragms and walls in CBC are for stabled wood whereas in SDPWS are for nailed wood.

11- Not reading and familiarizing oneself with the comments below key ASCE tables. Most tables are copied here with the required permissions from ASCE.

12-

Your Feedback Matters – Make Sure You Share It With Others

Good day,

As you reach the final pages of this book, I would like to express my sincere gratitude for choosing it as your guide to aid you in your journey toward success in the PE exam. I have poured countless hours into meticulously crafting the material in these pages.

Your opinion matters greatly in helping others discover the value of this resource. If you found this book beneficial, kindly consider leaving your feedback on the platform that you bought it from - like Amazon. Your words will not only acknowledge the hard work invested into producing this book but will also guide future readers in their quest for quality study materials.

Remember, your review is more than just feedback; it is a beacon for those seeking reliable resources. Your support can make a significant difference, ensuring that this book continues to assist aspiring professionals on their path to success.

Thank you for being a part of this journey, and I personally appreciate your commitment to sharing your experience with others.

Jacob Petro

PE ESSENTIAL GUIDES

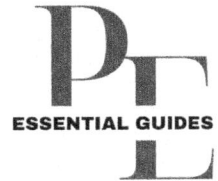

PE
ESSENTIAL GUIDES

Made in the USA
Coppell, TX
10 July 2025

51740554R00177